FATE AND EFFECTS OF OIL IN MARINE ECOSYSTEMS

Fate and Effects of Oil in Marine Ecosystems

Proceedings of the Conference on Oil Pollution Organized under the auspices of the International Association on Water Pollution Research and Control (IAWPRC) by the Netherlands Organization for Applied Scientific Research TNO Amsterdam, The Netherlands, 23–27 February 1987.

edited by

J. KUIPER

Wadden and North Sea Centre
De Koog (Texel), The Netherlands

W.J. VAN DEN BRINK

TNO Corporate Communication Department
The Hague, The Netherlands

1987 **MARTINUS NIJHOFF PUBLISHERS**
a member of the KLUWER ACADEMIC PUBLISHERS GROUP
DORDRECHT / BOSTON / LANCASTER

Distributors

for the United States and Canada: Kluwer Academic Publishers, P.O. Box 358, Accord Station, Hingham, MA 02018-0358, USA
for the UK and Ireland: Kluwer Academic Publishers, MTP Press Limited, Falcon House, Queen Square, Lancaster LA1 1RN, UK
for all other countries: Kluwer Academic Publishers Group, Distribution Center, P.O. Box 322, 3300 AH Dordrecht, The Netherlands

ISBN 90-247-3489-4

Copyright

PRINTED IN THE NETHERLANDS

Preface

A. Rörsch
Member of the TNO Board of Management

Like all living creatures man has from the very outset influenced the environment. Initially, the traces of human activity were hardly noticeable and so were their effects on the equilibrium of the ecosystem as such.
However, as soon as man learned how to use tools, he was able to influence his surroundings more drastically, and to proliferate more rapidly. As a matter of fact that is the time when things went wrong, because a process was started off which was to continue with ever-increasing speed and on an ever-increasing scale. The present condition of nature as a result of the activities of mankind is generally known.
Whether it is an accident with a nuclear plant or the vanishing of tropical rain forests, acid deposition or the pollution of soil, water and air, environmental disasters almost seem to be the order of the day. It is striking that with all these - more or less arbitrary - examples the provision of energy plays a role. In this respect one can add an even more important energy carrier to the list, namely: crude oil.
At the slightest mishap the transport of oil over sea and offshore oil production may have serious consequences for the marine ecosystem. From the papers in this book it becomes clear that oil spills at sea continue to affect the ecosystem for a very long time. That's why it is essential that policy-makers both in government and in industry should know exactly what happens to the ecosytems in case of a mishap. In this connection the research results presented during the TNO Conference on Oil Pollution dealing with the "Fate and effects of oil in marine ecosystems" are of major interest in that they provide policy guidelines with respect to risk assessment and risk management.

Environmental research is one of the main fields of activity of the Netherlands Organization for Applied Scientific Research, TNO. Within the framework of this research TNO also devotes a great deal of attention to the possibilities of dealing with the consequences of oil pollution at sea. By the organization of this conference and the publication of this book the Organization hopes to contribute to the solution of at least part of the problems stemming from human activity.

INTRODUCTION

J. Kuiper
Chairman
(TNO Marine Research Laboratory; Department of Biology,
Den Helder, The Netherlands*)

Considering the overwhelming amount of literature on the environmental fate and effects of oil and oil compounds, one could wonder what is the use of another conference or book about a number of aspects of oil in marine ecosystems. As far as the organizers of the TNO Conference on this subject are concerned, and also for the IAWPRC under whose auspices the conference is held, two reasons prevailed.

The first is the next North Sea Ministers Conference which is scheduled to be held in London, in November 1987. During that conference existing and future policies regarding the use (or should we say abuse?) of the North Sea will be discussed at the highest level by the states bordering that sea.
Like all environmental issues, oil pollution stems from the activity of man. But unlike pollution caused by, for instance, PCBs that end up in the sea from a wide variety of sources mainly via the atmosphere, oil pollution at sea mostly originates from various, most of the time well-known point-sources (ships, oil rigs, rivers). This makes these problems amenable to environmental management and relatively simple to solve. However, to be able to formulate an effective and realistic policy, policy-makers - including the ministers taking part in the aforesaid North Sea conference - will need an insight into the fate and effects of the pollutants in the marine ecosystem. The organizers believe that this TNO Conference on oil pollution will alleviate at least part of this need. Accordingly, I am convinced that, if the political will exists, most of the problems associated with oil pollution at sea could soon become history.

Until now research has yielded a huge amount of scientific data on the fate and effects of oil pollution in the marine ecosystem. Although one of the aims of this research is to furnish decision-makers with requisite information, many researchers have failed to indicate how data should be interpreted in the process of risk analysis and risk management. In other words: decision-makers are forced to judge the relevance of this information all by themselves. Oil-related environmental research is by definition applied

*) present address: Wadden and North Sea Centre, De Koog, Texel, The Netherlands.

research, in which the translation of results into guidelines for the authorities is essential. This brings us to the second reason why this conference was organized: an attempt to help bridge the gap between research and risk policy. That's why the conference programme also includes a panel discussion on risk analysis and risk management.
Apart from the value to be attached to research results in the process of risk analysis and risk management, the possible significance of the outcome of research projects for the field situation also plays a central role in the various papers and posters.
During the conference (and so in these proceedings) this basic information is presented in three main sections both as reviews on the state-of-the-art in the different areas of relevant research and as case studies; a fourth section is devoted to the use of existing scientific knowledge in the process of risk analysis and risk management.

Section 1 deals with the physical and chemical behaviour of oil in the marine environment and the biological processes that affect the fate of oil in marine ecosystems. Different aspects of the fate of oil in time and space under various conditions (including tropical conditions) are described.

Elevated concentrations in some (abiotic or biotic) compartments of ecosystems might result in ecological effects. Section 2 is devoted to the effects of oil and oil compounds on species, populations and ecosystems, both in the laboratory and in the field. Also, important information is given on the fate and effects of oil in tropical ecosystems, because information available in literature on oil under tropical conditions is remarkably scarce. Attention is paid to interesting new developments with respect to the possibilities of predicting ecological effects of contaminants by means of different types of model ecosystems (or microcosms). The effects of measures to abate oil pollution resulting from accidental spills, such as treatment with dispersants, are also reviewed in this section.

The third section gives the results of monitoring studies in abiotic and biotic compartments of marine ecosystems, as well as of the monitoring of biological and ecological effects of oil spills on species and on the entire ecosystem. Here we touch upon the heart of the oil pollution problem in the field situation; here the hypotheses and predictions based on laboratory tests and research on model ecosystems are tested in practice.

Finally, a number of specific cases of risk analysis and risk management based on existing scientific knowledge are discussed in the fourth section.

I would like to thank all authors who have been willing to share their research findings with us during the TNO Conference on Oil Pollution. Also, I am very grateful to T. Bakke, P. Bockholts, P.D. Boehm, W.J. van den Brink, J.N. Butler,

E. Dahl, B. Dicks, H. Farke, E.S. Gilfillan, L. Lijklema, J.G. Quinn, G.A. Sergy, Ms. F. Soudan and J. Widdows -all Members of the Scientific Committee- who have reviewed all the abstracts submitted, thus ensuring that a high scientific level be maintained.

I hope that the comprehensive information collected in this book, as well as the exchange of knowledge and views during the conference, will contribute to a better preservation of the marine environment housing healthy ecosystems.

CONTENTS

II. "Effects of oil on species, populations and ecosystems, both in laboratory tests and in the field (including model ecosystems) and effects of the treatment of oil spills" 87

SECTION I

PHYSICAL AND CHEMICAL BEHAVIOUR OF OIL IN THE MARINE ENVIRONMENT AND BIOLOGICAL PROCESSES AFFECTING THE FATE OF OIL IN MARINE ECOSYSTEMS

THE PHYSICAL FATE OF WEATHERED CRUDE AND EMULSIFIED FUEL OILS AS A FUNCTION OF INTERTIDAL SEDIMENTOLOGY

D.I. LITTLE

Oil Pollution Research Unit, Field Studies Council,
Orielton Field Centre, Pembroke, Dyfed, U.K.

ABSTRACT: The purpose of this series of field experiments was to study the initial behaviour and fate of contrasting oil types stranded on a wide range of sediment shorelines. Measurements were made of sediment grain size, moisture content, firmness and drainage parameters to assess their influence on oil infiltration. Oil was applied at 1 lm^{-2} (Nigerian crude) and 2 lm^{-2} (emulsified fuel oil) to a series of replicated field plots laid out at _ca._ mean high water of spring tides. Visual measurements of oil cover and infiltration were backed up by sediment hydrocarbon analysis carried out at intervals over a 1 year period. Residence times of the oils ranged from 3 days to >1 year. The emulsified oil tended to infiltrate only as far as permitted by antecedent moisture content and permeability and, on muddy shores, only _via_ large pores. Some plots were firmer as a result of oil application and this may increase oil residence time in a way analogous to the formation of "asphalt pavements". Oil uptake was reduced on low shear strength sediments, but this may have been an indirect effect of waterlogging. Drainage of the surface sediments was a prerequisite to significant oil infiltration, but water table movements did not encourage, by "tidal pumping", penetration deeper than _ca._ 2 cm. Finally, sediment % mud content may be used to predict the likely oil residence time, possibly because this sediment parameter is very sensitive to energy level, and also because mud particles retain surface-bound contaminants more strongly than do sands.

1. INTRODUCTION

The importance of sediments in determining the fate and effects of spilled oil lies in their ability to act both as a hydrocarbon sink and then as a pollution source to the biological communities they support. Quantitative data on sediment hydrocarbon concentrations and composition, together with their modification by oil spill cleanup activities, can facilitate the prediction and assessment of ecological damage (Gilfillan _et al._, 1985). They assist in the calculation of oil spill mass balances and, in addition, provide the experimental basis for realistic pollutant exposure regimes in ecotoxicological studies (Anderson _et al._, 1984). Oil stranded on sediment shorelines, usually at high water mark, behaves in a quite different way to the hydrocarbons which accumulate in fine-grained sediments as a result of, for example, oil biopackaging and sedimentation. Any model which attempts to explain the sedimentological fate of oil must take account of these differences, in addition to modeling hydrodynamic and physico-chemical variables such as currents and oil partitioning. The model should ideally predict the onshore, offshore and longshore transport and fate of the oil (Gundlach _et al._, 1985). To achieve this, the probability of stranded oil being resuspended as weathered oil droplets,

Kuiper, J., Van den Brink, W.J. (eds), Fate and Effects of Oil in Marine Ecosystems. ISBN 90-247-3489-4

which may or may not be associated with sediment particles, must be understood (Boehm *et al.*, 1985). Clearly the development of these models is a major task.

The concept of oil 'residence time' has provided an operational framework for oil spill response planning in the absence of detailed algorithms of shoreline-oil interactions. The residence time of stranded oil is influenced chiefly by the available levels of mechanical energy, mainly from wave action (Owens, 1976), but it is also dependent on oil type, volume and the initial probability of infiltrating the sediment. Once oil is in the sediment, its residence time will depend not only on redox conditions, microbial degradation and water-washing, but also on the rate at which infiltration and accretion take the oil below the depth of mechanical disturbance (Little *et al.*, 1981). Much will also depend on the locus of oil contamination, whether in the sediment matrix or its macropores (Howard and Little, in press). Intuitively, it may be suggested that the amount of oil infiltration will be determined by the firmness, texture, permeability, antecedent moisture content and drainage of the sediment. This study investigates the possible role of these factors in the infiltration of two contrasting types of weathered oil into a wide range of sediment shorelines. It also suggests mechanisms whereby stranded oil becomes temporarily removed from the marine ecosystem *via* infiltration and burial, or else is reworked and further diluted in the coastal zone.

2. METHODS

2.1. Experimental design and study shores (Table 1 and Fig. 1)

A series of field experiments was carried out in 1984 on seven shores in southwest Britain to compare the fate of single, moderate to heavy treatments of lightly-weathered crude oil and medium fuel oil 'mousse'.

TABLE 1. Experimental shores, abbreviations, treatment dates and sediment character.

Study shore	Abbreviation	Treatment date	Sediment character
Angle Bay	A	1/9/84	Sheltered sand, gravel and algae; well drained
Sandyhaven Pill	S	3-4/9/84	Sheltered fine sands and vegetated mud; poorly to well drained
Steart Saltmarsh	St	1/10/84	Moderately sheltered vegetated mud; moderately drained
Westdale Bay	W	15/10/84	Very exposed medium sand; well drained
Broadhaven	Br	29/10/84	Moderately exposed fine sand; well drained
Freshwater West	F	30/10/84	Very exposed fine sand; well drained
Blue Anchor Bay	B	15-20/11/84	Moderately exposed fine sand; poorly drained

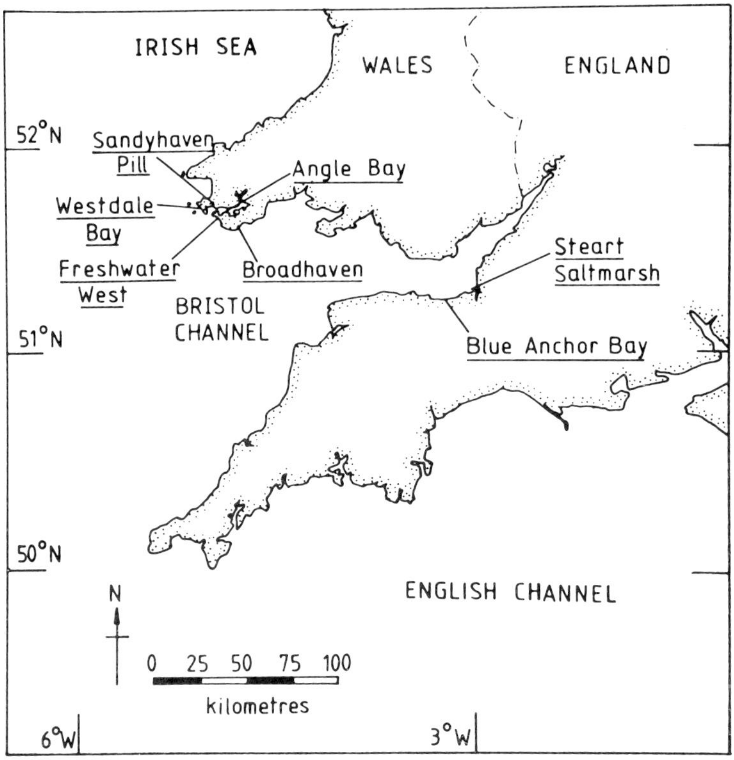

FIGURE 1. Study shore location map.

The sediment at the study sites varied from gravelly sand to vegetated silt and clay, and included three exposed sandy beaches of different morphology and fetch. The experiments were analogous to those carried out along an energy gradient of rocky shores and described by Thomas *et al.* (1983). The experimental design consisted of oil application to a set of eight 5 x 1 m plots marked out at *ca.* mean high water of spring tides on each shore. The oils were applied on the ebb tide during periods of spring to neap tidal cycles to simulate natural oil stranding and to allow maximum oil-sediment interaction before being flooded by the next spring tides.

2.2. Pollutants

2.2.1. Crude oil. Nigerian crude oil from the Forcados field was artificially weathered to simulate stranded oil. Fractional distillation to remove components with boiling points less than that of n-C_{12} (216°C) was carried out by Esso Petroleum Ltd. A capillary gas chromatograph trace of the aliphatic fraction is shown in Little *et al.* (in press). The viscosity was *ca.* 9 cs at 20°C. The oil was applied by hand-pumped sprayer at the rate of 1 lm^{-2}.

2.2.2. Mousse. This pollutant was prepared by Warren Spring Laboratory from a medium fuel oil, emulsified with 60% vv^{-1} seawater by agitation with compressed air. The water-in-oil emulsion had a similar aliphatic hydrocarbon boiling point range to the crude oil but its viscosity was *ca.* 53,500 cs at 20°C. The mousse was applied direct from 5 l cans at a rate of 2 lm^{-2}.

2.3. Hydrocarbon analyses

2.3.1. Percentage oil cover and visual observations. Surface oil cover was measured using a point frame. The presence, absence and nature of the oil (pooled, thin, sheen and stain) were recorded for 100 points per plot at intervals over a 1 year period. Trenches were dug to measure depth of oil infiltration, and colour transparencies were taken of the trenches in addition to overviews of each plot.

2.3.2. Sediment hydrocarbon analysis. Two or three sediment subsamples (10 cm x 10 cm x 5 cm depth) were taken from the plots at intervals from 4 to 388 days and then stored in aluminium trays at -20°C. All sampling utensils and containers were solvent-rinsed. Methods for hydrocarbon extraction, separation and gravimetric analysis are detailed by Baker *et al.* (1984).

2.4. Sedimentological analyses

2.4.1. Sediment movements. Erosion pins and dyed or artificial sediments were used to characterise energy levels at the sediment surface by acting as markers of sediment turnover and movement (King, 1951). The aim was to relate these data to changes in both oil cover and concentration through time. Dyed local sediment, graded coal dust or aluminium oxide powder, depending on sediment texture and colour, were placed into small trenches (10 cm long x 2 cm wide x 20 cm deep) or cylindrical core holes (6.5 x 20 cm) adjacent to the plots. The depth of clean sediment overlying the dyed sand was then monitored through time and the results related to the erosion pins. This combination does not permit the observer to decide whether accretion took place before or after disturbance of the pre-existing sediment surface. In many of these experiments, the oiled sediment itself acted as a visual marker so the problem did not arise. In addition, a beach profile and the height of each plot at intervals through time were determined by levelling.

2.4.2. Moisture content. A 50 g subsample of each homogenised sediment hydrocarbon sample was dried at 105°C until a constant mass was recorded. The dry mass was used in calculating the hydrocarbon concentration and also to characterise antecedent moisture conditions during the pre-tidal cover period at the beginning of the experiments.

2.4.3. Hydraulic conductivity. On each shore, between one and six undisturbed cores were taken in and around the experimental plots at the time of the oiling, using short (<20 cm) PVC core tubes (6.5 cm internal diameter). These were returned upright and field-moist to the laboratory. Air was removed from the core samples by gentle vacuum suction before testing. For sands and sandy silts, a constant head permeameter was used to determine hydraulic conductivity. For finer sediments, a falling-head permeameter was used (Ackroyd, 1969). These techniques permitted the measurement of water flow through the core under known hydraulic gradients, and thus may help to explain the penetration and behaviour of the oil.

2.4.4. Firmness. The unconfined compressive strength of the drained surface sediments was measured using a hand-held penetrometer equipped with accessory large-scale needle points (ASTM D1558) to permit measurement of penetration resistance in kg cm^{-2}.

2.4.5. Water table behaviour. Unpublished data from a previous study of water table and drainage of intertidal sediments (Little, 1984) were used in a comparison of the experimental plots. The method used involved transects of well-point piezometers from above high water to low water mark. The depth of the water table was recorded against time on spring and neap tides. Only data from August and September 1982 for stations adjacent to the present experimental plots were used. The total area over the water table curve against time, its initial rate of fall after emersion and the maximum depth reached, were all calculated from the piezometric data.

2.4.6. Grain size distribution. Except at Angle Bay, at least three bulk surface sediment samples (0-10 cm depth) were taken on each study shore at the beginning of the experiment. These samples were analysed using the techniques described by Little *et al*. (1984). The samples were split into sand and mud fractions by wet-sieving at 63 µm and further categorised by dry-sieving of the sand and gravel fractions. Electronic particle counting of the mud fractions was carried out when sufficent mud was present. The data were reduced by microcomputer using the method of moments (Krumbein, 1936).

3. RESULTS AND DISCUSSION

3.1. Hydrocarbons

Details of the visual observations, the % oil cover and total hydrocarbon (THC) concentrations on each shore through time are given by Little and Scales (in press). These showed that the range of oil residence times was from 3 days to >1 year depending on energy level. As expected, self-cleaning was significantly faster on more exposed, sandy beaches than on sheltered, muddy shores. Aspect, smaller fetch and slightly higher position on the shore help explain the slightly longer oil residence times at Broadhaven compared to the other sandy beaches; Westdale and Freshwater West. Broadhaven experienced a lower wave energy level and THC concentrations were reduced by mixing, dilution with new sediment and, possibly, by interstitial water-washing. This contrasted with the wholesale oil and sediment erosion at Westdale and Freshwater West.

Surface oil % cover remained relatively high until each tidal immersion, when some oil was laterally advected. Remaining oil was often buried by new sediments (wind, wave or tide-transported) and was, therefore, invisible from the surface. Mousse treatment resulted in more patchy oil distribution, but visible contamination lasted longer than was the case with the crude oil, particularly on pebbly or filamentous algae-covered sediment surfaces. The majority of both oil types remained concentrated within *ca*. 2 cm of the surface, although some migration of lighter components of the crude oil occurred; downwards into subjacent sediments and upwards by capillary rise into newly-deposited material. The ease of visual identification of the crude oil was often complicated by wet weather and the presence of diatoms or algae.

Blue Anchor Bay is a relatively high energy, poorly-drained, muddy sandflat. Most of the applied oil was removed by the first tidal cover. Similar behaviour was observed for both oil types on poorly-drained plots at Sandyhaven Pill. Rapid oil removal was due to hydrological protection provided by the high water table and subsequent tidal flushing. Natural cleanup was completed by sediment current ripple action under high energy conditions (Little _et al._, 1981, and in press). These poorly-drained plots were the only exceptions to the generally good drainage on the study shores selected. The upper beachface and backshore areas of sandy shorelines are often very well-drained. Such areas are also most likely to receive the greatest quantities of refloated and undispersed, stranded oil, and are beyond severe wave action except during the highest spring tides. Along with the high water mark of marsh environments, where vegetation also acts as an oil trap, these are the habitats where prolonged oil-sediment interactions are most likely to occur.

3.2. Sedimentology

3.2.1. _Sediment movements_. The movement of sediment is an important parameter if discrimination between oil infiltration and burial is required. These processes are clearly vital in two respects to correct interpretation of the fate of stranded oil. Firstly, the dilution and mixing of oiled sand with clean sand can be important natural cleaning processes. Because this paper is concerned with conditions antecedent to the oil stranding, subsequent movements of sediment are not treated in detail. Dilution of the oiled surface sediments by aeolian sand deposition followed by capillary rise of the oil on some sandy shores, were implicated by Little and Scales (in press) as a reason for fluctuating surface THC concentrations through time. This may also explain some of the variability in the graphs presented here. Secondly, under anaerobic conditions buried oil may degrade very slowly and may only be re-exposed by subsequent erosion. On Steart Saltmarsh, the accretion rate was relatively high compared to Sandyhaven vegetated plots. However, sedimentation at Steart generally decreased through time consistent with the shift from equinoctial tides after October 1984. Additionally, there was a gradient of increasing sedimentation along the shore. In beach and marsh environments sediment movements provide the link between oil stranding and the reworking of oiled sediment particles.

3.2.2. _Moisture content_. Sediment % moisture content was calculated for each sample for which sediment hydrocarbons were determined. This is in order to express THC concentration on a dry mass basis. It may be seen that sandy shores had lower moisture contents (1.3% to 17.8%) than muddy shores (23.6% to 64.4%). Some variation occurred within plots through time as a result of different drainage periods elapsing between previous tidal coverage and sampling for hydrocarbons. Fig. 2 is a plot of $\log_{10}$ transformed total hydrocarbon concentrations against $\log_{10}$ % moisture for the first 9 days of the experiments (generally pre-tidal cover). Logarithmic transformation of the data was carried out because it effectively 'stretched' the distribution of % moisture data in the highly-clumped 2.5% to 20% range. Wide scatter of the data is evident, but in the case of mousse-treated plots there is suggestion of a negative relationship. The higher the moisture content, the lower the oil concentration in the substrate. For emulsified or viscous oils, therefore, these data support the idea that antecedent moisture contents on intertidal beaches are particularly significant to oil penetration.

3.2.3. _Hydraulic conductivity_. The permeability of the surface sediments may be an important property for infiltration by stranded oils. The data range through five orders of magnitude, with the highest hydraulic conductivity being recorded at Broadhaven plot 3 (3.7×10^{-2} cm sec^{-1}). The remaining cores from Broadhaven and the cores from Freshwater West and Westdale were of a similar medium-grained sand, but

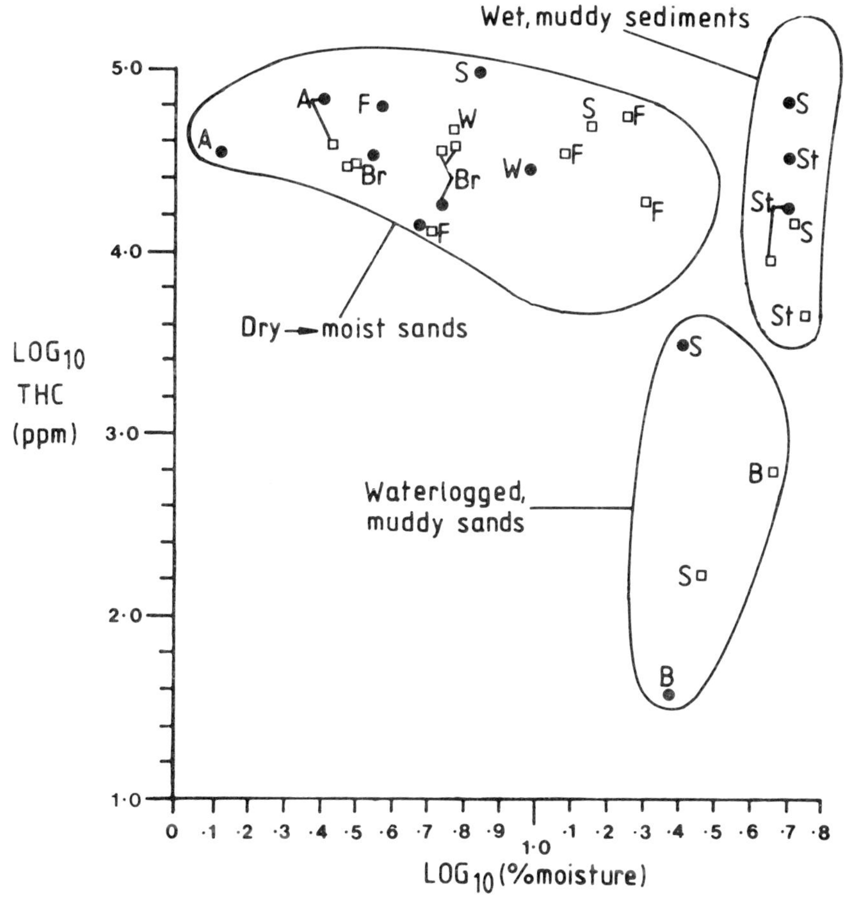

FIGURE 2. Plot of sediment % moisture against total hydrocarbon concentration (THC) at the first sampling visit (before tidal cover), ● = crude, □ = mousse. See Table 1 for shore abbreviations.

all were slightly less permeable. Hydraulic conductivity data for the Blue Anchor Bay sediments were between 1.9×10^{-3} to 5.3×10^{-3} cm sec^{-1}, the lowest value being recorded from a core which had two silt laminae completely occupying the core cross-section. The sandy sediments at Angle Bay were of a surprisingly low hydraulic conductivity (9.9×10^{-3} and 3.4×10^{-4} cm sec^{-1}), but it should be noted that these cores were sampled through the oil layer, with the mousse-treated plot giving the lower coefficient, perhaps because of the greater hydrophobic properties of the mousse compared to the crude oil. The sandy sediments at Sandyhaven Pill were all an order of magnitude less permeable than this, whilst values from the muddy plots ranged between 4.9 x 10^{-6} and 8.0 x 10^{-6} cm sec^{-1}. Steart saltmarsh yielded the least permeable core from the entire study (2.9 x 10^{-7} cm sec^{-1}), although the values for the other two cores from Steart were 3.2 x 10^{-5} and 5.8 x 10^{-5} cm sec^{-1}. These values were higher than those from the vegetated plots at Sandyhaven, an observation which may perhaps be explained by the sandy laminations in the Steart cores.

When these data were logarithmically-transformed and plotted against $\log_{10}$ THC concentration (days 1-9) no obvious overall relationship could be detected. Only in the case of the mousse-treated plots is it possible to discern the expected rise in THC concentration with increased hydraulic conductivity (Fig. 3). It is suggested that an improved relationship might have been observed had the unsaturated rather than saturated hydraulic conductivity been determined. This is because most of the wave-exposed beaches were oiled after swash tongues retreated from the marked out plots and may therefore already have been drained somewhat as the oil was applied.

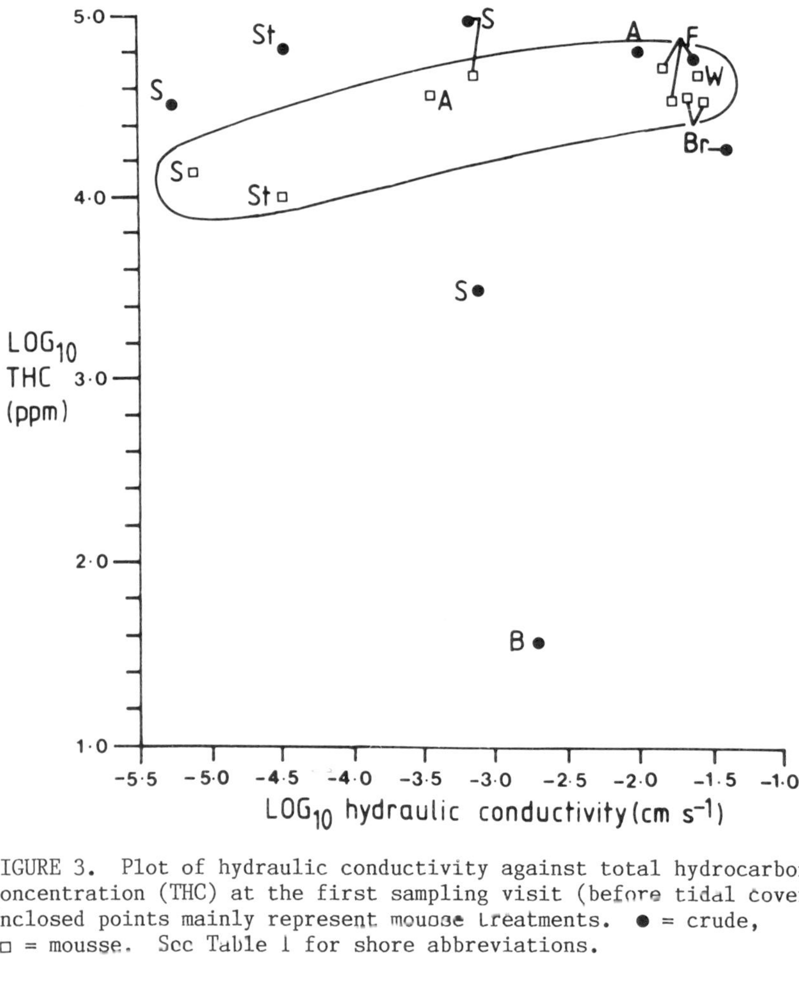

FIGURE 3. Plot of hydraulic conductivity against total hydrocarbon concentration (THC) at the first sampling visit (before tidal cover). Enclosed points mainly represent mousse treatments. ● = crude, □ = mousse. See Table 1 for shore abbreviations.

3.2.4. <u>Firmness</u>. The highest unconfined compressive strength values were recorded near plots 3 and 4 at Sandyhaven Pill (0.5–0.8 kg cm^{-2}). The remaining sediment firmness data show little variation, with widely contrasting sediments generally giving values of between 0.1–0.3 kg cm^{-2}. Slightly higher values than these were recorded in the oiled sediment at Angle Bay where the observations were made 4 days after mousse application and the tarry pavement was evidently relatively resistant to pressure by this time. In contrast, at the three sandy shores for which data are available, the oiled sediment firmness was frequently less than that of adjacent unoiled sediments. These readings were all made on the day of pollutant application so that the surface sediments may have been lubricated by the oils and thus were less resistant to pressure.

Fig. 4 illustrates the positive relationship which exists between the unconfined compressive strength and THC concentration. It would appear that weaker, uncompacted sediments do not promote oil uptake, and that

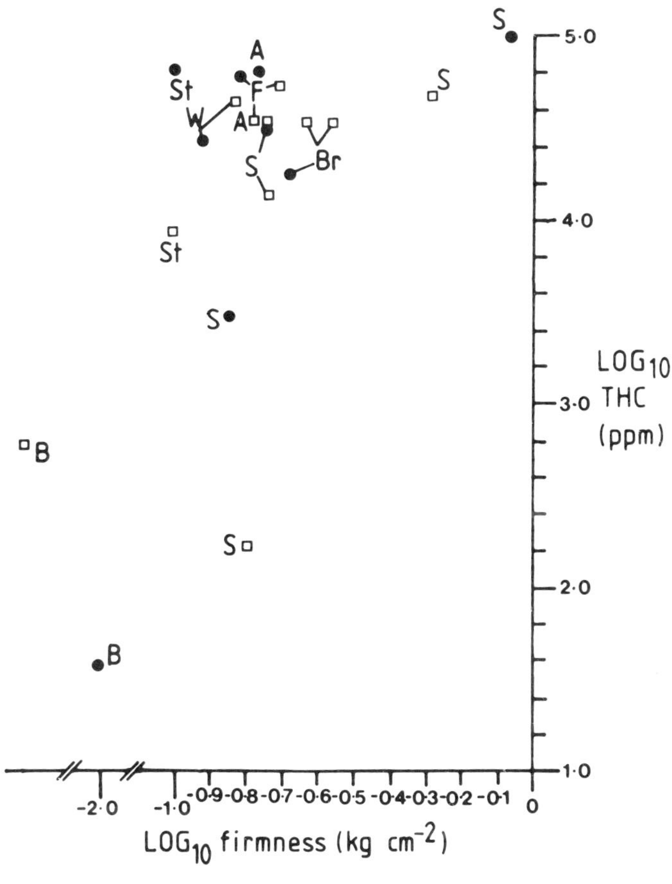

FIGURE 4. Plot of sediment unconfined compressive strength againsttotal hydrocarbon concentration (THC) at the first sampling visit (before tidal cover). Enclosed points mainly represent mousse treatments. ● = crude, □ = mousse. See Table 1 for shore abbreviations.

firmer sediments, particularly at plots 3 and 4 at Sandyhaven, do promote oil uptake. Although not measured quantitatively, plots with relatively low accretion rates at Steart Saltmarsh were also firmer than the others, and indeed had higher surface oil cover. The firmness of the sediment was itself influenced negatively by % mud and by % moisture content. The apparent positive effect of sediment firmness on oil infiltration may, therefore, be an indirect effect of dry, low % mud sediments encouraging oil infiltration, and of wet muds preventing it. It is difficult without further experimentation or the use of multiple regression analysis to separate these influences.

The potential importance of such data lies in the longer-term relationship between compressive strength and the probability of sediment erosion and transport. Oil pollution of sediments in some cases might increase sediment shear strength and thus retard sediment turnover and increase oil residence times. This phenomenon has been widely reported for gravelly shores (e.g. Owens, 1976), giving rise to the term "asphalt pavement". It has also been observed in sands during a study of the Sivand oilspill of September 1983. Levell (1976) commented on the increased stability of some plots in his experiments at Sandyhaven Pill.

3.2.5. Water table behaviour. Sandyhaven plots 5 and 6, and Blue Anchor plots 11 and 12 were more poorly-drained than the other plots. Fig. 5 illustrates the relationship between the maximum depth of the water table on spring tides and the initial (days 1-9) THC concentrations. Perhaps as a result of their very high water tables, Blue Anchor and Sandyhaven plots 5 and 6 showed lower oil concentrations. At sites where the maximum water table depth was > 0.2 m no further increases in THC concentration occurred with increasing sediment drainage depths. If anything, reductions in THC concentration occurred, for example at Sandyhaven on the muddy plots 1 and 2. The same is true when the initial ebb rate of fall of the water table was plotted against initial (days 1-9) THC concentrations. It would seem that drainage of the surface sediment layer is a prerequisite to oil penetration (Little *et al.*, 1981) but that repeated water table fluctuations may not necessarily lead to enhanced penetration or tidal pumping of oil, as suggested by Vandermeulen (1980). Enhanced penetration would be expected with inland spills from oil tanks where there is a head of oil. Large accumulations of oil in scour pits may perform the same function on beaches. Only where oil volumes are high and the depth to the water table small (3-4 cm) is water table behaviour likely to shape the oiled sand body in the intertidal zone. Some migration of lower molecular weight hydrocarbons occurred in sandy sediments, and this may be a function of sediment drainage parameters (McLaren, 1985).

3.2.6. Grain size distribution. The % mud (sediment particles < 63 µm in diameter) were plotted against THC concentration. Once again, the $\log_{10}$ transformed data were used because of the wide ranges in the data. However, this time the THC concentrations were plotted in temporal groupings, i.e. days 1-8 (first sampling visit and pre-tidal cover), days 9-87 (intermediate visits) and finally, the last sampling visit to each study shore. Figs. 6 to 8 illustrate the changes which occurred through time in the relationship between % mud and THC concentration. It should be noted that the % mud data are the same in each of the 3 THC data groupings, but it may be reasonably assumed that changes in the independent variable (% mud) on all the shores would be several orders of

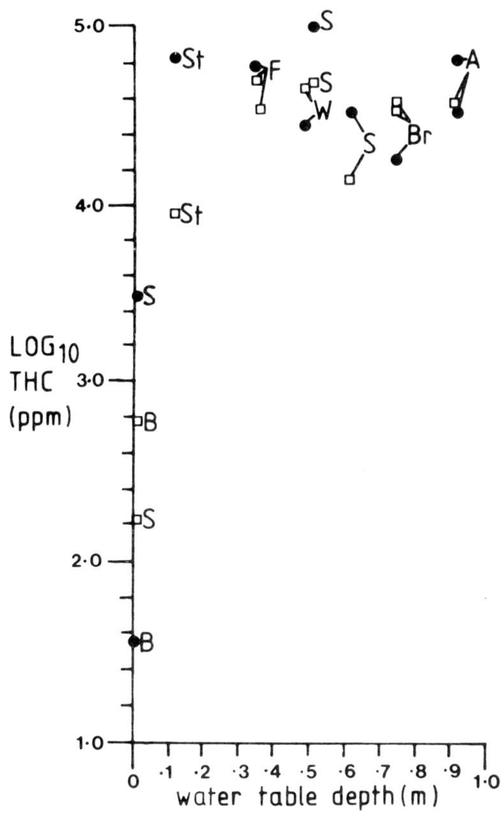

FIGURE 5. Plot of maximum depth to water table on spring tides against total hydrocarbon concentration (THC) at the first sampling visit (before tidal cover). Water table data from same station positions (Little, 1984). ● = crude, □ = mousse. See Table 1 for shore abbreviations.

magnitude less than the measured changes through time in the dependent variable (hydrocarbon concentrations).

At the beginning of the experiments (days 1-8), as would be expected, the % mud/THC relationship is a good reflection of the % moisture/THC relationship described above: the shores higher in mud (and moisture) content generally had lower and more variable sediment THC concentrations shortly after oil application (Fig. 6). The sandy shores with better drained sediment were tightly-grouped and, particularly for mousse-treated plots, showed comparatively little THC concentration variation. As with the % moisture/THC scatterplot, there is a much better fit for the sandy, better-drained sediments, with progressively more scatter in the data for muddy, wetter sediments. This scatter is presumably a function of the influence of other variables such as the existence of macropores as pathways for oil penetration (positive anomalies) and lower position on the shore in combination with high water tables (negative anomalies).

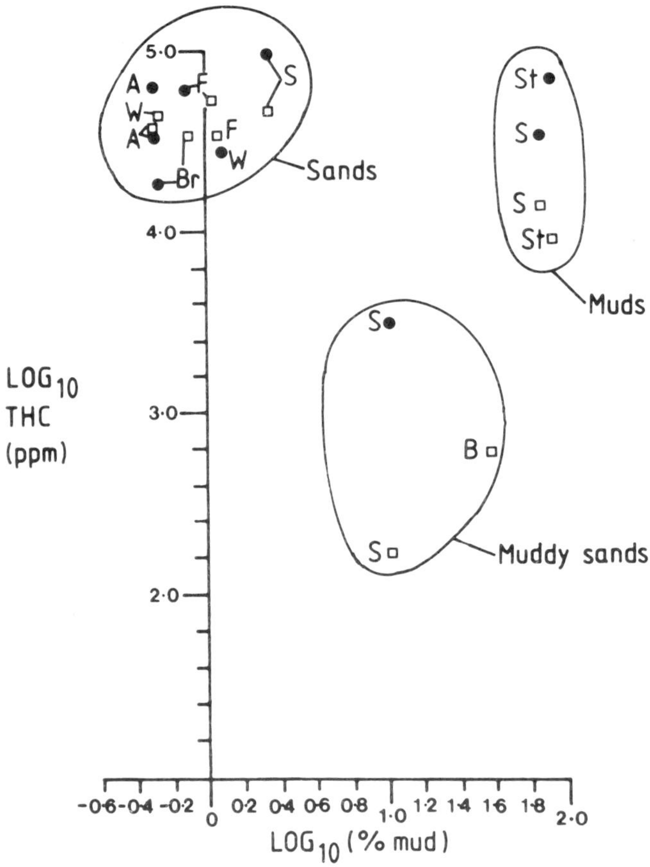

FIGURE 6. Plot of sediment % mud against total hydrocarbon concentration (THC) at the first sampling visit (before tidal cover). ● = crude, □ = mousse. See Table 1 for shore abbreviations.

Fig. 7 shows that between days 9 and 87 the tightly-grouped cluster of sandy shores has broken up, with only Broadhaven and one plot from each of Angle Bay and Sandyhaven still retaining extremely high THC concentrations. All the plots with 10% mud were still heavily-oiled. After 14 days, Freshwater West samples had been substantially cleaned while this process took ca. 2 months for Angle Bay. By the last sampling visit to each plot (Fig. 8) only Sandyhaven plot 1 (388 days), plots 2 and 3 (131-2 days), Broadhaven plot 3 (25 days) and plot 11 at Blue Anchor (3 days) remained as positive anomalies above an otherwise linear regression. Only one of these still-polluted plots was sandy. There were no negative anomalies below this line, and furthermore, several data points close to the line for both sandy and muddy shores agreed well with a previously-determined regression of background oil concentrations and mud content for Milford Haven sediments. These data suggest that

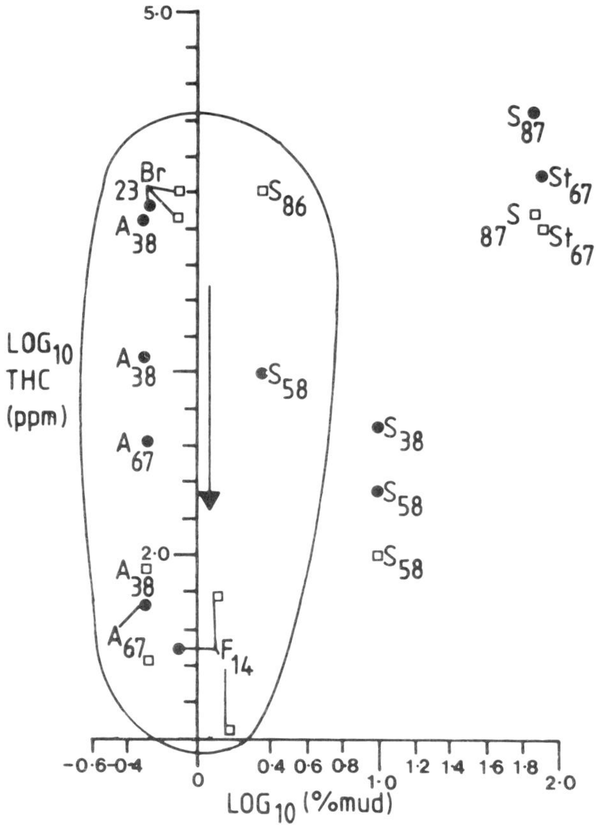

FIGURE 7. Plot of sediment % mud against total hydrocarbon concentration (THC) at the intermediate sampling visits (days 9-87). Some of the sandy group of stations visible in Fig. 6 have begun to lose their oil content (symbolised by the arrow) whilst the muddier groups of stations have not. ● = crude, □ = mousse. See Table 1 for shore abbreviations. Day on which each sample was taken is shown.

oil-coated sand particles lose their contamination by abrasion or export to the surf zone, whereas mud particles tend to retain some of their oil burden. Alternatively, the % mud acts as an analogue for energy level.

It may be suggested, therefore, that the length of time taken for any shore in this geographical area to attain the theoretical relationship shown in Fig. 8 is a good indication of the likely residence time of 'surplus' oil contamination. The residence time could, therefore, be predicted from knowledge of the sediment % mud content on shores not known to have been polluted or, after a major spill, where hydrocarbon analyses of all affected shores may be prohibitively expensive. This regression model is implicit in oil spill vulnerability indices, but numerical precision for comparable moderate to heavy oiling rates on a range of shores has been lacking.

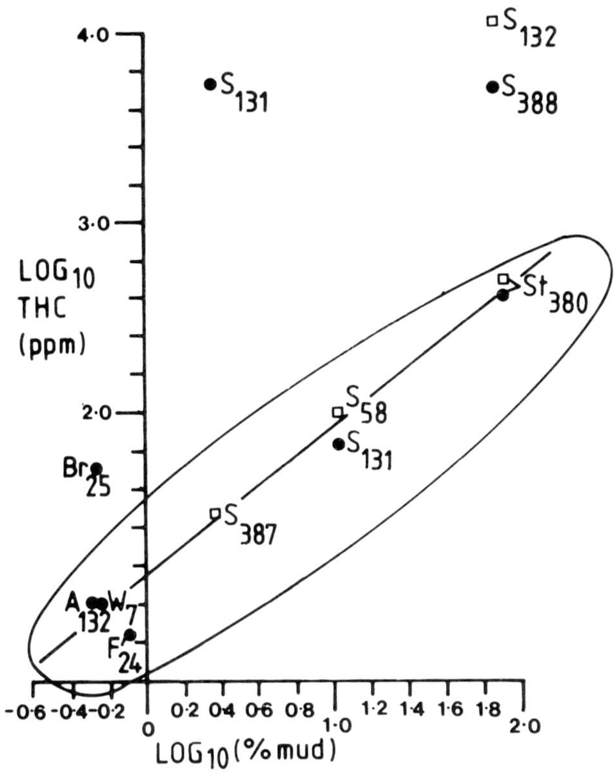

FIGURE 8. Plot of sediment % mud against total hydrocarbon concentration (THC) at the last sampling visits. Enclosed points represent those shores which have returned to 'background' THC concentration after the number of days shown. These points are expressed by the linear regression equation $y = 0.77x + 1.13$, $r = 0.96$ ($p < 0.001$). In combination with the number of days required to reach the line it may be used to predict the residence time of oil spills in this geographical area. ● = crude, □ = mousse. See Table 1 for shore abbreviations. Day on which each sample was taken is shown.

4. CONCLUSIONS

Variations in shoreline aspect and energy level are known to affect the residence time of stranded oil. The weather, the type of oil and the volume and degradation of the oil are unique to each real spill and thus tend to confound some of the generalisations about the physical fate of oil on shorelines. Oil retention is also considerably affected by burial under new sediments or by erosion, and both those are time and site-specific. In spite of these complications it would appear that sediment % moisture, % mud, firmness, drainage and hydraulic conductivity can all exert some influence on the initial infiltration of oil.

The tendency of mousse to infiltrate only as much as permitted by sediment antecedent moisture content and hydraulic conductivity was slightly more marked than with the crude oil. This was perhaps a result of the much higher viscosity of the mousse. On muddy shorelines mousse penetration was only significant in the presence of sediment macropores.

Some oil treatments increased the firmness of the surface sediment layers. Even where a true asphalt pavement does not form, it is suggested that increased firmness is a factor which can increase oil residence time.

Drainage of the surface sediment layer was a prerequisite to oil infiltration, but the speed, periodicity and maximum depth of water table fluctuations did not necessarily lead to enhanced oil infiltration by depth or volume. This would be expected, for example, with inland oil spills from tanks where there is a head of oil available for infiltration, but not on sediment shorelines except perhaps in scour pits.

Sediment % mud content may be used to predict the likely residence time of oil on shores not known to have been oiled. This is implicit in oil spill vulnerability indices, but the data permit more precise predictions for north temperate, clastic sediment shores. The data also support the view that oil-coated sand particles lose their contamination by abrasion or export to the surf zone, whereas mud particles retain their oil burden.

5. ACKNOWLEDGEMENTS

This experiment was financed by the Institute of Petroleum, BP International Ltd. and the Commission of the European Communities under Contract Number ENV.837.UK(H). Pollutants were prepared and supplied by Esso Petroleum Ltd. and Warren Spring Laboratory. Thanks are also due to the national and local organisations which co-operated over access to their sites, and to the many colleagues who helped in the field and laboratory.

REFERENCES

1. Ackroyd TNW: Chapter 8. Permeability. In: Laboratory Testing in Soil Engineering. Soil Mechanics Ltd, London. 233 pp, 1969.
2. Anderson JW, Vanderhorst JR, Keisser SL, Fleischmann ML and Fellingham GW: Recommended methods for testing the fate and effects of dispersed oil in marine sediments. In: Tom E Allen (ed), Oil Spill Chemical Dispersants: Research, Experience and Recommendations. ASTM STP 840, Philadelphia, PA. 224-238, 1984.
3. Baker JM, Crothers JH, Little DI, Oldham JH and Wilson CM: Comparison of the fate and ecological effects of dispersed and non-dispersed oil in a variety of intertidal habitats. In: Tom E Allen (ed), Oil Spill Chemical Dispersants: Research, Experience and Recommendations. ASTM STP 840, Philadelphia, PA. 239-279, 1984.
4. Boehm PD, Steinhauer W, Requejo A, Cobb D, Duffy S and Brown J: Comparative fate of chemically dispersed and untreated oil in the Arctic: Baffin Island Oil Spill Studies 1980-1983. Proceedings of the 1985 Oil Spill Conference (Prevention, Behaviour, Control, Cleanup). American Petroleum Institute, Washington DC. 561-569, 1985.
5. Gilfillan ES, Page DS, Hanson SA, Foster JC, Hotham J, Vallas D, Pendergast E, Hebert S, Pratt SD and Gerber R: Tidal area dispersant experiment, Searsport Maine: an overview. Proceedings of the 1985 Oil Spill Conference (Prevention, Behaviour, Control, Cleanup). American Petroleum Institute, Washington DC. 553-559, 1985.

6. Gundlach ER, Kana TW and Boehm PD: Modeling spilled oil partitioning in nearshore and surf zone areas. Proceedings of the 1985 Oil Spill Conference (Prevention, Behaviour, Control, Cleanup). American Petroleum Institute, Washington DC. 379-383, 1985.
7. Howard S and Little DI: Effect of infaunal burrow structure on oil penetration into sediments. Proceedings of the 1987 Oil Spill Conference (Prevention, Behaviour, Control, Cleanup). American Petroleum Institute, Washington DC. In press.
8. King CAM: The depth of disturbance of sand on sea beaches by waves. Journal of Sedimentary Petrology 21, 131-140, 1951.
9. Krumbein WC: Application of logarithmic moments to size-frequency distributions of sediments. Journal of Sedimentary Petrology 6, 35-47, 1936.
10. Levell D: The effect of Kuwait crude oil and the dispersant BP1100X on the lugworm *Arenicola marina* L. In: Baker JM (ed), Marine Ecology and Oil Pollution. Applied Science Publishers, Barking, London. 131-185, 1976.
11. Little DI: Water table and drainage characteristics of intertidal sediment. Report to British Petroleum and Commission of European Communities. FSC/OPRU/2/84, 1984.
12. Little DI, Abbiss TP, Baker JM, Rowland SJ and Tibbetts PJC: The fate and effects of dispersant-treated compared with untreated crude oil, with particular reference to sheltered intertidal sediments. Proceedings of the 1981 Oil Spill Conference (Prevention, Behaviour, Control, Cleanup). American Petroleum Institute, Washington, DC. 283-293, 1981.
13. Little DI, Staggs MF and Woodman SSC: Sample pretreatment and size analysis of poorly-sorted cohesive sediments by sieve and electronic particle counter. In: Parker WR and Kinsman DJJ (eds), Transfer Processes in Cohesive Sediment Systems. Plenum Press, New York. 47-74, 1984.
14. Little DI, Abbiss, TP, Baker JM, Dicks BM and Tibbetts PJC: The effects of chemical dispersants on the flux of stranded crude oil into fine intertidal sand. Rapports et Procès-Verbaux des Réunions, Conseil International Exploration de la Mer 186. In press.
15. Little DI and Scales DL: The physical fate of oil stranded on sediment shorelines. Proceedings of the 1987 Oil Spill Conference (Prevention, Behaviour, Control, Cleanup). American Petroleum Institute, Washington DC. In press.
16. Mclaren P: Behaviour of diesel fuel on a high energy beach. Marine Pollution Bulletin 16(5), 191-196, 1985.
17. Owens EH: Mechanical dispersal of oil stranded in the littoral zone. Journal of Fisheries Research Board of Canada 35, 563-572, 1978.
18. Thomas AD, Baker JM and Hall J: Rocky shores: Relationship between exposure grade and natural cleaning time of stranded oil. Report to Exxon Corporation, FSC(Slapton)/3/83, 1983.
19. Vandermeulen JH: Chemical dispersion of oil in coastal low-energy systems: salt-marshes and tidal rivers. In: Mackay D, Wells PG and Paterson S (eds), Chemical Dispersion of Oil Spills: An International Research Symposium. University of Toronto Institute for Environmental Studies, Toronto. 103-107, 1980.

WEATHERING OF MINERAL OIL IN MARINE ENVIRONMENT

N.ADLER
N.KOPRIVANAC
Faculty of Technology, University of Zagreb, Yugoslavia

INTRODUCTION

Petroleum hydrocarbons disseminated by men into the marine environment cause many pollution problems. The fate of a spill of crude oil or refined product in the marine environment is determined by spreading, evaporation, dispersion, emulsification, dissolution, sedimentation and various degradation processes (1). That is why analytical problems to identify the sources of hydrocarbon oil spills are serious(2).

An analysis of various marine samples of oil spread on the surface of water is often required because of spillage of crude oil or refined petroleum products. The object is to identify oil for legal purposes and to provide information for inspection authorities. There are no generally accepted techniques for their assay. A variety of analytical methods for hydrocarbon detection and determination are applicable to marine oil spill identification: gravimetry, colorimetry, IR and UV absorption spectrophotometry, UV fluorescence spectrophotometry, atomic absorption spectroscopy (AAS), inductively coupled plasma spectroscopy (ICP), liquid or gas chromatography and mass spectrometry (3,4).

The method and the procedure to be used in any particular situation depend on the nature of the problem to be solved, and the degree to which the chemical composition of the oil must be determined will define the analytical method applied. In the case of oil spill identification the chosen method must be sensitive, fast and reliable for the analysis of oil in sea water.

In about ten years of our experience with the identification

Kuiper, J., Van den Brink, W.J. (eds), Fate and Effects of Oil in Marine Ecosystems. ISBN 90-247-3489-4

of oil spills we have chosen among all the mentioned methods, IR and UV techniques for their simplicity, fastness and accuracy.

EXPERIMENTAL

The changes in hydrocarbon composition of three samples: crude oil (d_{15^oC}=0.870 g/cm^3), light gas oil (d_{15^oC}=0.832 g/cm^3) and heavy gas oil (d_{15^oC}=0.921 g/cm^3) during weathering processes were studied. Samples were spread and then taken during 15 sunny days (July, August) from the surface of the Adriatic sea by the coast near Zadar. The 2 mm of oil slick were spread on the depth of about 1 m (Fig.1). Average temperature of the sea was 22°C.

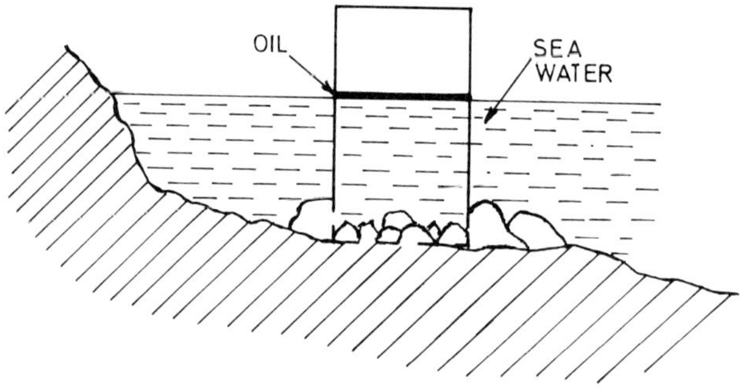

Fig.1.WEATHERING EXPERIMENT

1. Ratio of Infrared Absorbances - RIA

The samples were collected from the surface and analyzed on a Perkin-Elmer IR spectrophotometer, model 257.
The maximal absorbances were measured at the wavelength of 720, 810, 1375 and 1600 cm^{-1}. The following RIA were calculated: 720/1375, 810/1735, 810/720, 1600/1375, 1600/720 and 810 cm^{-1}/1600 cm^{-1} (3), proposed by Kawahara (5,6) and modified by Mattson (7,8).

2. Ratio of Ultraviolet Absorbances - RUA

UV-absorbances of the mentioned samples were measured on a Unicam SP 8000 Ultraviolet Recording Spectrophotometer using n-hexane (Carlo Erba) as a solvent. The UV absorbances were measured in the region between 200 and 400 nm. The peaks at 228 and 256 nm were observed and measured and their ratio of UV absorbances were calculated - RUA 228/256. The method was proposed and applied by Levy (9).

RESULTS AND DISCUSSION

RIA. The maximal absorbances of IR spectra at the 720,810, 1375 and 1600 cm^{-1} which correspond to the hydrocarbon structure in paraffinic, naphtenic, methyl and aromatic structure were measured for all samples. Table 1 shows their ratios. RIA shows the best correlation at 810/720 cm^{-1} - RIA_1 and 810/1375 cm^{-1} - RIA_2. During weathering (W) of crude oil, light and heavy gas oil a regular slow decrease occurs (Fig. 2 and 3).

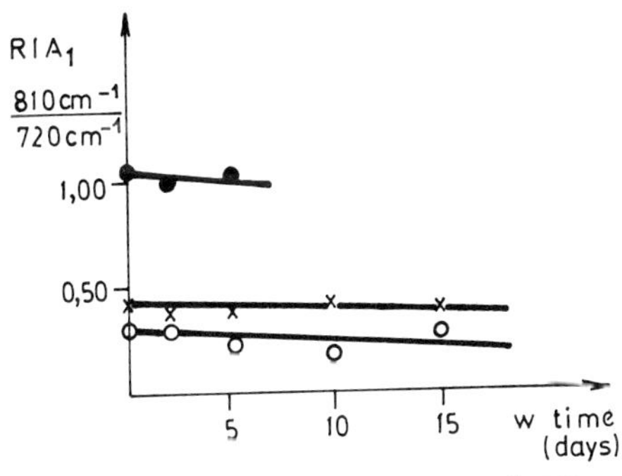

Fig. 2. CORRELATION OF RIA 810cm^{-1}/720cm^{-1} AND TIME OF THE EXPOSURE (W)

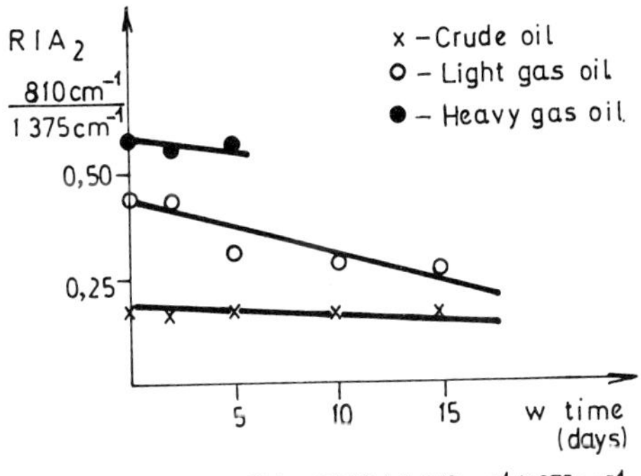

Fig.3. CORRELATION OF RIA 810cm^{-1}/1375cm^{-1} AND TIME OF THE EXPOSURE (W)

This correlation proved to be linear with negative coefficient "a":

$$RIA = -aW+b$$

Table 2 shows the relevant equations, i.e. the values of coefficients a and b for the two ratios and for each sample and

TABLE 1 - Ratios of IR absorbances - RIA

Sample	Time (days) W	$\frac{720cm^{-1}}{1375cm^{-1}}$	RIA$_2$ $\frac{810cm^{-1}}{1375cm^{-1}}$	RIA$_1$ $\frac{810cm^{-1}}{720cm^{-1}}$	$\frac{1600cm^{-1}}{1375cm^{-1}}$	$\frac{1600cm^{-1}}{720cm^{-1}}$	$\frac{810cm^{-1}}{1600cm^{-1}}$
Crude oil	0	0.462	0.179	0.387	0.328	0.709	0.545
	2	0.571	0.174	0.375	0.515	0.906	0.413
	5	0.476	0.174	0.366	0.396	0.833	0.440
	10	0.716	0.146	0.368	0.716	1.000	0.368
	15	0.546	0.171	0.344	0.421	0.771	0.407
Light gas oil	0	1.518	0.444	0.292	0.685	0.451	0.648
	2	1.350	0.421	0.311	0.807	0.597	0.521
	5	1.323	0.308	0.233	0.632	0.477	0.488
	10	1.513	0.305	0.201	0.402	0.266	0.758
	15	1.244	0.285	0.219	0.774	0.622	0.368
Heavy gas oil	0	0.482	0.571	1.185	0.553	1.148	1.032
	2	0.528	0.547	1.035	0.754	1.428	0.725
	5	0.512	0.560	1.095	0.585	1.142	0.958

Table 3 their standard deviations. It is evident that three samples can be identified inspite of time of the exposure.

This method represents modified method of the one proposed by Kawahara(6). Mattson (8) in his study proposes infrared fingerprinting of mineral oils as a promising method in the identification of oil spills. However, he points out that additional studies with natural aging must be carried out in order to determine the extent to which the method can be applied. All our experience in connection with natural aging have shown that fingerprint method is not reliable. On the contrary, the ratio of infrared absorbances is decreasing so slowly that the RIA_1 and RIA_2 can serve as a tool in order to identify the source of marine pollution by mineral oil, regardless of the age of oil spill. According to our previous experiments too (10), infrared fingerprinting alone cannot be applied if the oil slick is weathered.

TABLE 2 - Linear relationship of time (W) and RIA for two ratios

Sample	RIA = -aW + b	
	$810\ cm^{-1}/720\ cm^{-1}$	$810\ cm^{-1}/1375\ cm^{-1}$
Crude oil	$RIA_1=-0.7.10^{-2}W+0.38$	$RIA_2=-1.2.10^{-2}W+0.18$
Light gas oil	$RIA_1=-2.4.10^{-2}W+0.30$	$RIA_2=-4.3.10^{-2}W+0.44$
Heavy gas oil	$RIA_1=-4.5.10^{-2}W+1.15$	$RIA_2=-0.6.10^{-2}W+0.56$

TABLE 3 - Standard deviation for two key ratios: 1-$810\ cm^{-1}/720\ cm^{-1}$ 2-$810\ cm^{-1}/1375\ cm^{-1}$

Sample	Ratio	S( ) cm^{-1}
Crude	1	$\pm$ 0.013
oil	2	$\pm$ 0.040
Light	1	$\pm$ 0.046
gas oil	2	$\pm$ 0.075
Heavy	1	$\pm$ 0.075
gas oil	2	$\pm$ 0.085

Other methods such as measurement of physical properties of oil (refractivity index, density and average molecular weight) and structural group analysis (SGA), concentration of carbon in aromatic, naphtenic and paraffinic structure (% C_A ,% C_N , % C_P) and the same structure analysis on the basis of IR spectra, showed the regular changing in the course of ageing. In comparison with RIA, these methods are not accurate enough and take more time.

The values of RIA_1 and RIA_2 (Table 1) are different enough from sample to sample making it possible to distinguish one oil from the other, regardless of the age of oil spill. The lines on Figs. 2 and 3 illustrating time (W) and RIA correlation do not cross in 15 days. The same results were obtained when different crude oils were weathered.

RUA. The UV absorbance at 228 and 256 nm were measured and their ratio was calculated. Table 4 shows the ratio of UV absorbances (RUA) for all samples during the time of weathering.

TABLE 4 - Ratio of UV absorbances - RUA

Time days W	RUA 228/256		
	Crude oil	Heavy gas oil	Light gas oil
0	1.707	1.000	2.818
2	1.533	1.000	2.888
5	1.538	1.007	2.724
10	1.465	0.986	2.428
15	1.465	0.962	1.922

The coefficients a' and b' as well as the standard deviation obtained by polynomial regression are shown in Table 5.

TABLE 5 Linear relationship of time (W) and RUA

Sample	RUA=-a'W+b'	Standard deviation S(σ) nm
Crude oil	RUA=-0.01W+1.63	± 0.052
Heavy gas oil	RUA=-1.98W+0.99	± 0.012
Light gas oil	RUA=-0.06W+2.92	± 0.092

The RUA 228/256 values decrease slowly, so that the difference among these samples is evident, even when the oil slick is spread on the surface of the sea during 15 days. Figs. 4, 5 and 6 illustrate the correlations shown in Table 5. Our experience in several cases confirm Levy's findings (9), that RUA 228/256 can serve as a tool to identify the source of oil pollution, regardless the ageing of spread oil.

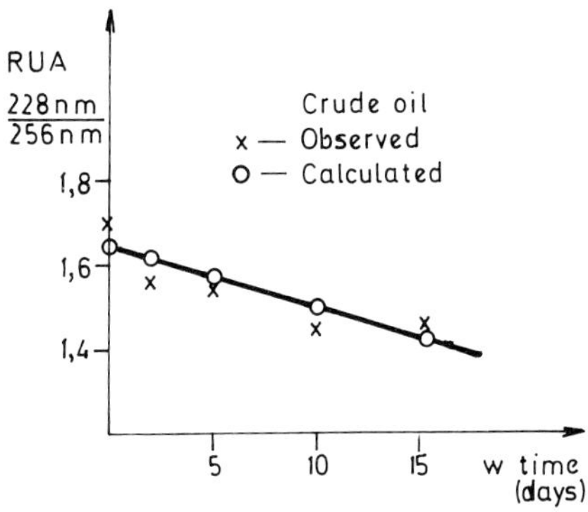

Fig. 4. CORRELATION OF RUA AND TIME OF THE EXPOSURE

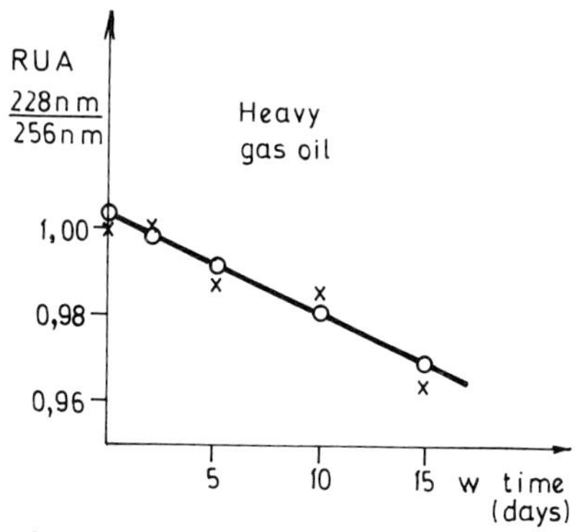

Fig. 5. CORRELATION OF RUA AND TIME OF THE EXPOSURE

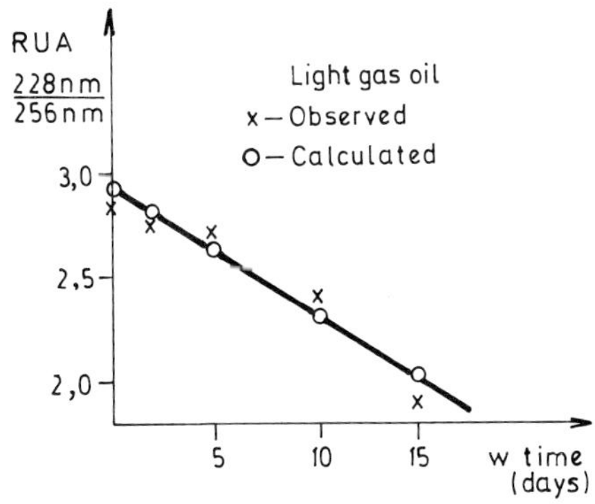

Fig. 6. CORRELATION OF RUA AND TIME OF THE EXPOSURE

The question to be mentioned in the case of UV measurement in general is, that this method, as well as the fluorescence spectroscopy, can be applied only in case when oil contain the aromatic and/or unsaturated hydrocarbons, which is not always obvious.

CONCLUSION

The characterization of oil spills on surface waters using the ratio of infrared absorbances (RIA) can be usefully applied for classifying petroleum pollutants in water, regardless of wheather the oil slick was weathered or not. The smallest influence of weathering on RIA was found for the key ratio 810 cm^{-1}/720 cm^{-1} and 810 cm^{-1}/1375 cm^{-1} i.e. the ratio of maximal absorbances which correspond to the hydrocarbons in naphtenic to paraffinic structure and naphtenic to methyl structure (RIA_1 and RIA_2). The correlation of RIA with the time of exposure (W) was found to be lincar with negative coefficient "a" for both ratios:

$$RIA = -aW+b$$

The ratio of ultraviolet absorbances measured at 228 and 256 nm (RUA 228/256) during weathering decrease slowly. The correlation between time and RUA is linear and has the same general form, i.e. RUA = -a'W+b'. The coefficient a' and b' are different enough for different samples that the identification of oil is possible, regardless how long the oil spill is spread on the surface of the sea.

LITERATURE

1. J.A.C.M. van Oudenhoven et al., Characteristic of Petroleum and its Behaviour at Sea, Concave Report, No 8, Den Haag 1983.
2. E.D.Goldberg, The Health of the Oceans. The Unesco Press, Paris 1976.
3. ANON., Research Activities Related to Oil Pollution Incidents, Cooperative Research Report No 107, International Council for the Exploration of the Sea Palaegrade, 2-4,DK-1261 Copenhagen K, Denmark, July 1981.
4. R.A.Stuart and R.D.Branch, Instrument News, The Perkin-

Elmer Co. 21(197o), No 2, 1-9.

5. F.K.Kawahara, Environ.Sci.Technol., 3(1969) 150-153.
6. F.K.Kawahara, D.G.Ballinger, Ind.Eng.Chem.Prod.Res.Develop., 1970.,9,553.
7. J.S.Mattson, H.B.Jr.Mark, R.L.Kolpack, C.E.Schutt, Anal. Chem.,42(1970)234-238.
8. J.S.Mattson, Anal.Chem. 43(1971)1872-1873.
9. E.M.Levy, Water Research,6(1972)57-69.
10. N.Adler,D.Božić, Nafta(Zagreb),30(1979)229-235.

Droplet Size Distribution of Naturally Dispersed Oil

Gerard A.L. Delvigne
Delft Hydraulics Laboratory
P.O. Box 177, 2600 MH Delft
The Netherlands

Abstract

The mechanical action of breaking waves and turbulence cause oil to break up into small droplets and diffuse in the water column. The droplet size is important in view of the dispersion stability, the interaction with marine life, and the uptake by sediment.
Laboratory measurements were performed on the droplet size distribution in various conditions. Turbulence was generated in the laboratory test facility by an oscillating grid, at the same energy levels as turbulence in the surf zone, coastal zone and deep-sea zone.
The droplet size distributions observed were strongly dependent on the turbulence energy level ($d \sim E^{-0.50}$) and on the duration of the turbulent state. Other distinct parameters were the oil type, weathering state and temperature, all being reflected in the single parameter of viscosity ($d \sim \nu^{0.34}$). The droplet size was independent of the salinity of the water, and the oil concentration. The droplet size distribution was similar for droplets generated from submerged oil lumps and those generated from a surface layer.

Introduction

The mechanical action of breaking waves and turbulence in the water flow causes spilt oil to break up into small droplets and the droplets to diffuse in the water column. The droplet size distribution of the dispersed oil is an important parameter because it affects:
- the stability of the dispersion (or oil-in-water emulsion) and consequently, the concentration of oil in the water column,
- the uptake of oil by marine life, e.g., the ingestion of oil droplets by zooplankton,
- the absorption of oil by suspended particulate matter and the subsequent sedimentation of oil on the sea floor. The persistence of sedimented oil is generally relatively long.

The present paper deals with laboratory measurements on the droplet size distribution of dispersed oil in a turbulent ambient. The turbulence level is related to the field turbulence in the surf zone, coastal zone and deep-water zone. The droplet size distributions were determined as a function of the following parameters:
- turbulence energy level and turbulence duration
- oil type, weathering state, and oil concentration
- temperature and salinity of the water.

Kuiper, J., Van den Brink, W.J. (eds), Fate and Effects of Oil in Marine Ecosystems. ISBN 90-247-3489-4

This study on the droplet size distribution of dispersed oil forms part of a study on the interaction of oil with suspended particulate matter in ocean conditions.

Dispersion of Oil: State-of-the-art

Spontaneous dispersion, or spontaneous emulsification, and mechanical dispersion can be distinguished as two different processes.
Spontaneous dispersion takes place under the influence of surface-active compounds, either within the oil itself (sometimes with increased concentration by the addition of chemical dispersants) or as a result of dissolved organics in the water phase. Spontaneous dispersion forms very small oil droplets (micelles), of the order of 1 µm, resulting in a very stable dispersion. The droplets of mechanically dispersed oil are, in contrast, at least one to several orders of magnitude larger than the micelles. The stability of this type of dispersion depends on the turbulence level of the ambient water.

The mechanical break-up of oil is dependent on the viscosity and surface tension of the oil. Both parameters are influenced by weathering processes, especially evaporation of light oil components and emulsification. The dispersion stability of oil depends on the terminal velocity W of the oil droplets and consequently on the density difference, $\Delta\rho$, between the droplets and the ambient water. $\Delta\rho$ decreases with evaporation and emulsification.

The sea state is an important parameter which influences the formation of oil droplets from a coherent surface slick and the initial penetration of the droplets in the water below the slick. These dispersion phenomena can be caused by the kinetic energy of breaking waves. The sea state also influences the macroscale vertical transport of oil droplets in the water, caused by turbulent diffusion, or by advection due to general sea currents or local inhomogeneous currents and circulation.
Aravamudan et al. (1982) compiled the results of theoretical studies by Raj (1977) and Milgram et al. (1978) producing a mathematical model for oil droplet formation and droplet distribution in an ambient flow with breaking waves. The model calculates a largest size, d_{max}, of the droplets formed, and a critical small droplet size d_{crit} below which break-up will not occur. d_{max} and d_{crit} depend on the wave frequency ω, the thickness of the oil layer h, the oil viscosity ν and the oil surface tension σ. The initial intrusion depth of a droplet, caused by breaking waves, is of the order of 1 metre; the final depth due to turbulence is calculated in the model from the turbulent growth of a breaking wave wake.
Johansen (1982) and Mackay et al. (1982) calculated the vertical spread of oil droplets beneath an oil slick with the help of seastate-dependent diffusion processes in the turbulence diffusion equation (ignoring advection by vertical currents) given by:

$$\frac{\partial C}{\partial t} = W \frac{\partial C}{\partial t} + \frac{\partial}{\partial z} (\varepsilon_z \frac{\partial C}{\partial z}) \quad (1)$$

with C = oil concentration
W = terminal velocity of oil droplet in calm water
ε_z = vertical diffusion coefficient
Equation (1) indicates that the vertical diffusion and dispersion stability strongly depend on W (and consequently on the droplet size) and ε_z,

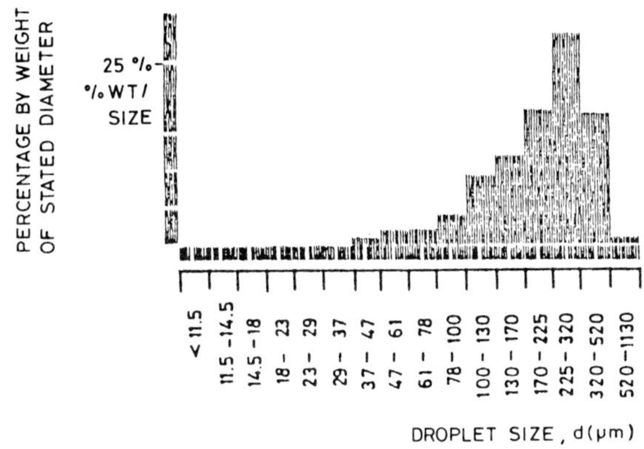

Figure 1. Droplet size distribution for oil dispersed by a breaking wave. Laboratory experiment. (Delvigne, 1984)

while ε_z, the vertical diffusion coefficient, in turn depends on all turbulence-generating mechanisms including currents, waves, wind and density gradients.

Field data on dispersed oil droplet size distributions are very scarce. Stable oil dispersions with a mean droplet size around 20 to 40 μm have been found under vigorous wave and current action (Forrester, 1971; Berridge et al., 1968). Delvigne (1984, 1985) measured the droplet size distribution of oil dispersed by breaking waves in a laboratory flume. A typical distribution is given in Figure 1. Milgram et al.'s calculation method for the critical droplet size in laboratory conditions results in $d_{crit} \approx 50$ μm.
Field data on oil concentration in the water range from 500 ppm in the near-shore zone in the case of a large oil spill, to 0.01 ppm for test spills.

The dispersion of spilt oil and processes dealing with dispersed oil (uptake by marine life, interaction with suspended particulate matter) are influenced by many ocean parameters, e.g. temperature, salinity, and turbulence conditions due to wind, currents and waves. Temperature and salinity conditions are relatively easy to determine. Turbulence conditions can be characterized by the energy dissipation rate E and the vertical diffusion coefficient ε_z as the vertical spread of droplets. Delvigne et al. (1986) surveyed the derivations of E and ε_z in the following ocean zones:

- surf zone, with breaking waves as the dominating mechanism,
- coastal zone, where turbulence due to spilling ("whitecapping") and bottom shear are equally important, and
- deep-water zone, where turbulence due to whitecapping is dominant.

Laboratory facility

Measurements on dispersed oil droplet size distribution in turbulent ambient water have been performed in the laboratory facility shown in Figure 2. The height of the column was 4 m, the internal diameter 0.3 m. The os-

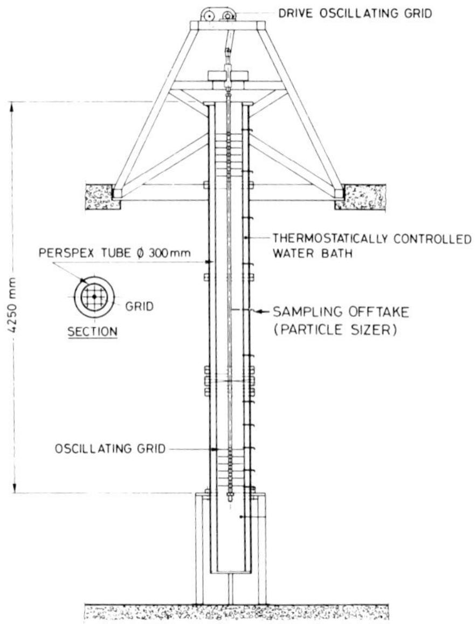

Figure 2. Laboratory facility.

cillating grid generated homogeneous turbulence. It is well-known that the turbulence spectrum in laboratory conditions is very different from field conditions as far as the large-eddy part of the spectrum is concerned, but the small-scale turbulence structures are very similar if the energy dissipation rates are equal. The small-scale turbulence is especially responsible for the breaking up of oil parcels into small droplets.

The energy dissipation rate E in the column could be varied between 0 J/m^3s (stagnant grid) and 3500 J/m^3s. Typical field values for E in different situations are:

- deep-sea $E = 10^{-4}$ to 10^{-2} J/m^3s,
- estuary $E = 0.1$ to 1.0 J/m^3s,
- surface layer $E = 1.0$ to 10 J/m^3s,
- breaking wave $E = 10^3$ to 10^4 J/m^3s,

Large rising oil droplets ($d \geq 5$ mm) were introduced into the column near the bottom. The grid was switched on when the droplets were distributed more or less homogeneously in the column. In a few cases, the oil was introduced as a thin surface layer to check for differences in droplet size distribution with surface or subsurface spills.

The droplet size distribution was measured with a laser-beam particle sizer with continuous flow-through water sampling.

Measurements

Oil droplet size distributions were measured as a function of parameters with the following values:

- oil type:
  - . PB_0 - Prudhoe Bay Crude Oil, fresh
  - . PB_3, PB_{10} - Prudhoe Bay Crude, weathered by evaporation over a period of 3 and 10 days
  - . Eko - Ekofisk Crude Oil, fresh

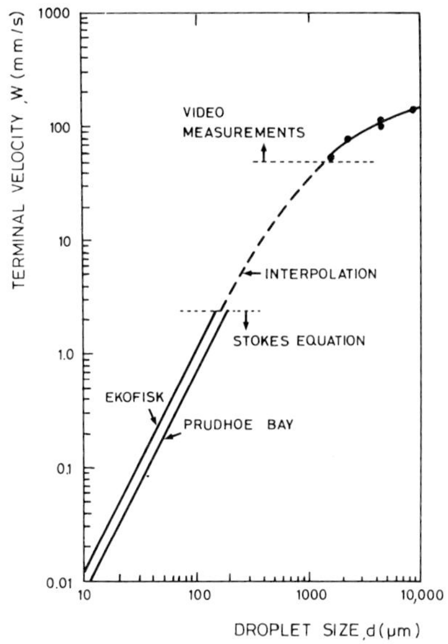

Figure 3 Terminal velocity W versus droplet size d.

- grid frequency, indicated by the energy dissipation rate E (0 - 3500 J/m^3s)
- duration of grid oscillation:
  - . t_E = 5 s (simulation of a single breaking wave)
  - . t_E = 600 s (steady state)
- temperature T = 20, 12 and 4 °C
- water salinity S = 30 $^o/oo$ (seawater), 10 $^o/oo$ (brackish) and 0 $^o/oo$ (fresh water)
- oil concentration in the column C = 20, 50 and 150 ppm
- oil introduction: surface and subsurface

The tests carried out are summarized in Table 1.

The oil viscosity ν is an important parameter for the break-up of oil droplets. The oil viscosities of the oil types tested are given in Table 2.

Droplet sizes with d < 1100 μm were measured with the laser-beam particle sizer. Larger droplet sizes were determined from video measurements by observing the terminal velocity W of the droplets. The W-d relationship, see Figure 3, was derived from the Stokes equation for small droplets, video observations of W and d for very big droplets, and an interpolation for intermediate sizes.

Results

Typical distributions indicated by the particle sizer are given in Figures 4a and 4b. The figures show the time-dependent droplet size distributions for Ekofisk oil, with E = 3500 J/m^3s, T = 20 °C, S = 30 $^o/oo$, C = 50 ppm, subsurface spill. Figure 4a is for an oscillation duration t_E = 5 s and Figure 4b for t_e = 10 min. Figure 4b indicates that a steady distribution is reached after about 5 minutes, because the distributions at 6, 8 and 10 minutes after the start of the grid oscillation are quite similar. The slight shift to smaller droplet sizes with time, shown in Figure 4a (t_E =

Table 1. Summary of tests

Legend — Turbulence duration, t_E: ● t_E = 600 s; ◐ t_E = 20 s; ○ t_E = 5 s; · t_E = 0 s

oil type	weathering state; days of evaporation	temperature, °C	salinity, $^0/_{00}$	concentration, ppm	1. subsurface spill 2. surface spill	Energy dissipation rate E (J/m^3s) 3500	1480	440	185	55	7	.85	0
Eko-fisk	Eko_0	20	30	50	1	● ○	● ○	● ○	● ○	● ○	◐	◐	·
	Eko_0	12	30	50	1	●		●				◐	·
	Eko_0	20	30	50	2		●	●					
Prud-hoe Bay	PB_0	20	30	50	1	● ○		● ○		● ○	◐	◐	·
	PB_0	12	30	50	1	●		●			◐		
	PB_0	4	30	50	1	● ○	● ○	● ○	● ○		◐	◐	
	PB_0	20	30	20	1	●		● ○					
	PB_0	20	30	150	1	● ○		● ○					
	PB_0	20	10	50	1	●		●					
	PB_0	20	0	50	1	●		●				◐	
	PB_3	20	30	50	1	● ○		● ○				◐	
	PB_{10}	20	30	50	1	● ○		● ○				◐	

Table 2. Oil viscosity

Oil type	Oil viscosity ν (cSt) at temperatures of 20°C	12°C	4°C
Eko	8.0	10.4	12.3
PB_0	92	124	233
PB_3	122		
PB_{10}	≈ 220 (non-Newtonian)		

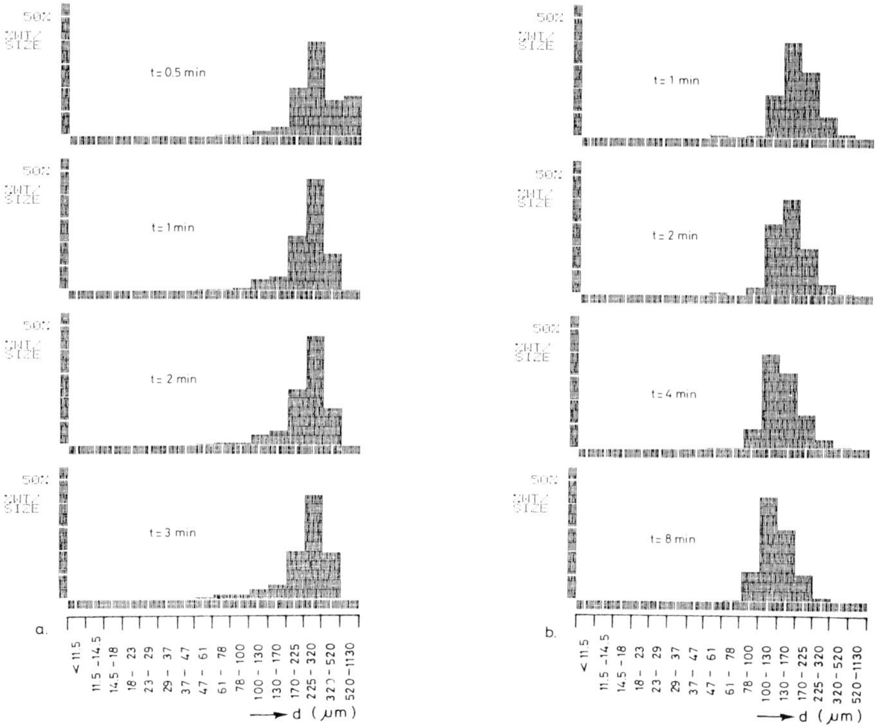

Figure 4. Droplet size experiments with different turbulence duration, t_e = 5 s (a), and t_E = 600 s (b). (PB_0, E = 3500 J/m^3 s, t = time after start of turbulence by grid (oscillation).

5 s), is due to the surfacing of relative large droplets.

The measurements consisted of the droplet size distributions in size ranges as indicated in Figure 1. Most of the distributions are compiled in Figures 6 and 7 by the smallest droplet size observed, d_{min}, the biggest size d_{max}, and the mean size d_{av}.

The following results were obtained from the measurements:

Salinity S. Measurements with S = 30 ^{o}/oo (seawater), S = 10 ^{o}/oo (brackish water) and S = ^{o}/oo (fresh water) where performed. Examples are given in Figure 5a. The droplet size distribution appeared to be independent of the salinity.

Concentration C. Measurements with oil concentrations C = 20, 50 and 150 ppm show similar droplet distributions. Consequently, the effect of the coagulation of oil droplets is not important in this concentration range. Measurement examples are given in Figure 5b. (The narrow distribution for C = 20 ppm is apparently due to the low concentration).

Turbulence level. Measurements on the effects of the turbulence level have

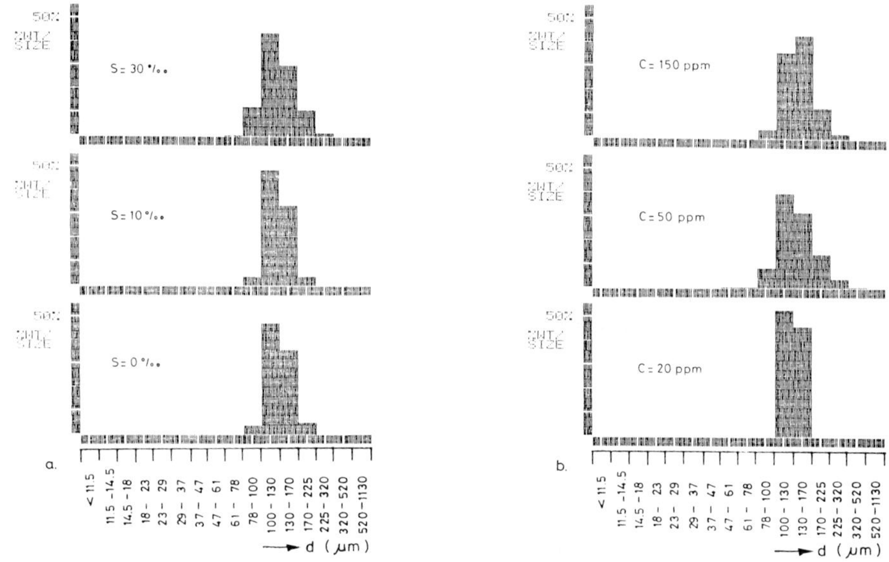

Figure 5. Droplet size experiments with different salinities (a) and oil concentration (b). (PB_0, E = 3500 J/m^{3}s, t_E = 600 s).

been plotted in Figure 6 (Ekofisk Crude) and Figure 7 (Prudhoe Bay Crude); d_{max}, d_{min}, and d_{av} are shown.

The figures show the difference in size distribution for $t_E = 5$ s (equivalent to a single breaking wave), and the equilibrium at $t_E > 300$ s.

The droplet sizes of the low-viscous Ekofisk oil are considerably smaller than the sizes found with Prudhoe Bay oil.

The figures suggest a linear relationship between log d (with d = d_{max}, d_{av} and d_{min}) and log E, in the high turbulent energy (E) range. This is shown in the Figures 6 and 7 by straight lines. All lines appear to have the same slope. This relationship cannot be extended apparently to the low turbulent energy range, probably because the break-up of big oil droplets is not only due to turbulence, but also to the shear which develops on oil droplets rising in the ambient water. Clearly, this effect is most significant for low-turbulent water and big oil droplets. The break-up of droplets is due only to the rising velocity in the experiments in which E = 0.

Temperature. The temperature affects oil viscosity and consequently the break-up of droplets. The effect of temperature on the measured droplet size distribution was very clear; Figure 8 shows the effect on the mean droplet size d_{av} of PB_0, for different energy levels, E.

Weathering state. Figure 9 shows the d_{av} - E relationship for PB_0, PB_3 and PB_{10}, indicating weathering of Prudhoe Bay oil by evaporation over a period of 0, 3 and 10 days. Weathering increases the viscosity, see Table 2, and leads to bigger oil droplets.

Surface and subsurface oil. The bulk of the measurements were performed

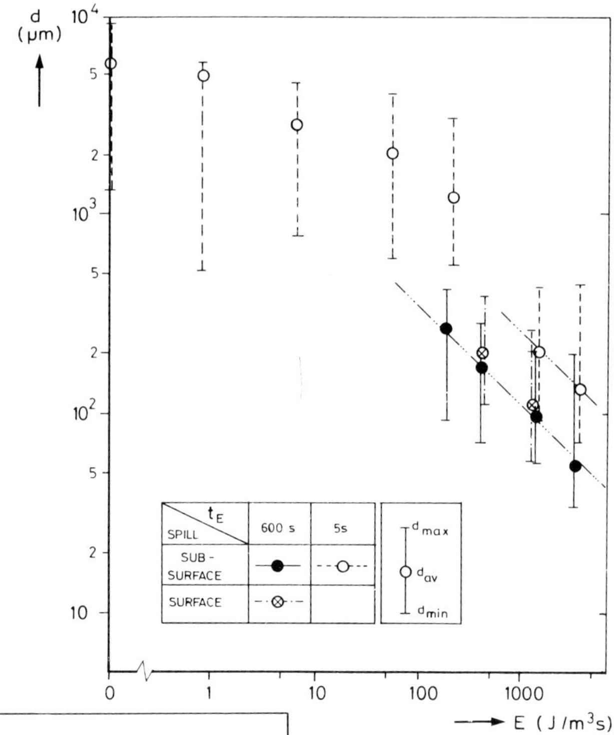

Figure 6. Droplet size distribution versus turbulent energy dissipation rate for Ekofisk Crude.

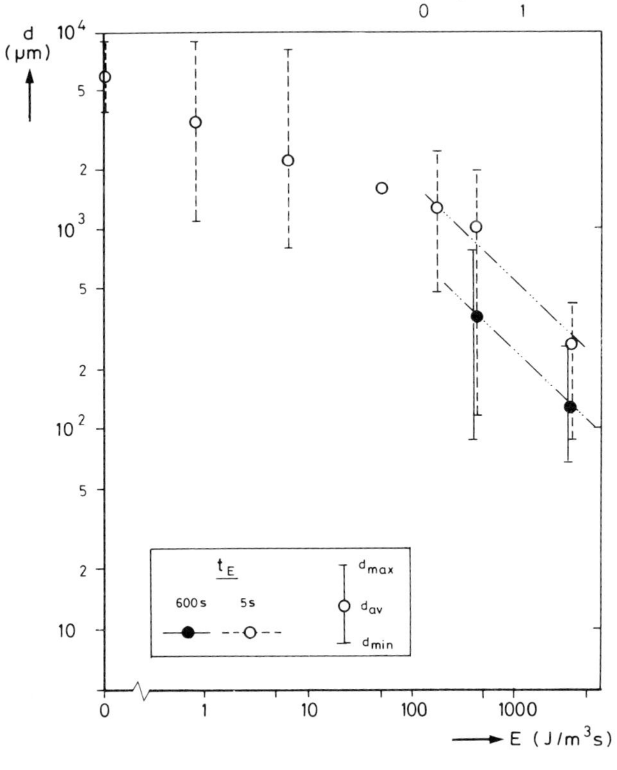

Figure 7. Droplet size distribution versus turbulent energy dissipation rate for Prudhoe Bay Crude (PB_0).

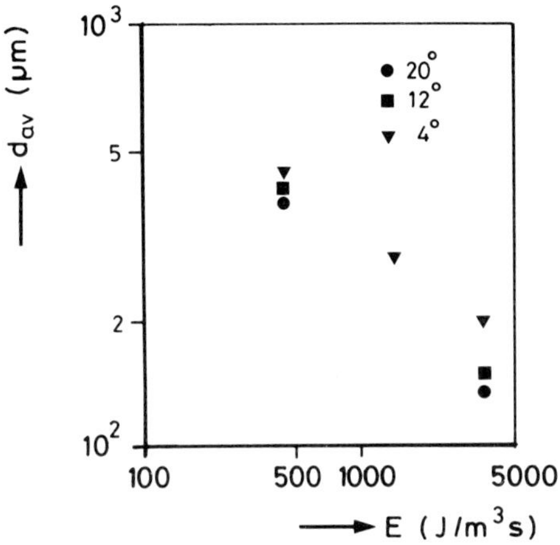

Figure 8. Mean droplet size versus turbulent energy dissipation rate for different temperatures. (PB_0)

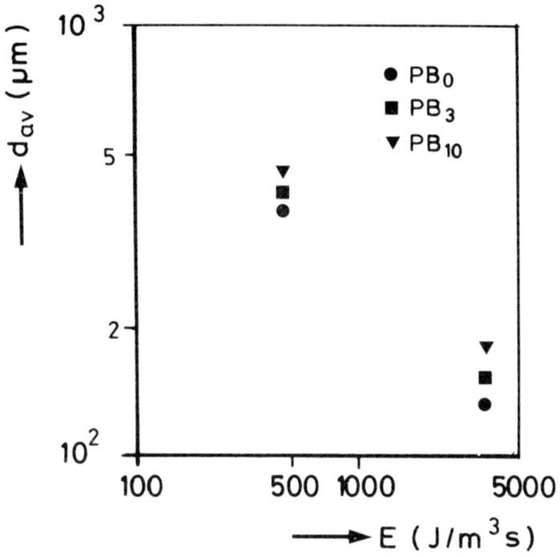

Figure 9. Mean droplet size versus turbulent energy dissipation rate for different oil weathering states.

with oil parcels introduced into the column near the bottom. Two experiments were carried out with thin oil layers (thickness 50 - 100 μm) on the water surface. Figures 6 shows almost identical droplet size-energy relationships for subsurface and surface oil. (Note that the oil droplet sizes are bigger than the thickness of the surface layer in contradiction to the Aravamudan et al's (1982) assumption on the maximum droplet size).

The results of these experiments indicate that the droplet sizes are independent of oil concentration, water salinity and surface or subsurface oil introduction. The droplet size distributions depend on turbulent energy, temperature, oil type and oil weathering state, the latter three parameters strongly influencing the oil viscosity. Figure 10 relates droplet size d, turbulent energy dissipation rate E, and oil viscosity ν. Obviously, the viscosity ν is a significant parameter for the d-E relationship.

Conclusions

The following conclusions were drawn from the laboratory experiments on the effect of turbulence on the droplet size distribution of dispersed oil:

1. The droplet sizes increase with increasing oil viscosity ν, see Figure 10. In high-turbulent water:

$$d = c_1 \nu^{0.34} \tag{2}$$

   where d_{min}, d_{max} and d_{av} can be substituted in Equation (2). The proportionality constant c_1 depends on turbulence intensity (E) and turbulence duration t_E.
2. The droplet sizes are independent of oil concentration, water salinity

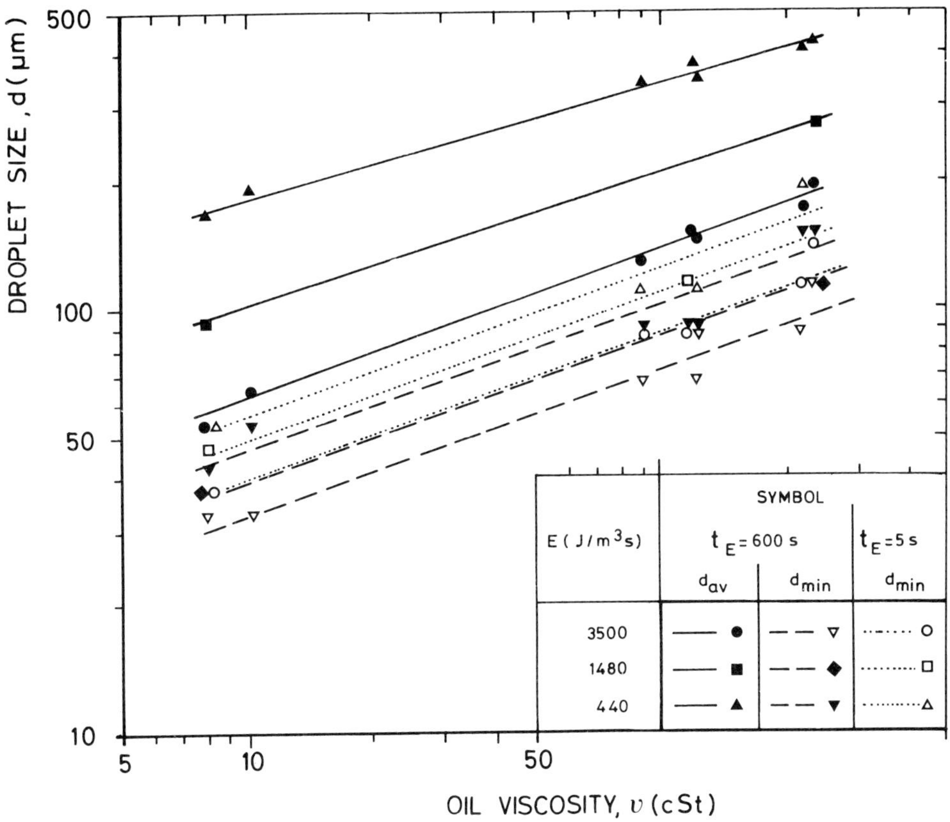

Figure 10. Droplet size distribution versus oil viscosity, for different values of turbulent energy dissipation rate E and turbulence duration t_E.

and whether the oil is introduced in the water from the water surface, or as submerged oil parcels. The droplets depend on oil type, weathering state and temperature only as far as the viscosity is concerned (Equation 2).

3. The droplet sizes decrease with increasing turbulence, see Figures 6 and 7. In high-turbulent water:

$$d_{av} = c_2 E^{-0.50} \qquad (3)$$

with E = turbulent energy dissipation rate. The proportionality constant c_2 depends on oil type, viscosity, ν, and turbulence duration, t_E.

4. The droplet sizes in low-turbulent water are smaller relative to Equation (3), because of the additional break-up of big droplets due to terminal velocity.

5. Droplet size equilibrium is attained at about t_E = 5 min. Consequently, the droplet size distribution obtained in breaking waves depends on a time parameter.

Note: After completion of this paper, experiments were performed with a different application of the particle sizer, and with microscopic observation of oil droplets. The experiments show considerable smaller oil droplets, down to 1 μm, than indicated by d_{min} in the preceding sections. The concentration of these small droplets decreases sharply with decreasing diameter. Then d_{min} in this paper indicates the smallest droplet size with a detectable concentration by the used measuring technique relative to the high concentration of bigger droplets.

Acknowledgments The study was commissioned by the U.S. Department of the Interior, Minerals Management Service, Anchorage, Alaska, as part of a research program on the transport of spilt oil and the interaction of oil and suspended particulate matter. (OCS Study MMS 86-0029).

References

1. Aravamudan, K., Ray, P., Ostland, J., Newman, E. and Tucker, W. (1982). Break-up of oil on rough seas - simplified models and step-by-step calculations. Report No. CG-D-28-82, U.S. Coast Guard, Washington, D.C., U.S.A.

2. Berridge, S.A., Dean, R.A., Fallows, R.G. and Fish, A. (1968). The properties of persistent oils at sea. J. Institute of Petroleum 54, p. 300.

3. Delvigne, G.A.L. (1984). Laboratory facility for the simulation of the natural and chemical dispersion of oil. Report No. M1933-2, Delft Hydraulics Laboratory, Delft, The Netherlands.

4. Delvigne, G.A.L. (1985). Experiments on natural and chemical dispersion of oil in laboratory and field circumstances. Proceedings of the 1985 Oil Spill Conference, American Petroleum Institute, Washington, D.C., U.S.A.

5. Delvigne, G.A.L., Roelvink, J.A. and Sweeney, C.E. (1986). Research on vertical turbulent dispersion of oil droplets and oiled particles-literature review. Report No. Z75-1, Delft Hydraulics Laboratory, Delft, The Netherlands.

6. Forrester, F.D. (1971). Distribution of suspended oil particles following the grounding of the tanker Arrow. J. Marine Res. 29, pp. 141-170.

7. Johansen, O. and Audunson, T. (1982). Drift of submerged oil at sea. Report No. P 319/1, Continental Shelf Institute, Trondheim, Norway.

8. Mackay, D., Shiu, W.J., Hossain, K., Stiver, W., McCurdy, D., Paterson, S. and Tebeau, P.A. (1982). Development and calibration of an oil spill behavior model. Report No. CG-D-27-83, U.S. Coast Guard, Washington, D.C., U.S.A.

9. Raj, P.K. (1977). Theoretical study to determine the sea state limit for the survival of oil slicks on the ocean. Report No DOT-CG-D-90-77, U.S. Coast Guard, Washington, D.C., U.S.A.

HYDROCARBON BIODEGRADATION IN MARINE SEDIMENTS : A BIOCHEMICAL APPROACH.

TH. HÖPNER, H. HARDER, K. KIESEWETTER AND B. TEGELKAMP.

FACHBEREICH BIOLOGIE DER UNIVERSITÄT, POSTFACH 2503,
D-2900 OLDENBURG.

1. INTRODUCTION

In an oil pollution sensitivity index (1 = low, 10 = high sensitivity) of coast types tidal flats have been ranked in position 9 (2). It is expected that the high-water shoreline will be heavily polluted after a near-coast tanker accident, while large tidal flats will be moderately, but ecologically critically, affected (1). Oil will penetrate into deeper sediment layers, thus enhancing toxicity against benthic organisms and favouring transition into anoxic zones, where biodegradation proceeds slowly, if at all (6). While mechanic oil removement is possible on sandy areas, it is mostly excluded at muddy flats which are frequent at the German coast. In these cases biodegradation is a main mechanism of restoration. To include biodegradation into an oil pollution management it is necessary to know biodegradation rates under the present conditions and support of biodegradation is desirable. This report deals with hydrocarbon biodegradation in tidal sediments and draws conclusions on practical measures of oil biodegradation support. Results have been obtained from in-situ experiments at tidal flats (4), from laboratory experiments using fresh sediments (5), as well as from laboratory experiments using a standardized and inoculated sediment mixture.

During these studies we made the experience that the silty sediment is a reaction medium of a high biocatalytic activity. The relatively high aerobic degradation rates which we attained in our laboratory studies encouraged us to start also experiments in anaerobic degradation. We present first results and compare them with aerobic ones.

2. IN SITU EXPERIMENTS

Tidal flat areas (12 m^2, mud, Jadebusen) were polluted with 300 g Statfjord oil by spreading out an oil sand mixture. We started the experiments in August 1983 and May 1984 and followed oil biodegradation and elimination, respectively, over several months with single measurements up to summer 1986. We drew cores, eluted lyophilized subsamples with tetrachlorocarbon and recorded the IR-extinction in the range of the hydrocarbon peak. The marked difference between the shapes of the curves (Fig. 1 and 2) was mainly a consequence of the erosion-conditions which were stronger after August 83 than after May 84, while in 1984 sedimentation prevailed. Addition of magnesium-ammonium-phosphate as an example of nutrient support had no effect (Fig. 1). Later we examined this experience in labo-

Kuiper, J., Van den Brink, W.J. (eds), Fate and Effects of Oil in Marine Ecosystems. ISBN 90-247-3489-4

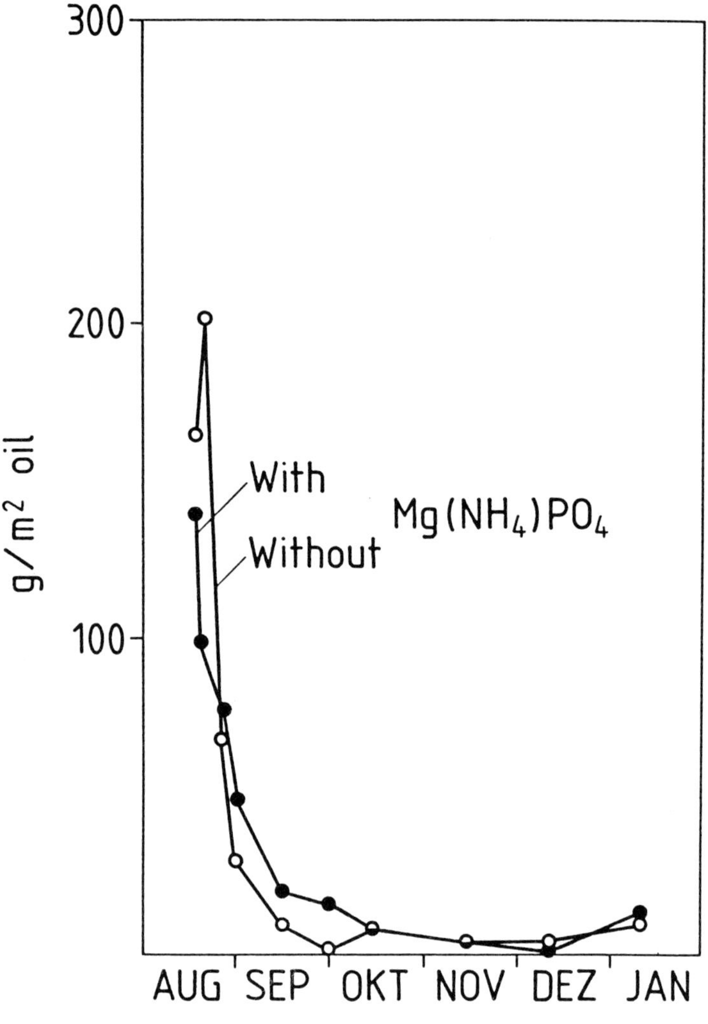

Fig. 1. Decrease of oil concentration (in g/m^2) after pollution of a mudflat with 300 g/m^2. o—o oil alone, ●—● oil plus 5 moles/m^2 $Mg/NH_4)PO_4$ as a constituent of the sand-oil-mixture. IR-assay. First sample was drawn one day (two tides) after application of oil.

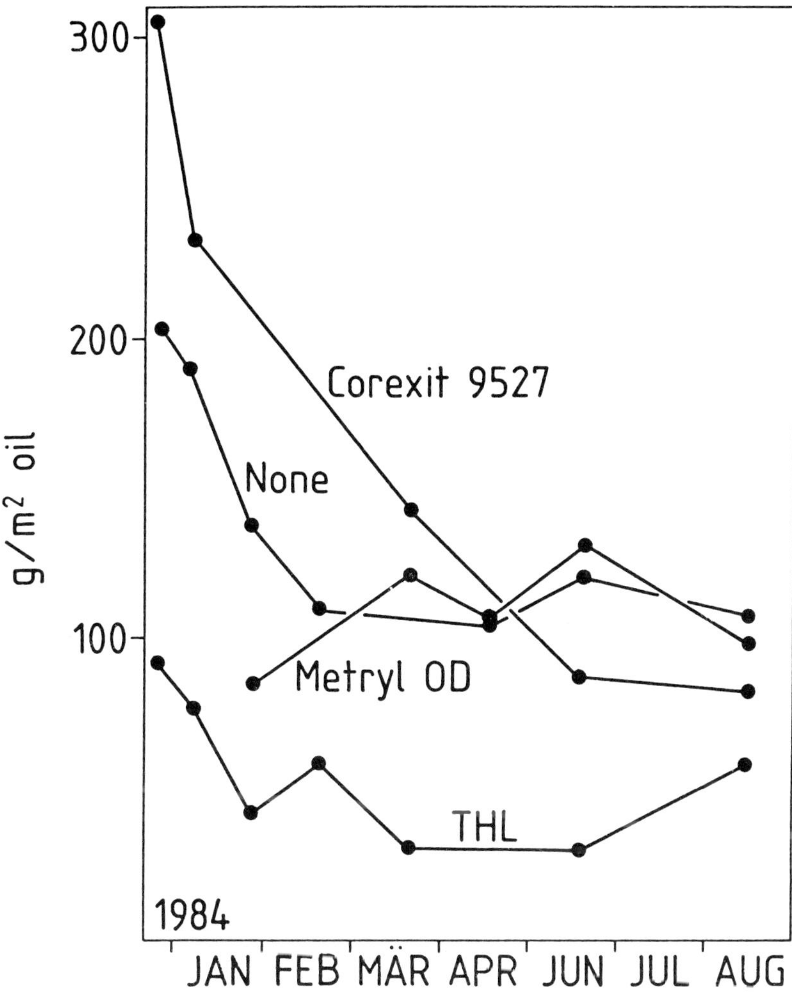

Fig. 2. As Fig. 1, but no nutrients. Addition of dispergators as indicated. For dispergator-concentrations see text. First sample as in Fig. 1.

ratory experiments where the nutrient demand could be clearly proven. Of the dispergators added Metryl OD (30 g per m^2) had no effect, Corexit 9527 (same) inhibited , anionic Trehalose lipide (THL, 3 g per m^2) enhanced degradation/elimination (Fig. 2). In both experiments the oil containing layer had been buried by some centimeters of mud. The upper border of the oil became identical with the reduction horizon indicating that oxygen had been exhausted and that further oxygen access proceeded only from above. Settling by benthic organisms was examined and recorded. The oil layer became perforated within the next months, but the oil content of the remaining anaerobic oil-"islands" has almost been constant till today. The oil is still fluid even after three years, it smells like the raw oil, and it seems by no means "aged".

We concluded that the sedimentation-erosion-conditions at tidal flats are by far the most important influences on the fate of an oil spill. If eroding conditions prevail, oil will be eliminated rapidly, and combating activities are dispensable as long as only the state of the affected tidal flats is concerned. In the case of need these conditions can be examined easily and quickly. This is the first, most important (and simple) measure after an oil spill of tidal flats.

Problems come in if affected tidal flats are of low sediment mobilitiy or moreover, if sedimentation prevails. In the latter case the oil will be buried and excluded from oxygen access. The field studies showed that an anaerobic oil degradation did not contribute to the observed decrease and that degradation was exclusively dependent on oxygen access. Under field conditions an influence of nutrients was not visible. We assumed that the remineralization rate attained the nutrient consumption rate as long as the degradation was oxygen limited. We decided to examine these conclusions in laboratory experiments.

3. LABORATORY EXPERIMENTS WITH FRESH SEDIMENTS

The aim of the experiments was to examine aerobic biodegradation rates as a first step to follow rates under various conditions of oxygen supply. For these examinations fresh sediments from the above treated areas were subjected with some seawater so that they could be shaken. The standard upset contained 55 g sediment (dry weight) and 30 ml water in a 250 ml Erlenmeyer flask. In addition to Statfjord oil, hexadecane was used as a model hydrocarbon. If nutrients (nitrate, orthophosphate) were added, the recommendations of GIBBS (3) (60 mg nitrate-N and 6 mg phosphate-P per gram hydrocarbon) were followed. Incubation time was mostly 10, 20, 30 and 40 days at 16° C. Oil content was followed as above by tetrachlorocarbon extraction and IR assay, which means, that the total degradation to carbon dioxide (or other small products) was followed.

There was no significant biodegradation in a sandy sediment even after nutrient addition (Fig. 3). However, in muddy sediments degradation of hexadecane (Fig. 4) and oil (Fig. 5) started after a lag phase of 10 to 20 days and attained a high rate in the presence of nutrients, whereas the rate was low without nutrients. After sterilization the degradation rate was zero (Fig. 6 and 7). An influence of dispergators on hexadecane

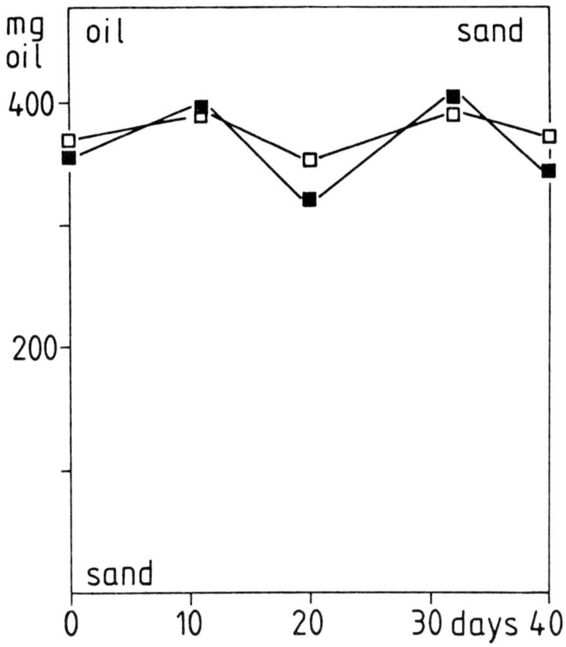

Fig. 3. Behaviour of Statfjord oil in an aerobic shaking-culture of fresh sandy sediment. □—□ without nutrient addition. ■—■ with nutrients. For composition of the cultures see text.

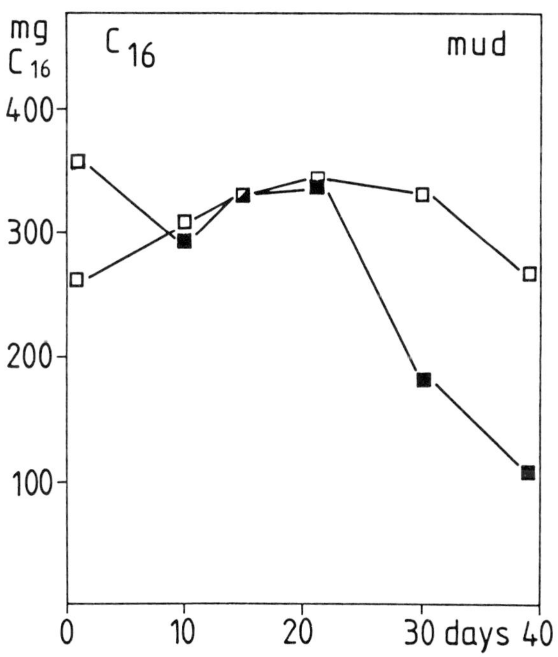

Fig. 4. Hexadecane degradation in a fresh muddy sediment. □—□ without nutrients. ■—■ with nutrients.

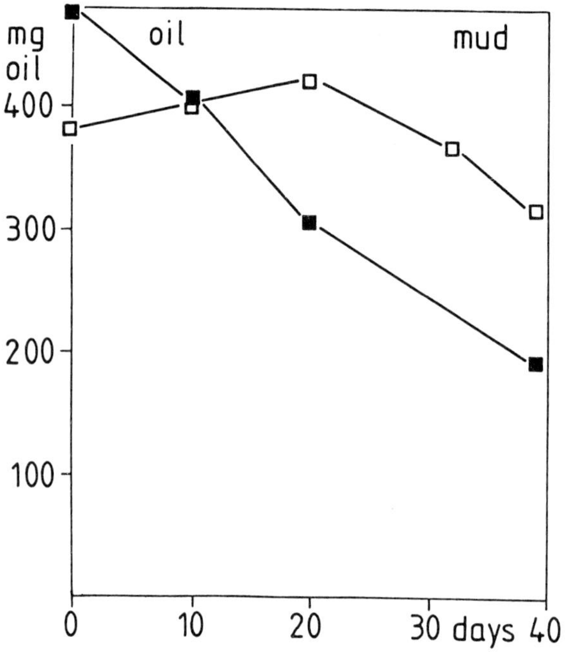

Fig. 5. Statfjord oil degradation in fresh mud. Symbols as in Fig. 4.

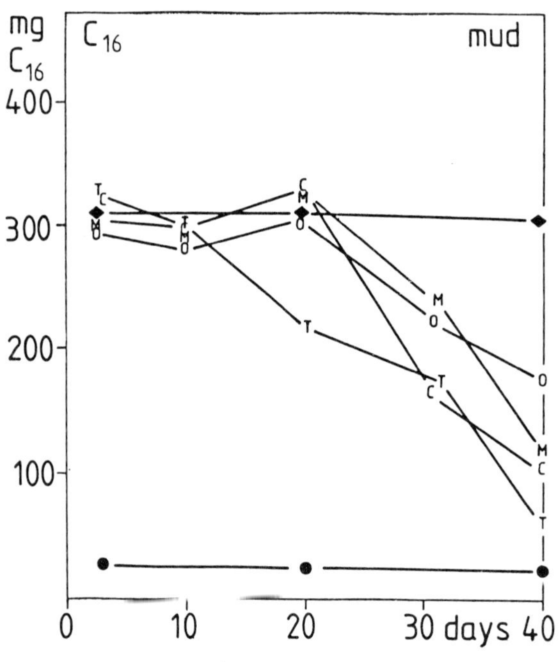

Fig. 6. Hexadecane degradation in fresh mud in the presence of nutrients. ◆—◆ sterilized. C-C-C with Corexit 9527. M-M-M with Metryl Od. T-T-T with THL. O-O-O without dispergator. ●—● no hexadecane, no dispergator. For concentrations see text.

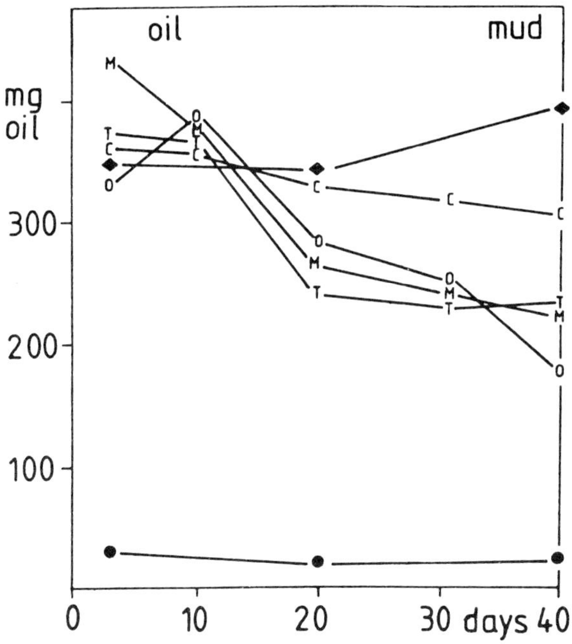

Fig. 7. Statfjord oil degradation as in Fig. 6. Same symbols as in Fig. 6.

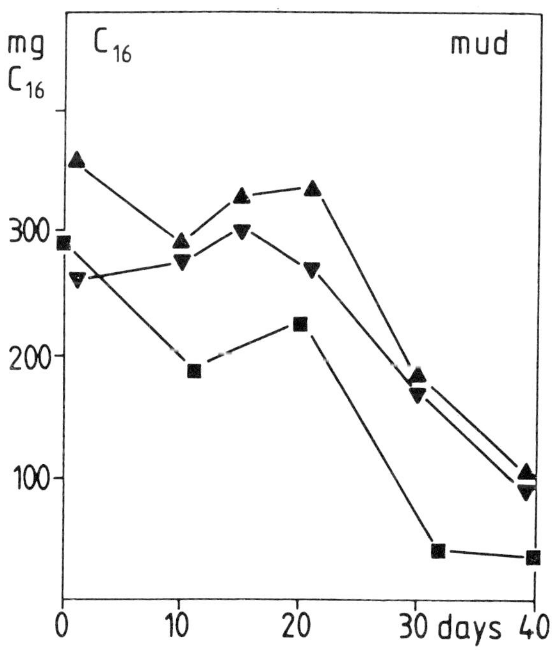

Fig. 8. Hexadecane biodegradation in the presence of nutrients in fresh mud and in sterilized mud after addition of bacteria. ■—■ original fresh mud. ▼—▼ mud sterilized, addition of 10^4 bacteria (total colony forming units). ▲—▲ same, but 10^6 bacteria.

degradation was not visible, however, THL seemed to abolish the lag phase (Fig. 6), and inhibition by Corexit 9527 could be observed in the case of oil (Fig. 7). It was not possible to enhance a degradation rate by adding an adapted inoculum (Fig. 8), there was only a tendency to shorten the lag phase.

We concluded that the adsorption of the hydrocarbons (and perhaps of the bacteria) to the large surfaces of the silt was the deciding reason for the high rates, because there was no degradation in sand. The silt used had a specific surface of 2 - 3 m^2 per gram as measured by the BET-method. The effect of nutrients was only visible, if more than 100 mg hydrocarbon were added. Up to this amount the endogenous nutrient content of the mud was sufficient. In the fresh sediment samples a competent population of bacteria seemed always to be present.

The biodegradation rates obtained are remarkable. In sea water and under conditions reflecting open sea states GIBBS (3) found 30 g per m^2 and year as a guideline, a value which has been found to be likely after various outdoor observations. Brought to the same dimensions, about 70 kg per year were degraded in one m^3 sediment, which obviously is not an appropriate comparision (because the degradation rate per surface unit is deciding), but throws light on the high biodegrading capacity of an aerobic fine grained sediment.

In the progress of the experiments the varying properties of the sediment samples and the different sizes of the lag phases became tedious. So we changed the experimental basis and turned to a standard system with constant and reproducible properties.

4. LABORATORY EXPERIMENTS WITH A STANDARDIZED AND INOCULATED SEDIMENT MIXTURE

A large quantity of muddy sediment from the site of the above mentioned samples was dried, sieved, homogenized and stored. For each shaking culture 55 g of this material were used. Seawater and distilled water, respectively, were added to attain the previous conditions. To prepare an adapted inoculum, biologically active surface-sediment samples were drawn from an indoor mudflat model (2 m^2) which we have been keeping with its full zoobenthic and phytobenthic populations in a glass house for 16 months. These samples were preincubated with hydrocarbons under the conditions of the degradation experiments and subsamples were used to inoculate the incubation upsets.

With this system the biodegradation rates were confirmed (Fig. 9). The lag phase phenomenon was abolished. A dependence on the hydrocarbon concentration was clearly visible. The nutrient dependence was also confirmed. Oxygen supply was found to be crucial as can be seen in Fig. 10, where the highest rate was obtained under continuous oxygen saturation by vigorous shaking (oxygen concentration not yet controlled), while a low rate was obtained in a closed flask filled with nitrogen. The intermediate result belongs to a closed shaken flask, in which only the present ca. 200 ml air were available. Nutrients had been added as given above.

It is well known that in fresh natural sediments a hydrocarbon degradation is possible. The bacterial populations exist because natural or anthropogenic hydrocarbons are always pre-

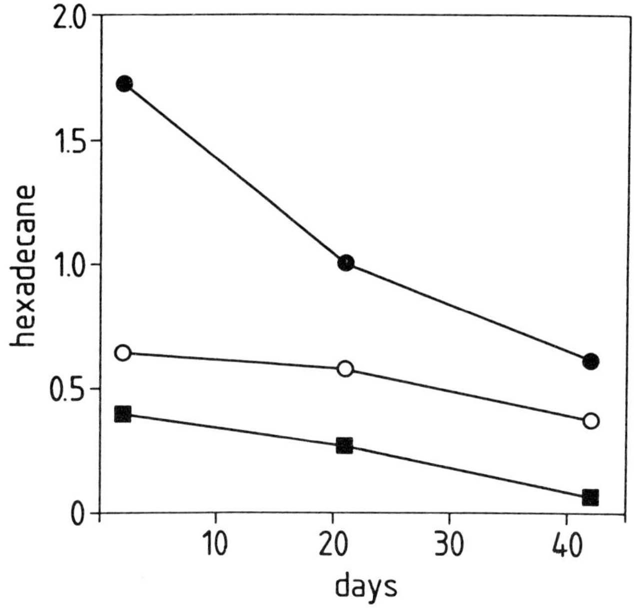

Fig. 9. Hexadecane degradation in the standardized and inoculated sediment mixture in the presence of nutrients. Starting amount of hexadecane: ■—■ 386 mg. ○—○ 773 mg. ●—● 1546 mg.

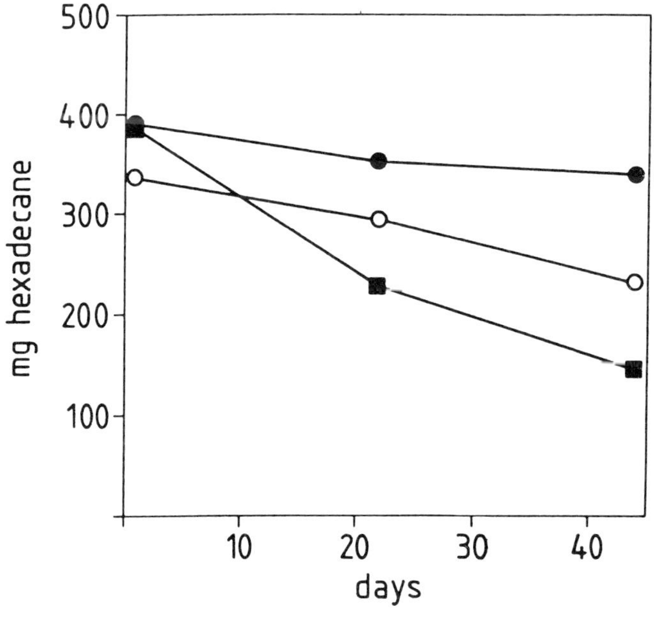

Fig. 10. Aerobic (■—■), anaerobic (●—●) and oxygen limited (○—○) degradation of 386 mg hexadecane in the upset as Fig. 9. For specifications see text.

sent at least in traces. As we could show, the specific surface of the sediment influences the degradation rate. Initially we were surprised that it was so easy to reproduce these results in a standard sediment system. Our laboratory upset in fact is a small fermenter with the special feature, that we do not work in a homogeneous liquid system, but in a biphasic system in which the water insoluble substrate is not in the state of emulsified droplets, but in the state of being adsorbed to a solid surface. The ratio between bacterial biomass and degradation rate is in the range of the ratio in an optimized biotechnological fermenter system, as preliminary calculations showed.

The high rates encouraged us to follow biodegradation by oxygen consumption in a Warburg Apparatus. The standard upset was reduced to 20 g sediment. Incubation at 16° C was done at a shaker in a climate room, and the Warburg flask was connected to the manometer at different times after the start of incubation. Fig. 11 shows a three-hours-observation on the 5th day after the start of incubation. The nutrient influence was visible. The upsets without hexadecane showed a slow oxygen consumption, because the inoculum contained a small amount of hexadecane. If the hexadecane degradation rate was computed according to the combustion equation, the rates stated above were confirmed. The time-saving manometric technique will be preferentially used in the future.

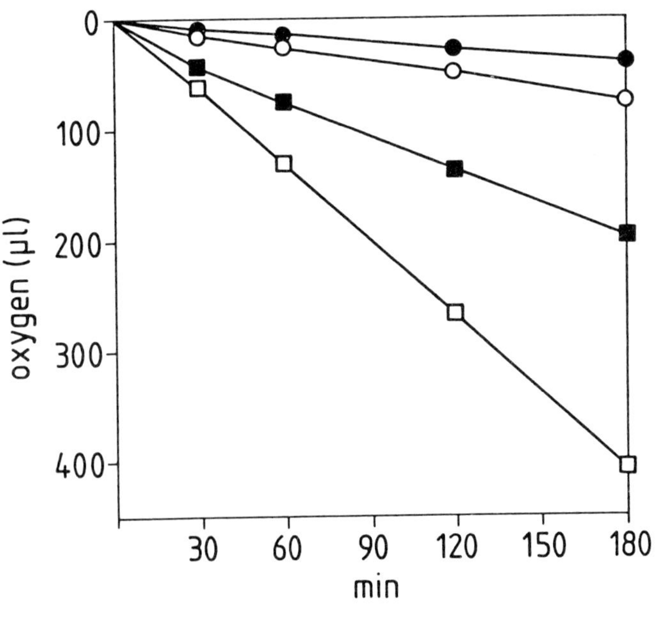

Fig. 11. Manometric assay of hexadecane degradation in a Warburg apparatus. □-□ complete, 20 g sediment, 10 ml seawater, 773 mg hexadecane, 100 mg nitrate-N, 10 mg phosphate-P, inoculum. ○—○ same, but no hexadecane. ●—● without hexadecane and nutrients. ■—■ with hexadecane, without nutrients. For further specifications see text.

5. FIRST ATTEMPTS TO STUDY ANAEROBIC DEGRADATION

It is commonly accepted that the first step of a biodegradation of an aliphatic hydrocarbon is the introduction of a first oxygen atom into the substrate. This reaction converts the hydrocarbon molecule into an alcohol. The subsequent two reactions convert the alcohol into an aldehyde and a fatty acid. The first step is catalyzed by an oxygen dependent mixed function hydroxylase. No reaction is known in which the oxygen is replaced by an electron acceptor (resp. oxygen donor) besides molecular oxygen itself. The subsequent reactions can be catalyzed by an alcohol dehydrogenase and an aldehyde dehydrogenase using electron acceptors like NAD^+ or $NADP^+$. In the case of aromatic hydrocarbons the first step is catalyzed by an oxygen dependent mono- or dioxygenase. After a reduction reaction which converts the mono- or dioxygen adduct into a mono- or diol, the subsequent steps can be independent of oxygen, too. This means that according to the present knowledge an absolutely oxygen independent degradation cannot be expected. Final electron acceptors of the subsequent reactions can be sulfate (which always is present in a marine system) or nitrate (which is unlikely under anaerobic conditions).

The result of this consideration is that in the absence of oxygen a biodegradation is not possible, but that a combined oxygen and nitrate (sulfate) dependent reaction needs much less oxygen. It has to be evaluated experimentally whether a nitrate (sulfate) dependent degradation can be achieved in the presence of a constant low oxygen concentration or whether oxygen has to be admitted discontinuously in appropriate intervals and amounts. We started experiments to study anaerobic reaction rates and to find practicable conditions. In these experiments only hexadecane has been used. Up to now we have no experiences with aromatic compounds. Some first results are presented here.

As Fig. 10 showed, degradation in the "absence" of oxygen is low or zero. We have to stress the fact that this means also, that there was no or only low degradation by a sulfate-dependent reaction, because sulfate was always present as a constituent of the seawater (ca. 23 mmol l^{-1}). We did not have any success to accelerate degradation by addition of nitrate in the absence of oxygen as well as in the presence of limiting oxygen. From a biochemical or cellular standpoint it is doubtful whether this accleration could be expected. Oxygen limitation can mean, that the appropriate amount of the present hydrocarbons is converted to the alcohol, which provides the substrate for a subsequent nitrate dependent degradation. But it can also mean, that a small part of the hydrocarbon is fully degraded. In the case of hexadecane we are presently convinced that the latter consideration is true, because the preliminary gas chromatographic assays showed no free intermediates of the degradation. The first alternative has to be based on the assumption that products of the oxygenating reactions (alcohols, aldehydes, fatty acids) leave the bacterial cells and become accessible to a nitrate dependent degradation. This is unlikely at least in the case of aliphatic hydrocarbons. But it cannot be excluded in the case of aromatic hydrocarbons. Our present experiences are restricted to aliphatics.

Before drawing final conclusions we wanted to prove that denitrification reactions on principle worked in our standard laboratory system. Our first substrate was glucose. We found a stoichiometric nitrate decay with a rate of 1 mmol per day in 100 g dry sediment. So we were shure that denitrification took place if a suitable substrate was used.

To elucidate the hydrocarbon dependent denitrification step by step we jumped across the questionable first oxygenation reaction and used its products as substrates, which means in the special case of hecadecane degradation, that we examined cetyl alcohol and palmitic acid. We used the stoichiometric amount of nitrate according to the denitrification reaction, and in a second upset only one sixth of the stiochiometric amount. The results (Figs. 12 and 13) were according to the expectations in the case of the fatty acid. The substrate was fully degraded if the stoichiometric amount of nitrate was present, while the reaction stopped after consumption of the limiting nitrate amount. The reaction rate was about 1.1 mmol nitrate per day in 100 g dry sediment, this means, it was the same as with the substrate glucose. In contrast to this, the result was surprising in the case of the alcohol: there was no or only a low and delayed degradation.

In the moment we cannot offer a reasonable explanation why the classical fatty acid degradation works pretty well in our anaerobic system, while the alkohol and aldehyde dehydrogenation do not work or work very slowly. Both types of reactions are dependent on oxidized coenzymes. The biochemical approach to hydrocarbon biodegradation conditions seems promising and we shall follow it consequently.

6. CONCLUSIONS

The results show that aerobic muddy sediments are highly active in hydrocarbon biodegradation. As long as a rapid aerobic degradation proceeds, the nutrient support can be rate limiting. The rapid oxygen consumption connected to this activity has the consequence that oxygen very soon becomes rate limiting. Under this condition a rate limitation by nutrients is not visible.

At least the first step of hydrocarbon biodegradation, the initial hydroxylation, is obligatorily oxygen dependent. Even if nitrate or sulfate serve as electron acceptors in sediments, these reactions are restricted to the degradation of the first oxygen containing intermediates (alcohols, phenols, carbonic acids). A biodegradation absolutely independent of oxygen cannot be expected.

It is commonly known that in anoxic sediments a hydrocarbon degradation is extremely slow or absent. In the case of the unbranched aliphatic hydrocarbon hexadecane we showed, that the rate limiting step is the initial hydroxylation and/or the subsequent conversion of the first product to the fatty acid. Nitrate is a suitable electron acceptor for anaerobic fatty acid degradation. Its conversion to dinitrogen has been derived from the stiochiometry of the reaction, but the identification of the gaseous products is still in preparation. Fatty acid degradation using sulfate as electron acceptor cannot be exclu-

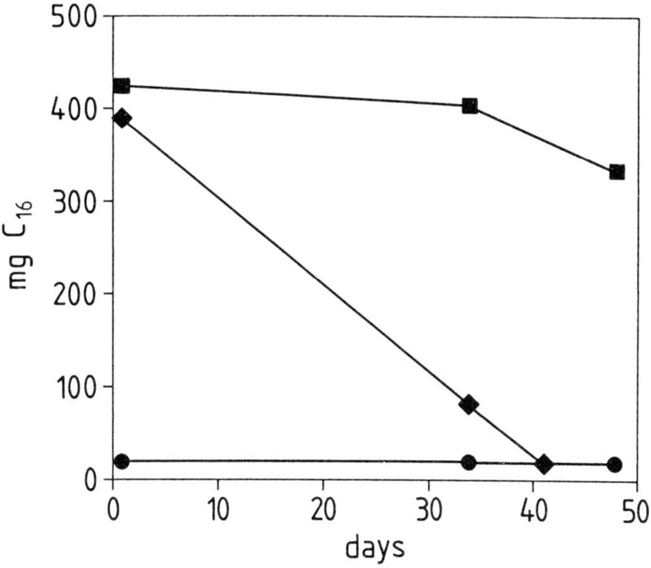

Fig. 12. Degradation of palmitic acid (◆—◆) and cetyl alcohol (■—■) in the presence of 340 mg nitrate-N and 10 mg phosphate-P. ●—● no substrate addition.

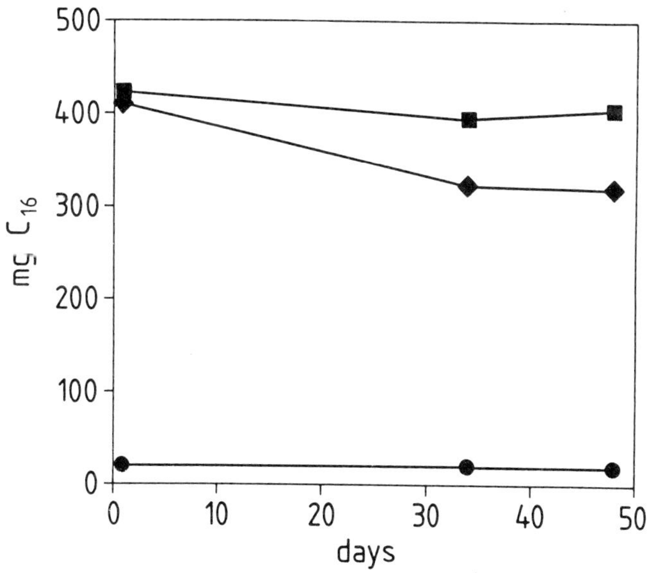

Fig. 13. Same as Fig. 12, but only 57 mg nitrate-N.

ded, but if at all, it proceeds very slow compared to the denitrification reaction.

These experiences are restricted to aliphatic hydrocarbons and it is necessary to examine the behaviour of various aromatic compounds and of raw oil and raw oil products. Nevertheless the anaerobic experiments confirm the common experiences of the generally low rates and give some insight into the reasons. For practical applications some conclusions can be drawn:

6.1. In muddy sediments the apparent hydrocarbon biodegradation rate is limited by oxygen. The first means to support biological degradation is to overcome the oxygen limitation. Mechanical treatment (e. g. mechanical or hydraulic dredging by means of a shallow-drought vessel) of oil containing sediments seems to be the first measure which should be examined.

6.2. Hydrocarbons in an adsorbed state are more easily degradable than in the state of a swimming layer, of droplets or of an emulsion. To support biodegradation an oil containing sediment should be intensively harrowed to achieve a thorough partition of the oil in the upper sediment layer. This suggestion is in some respects in contradiction to customary opinions. At the same time the treatment is a means to support the sediment with oxygen.

6.3. When the oxygen limitation has been removed, nutrient limitation is the next bottleneck. Support of biodegradation by nutrients is only effective in the absence of oxygen limitation. Under this condition the biodegradation rate can be enhanced manifold by addition of nutrients. As we could show in coinciding experiments, soluble nutrients artificially introduced into the sediment have a half life of several months in an undisturbed sediment. Since a sediment agitation should be repeated to maintain the oxygen support, a rapid loss of added nutrients can be expected, if easily soluble materials are used. Slowly available preparations are presumably more economical and effective.

6.4. Treatment with dispersants of oil after its arrival at tidal flats is not recommendable.

6.5. Biodegradation at or in tidal sediments cannot be supported by spreading out hydrocarbon degrading bacteria.

6.6. The open questions from the field experiments led us to laboratory examinations. It could be shown that laboratory experiments using a standard sediment system provide a hopeful access to the open questions. They demonstrated, that the sediment is a very active biocatalytical reaction medium for aerobic and anaerobic biodegradation of organic carbon. Degradation conditions can be worked out and optimized in the laboratory, and technical and chemical measures can be conceived to approach and maintain these conditions at the tidal flats. We assume that this kind of combining laboratory and field methods provides access to reasonable oil spill management procedures.

ACKNOWLEDGEMENTS

This work was supported by the German Federal Minister of Research and Technology (Grants 01 ZV 131 and 01 ZV 85070) and by the Commission of European Communities. The skillful assistance of B. Kürzel, C. Orliczek and I. Wolke is gratefully acknowledged.

REFERENCES

1. Breuel A: Oil Spill Cleanup and Protection Techniques for Shorelines and Marshlands. Noyes Data Corp. Park Ridge, New Jersey, 1981.
2. D'Ozouville L, Berné S, Gundlach ER et Hayes MO: Évolution de la Pollution du Littoral Breton par les Hydrocarbures de L'Amoco Cadiz entre Mars 1978 et Novembre 1979. In: Amoco Cadiz, Conséquences d'une Pollution Accidentelle par les Hydrocarbures. Paris, 1981. pp 55-78.
3. Gibbs CF: Quantitative Studies on marine biodegradation of oil. I. Nutrient limitation at 14°C. - Proc. Roy. Soc. London 188:61-82, 1975.
4. Höpner Th, van Bernem KH, Brüggemann W, Golombek P, Kant U, Kiesewetter K, Michaelsen M, Neugebohrn L, Ramm G, Suda R und Wonneberger K: Untersuchungen mit dem Ziel des Vorschlags biologischer Maßnahmen bei Ölverschmutzungen der Wattoberfläche und des Deichsvorlandes. Abschlußbericht des Vorhabens 01 ZV 131 an den Bundesminister für Forschung und Technologie. Oldenburg 1985.
5. Kant U, Kiesewetter K, Michaelsen M and Höpner Th: Estimation of Hydrocarbon Biodegradation Velocities in Tidal Sediments under Standard Conditions. Final Report to the Commission of European Communities. Oldenburg 1985.
6. Kuiper J, van het Groenewoud H, Hornsman G, de Wilde P, Berghius E, Kok A, Wolf W and Schoonheden H: A Study of Marine oil pollution in outdoor model ecosystems representing a tidal flat. Final Report. TNO, 2600 AE Delft, The Netherlands, 1982.

A SIMPLE REMOTE SENSING SYSTEM FOR THE DETERMINATION OF DISPERSANTS EFFECTIVENESS

R. H. Goodman and H. M. Brown
Esso Resources Canada Limited
Calgary, Canada

ABSTRACT

This paper describes the development of a simple remote sensing system for the detection of oil on water. Signals from sensors operating in the ultraviolet and infrared are combined in a computer-based image processing system to produce information on the area of thick and thin portions of the slick, from which the effectiveness of dispersants can be calculated.

The design parameters, the selection of sensors, and the integration of these sensor signals into a user friendly display is discussed. The use of this UV/IR system in field oil spill experiments and in an actual spill situation is described.

1. INTRODUCTION

The measurement of the effectiveness of a dispersant and even its definition (Brown et al, 1987) are still the subject of controversy after more than a decade of research on the subject. The concept of a dispersant is simple indeed; to so enhance the natural ocean dispersion processes that the oil will be dissipated into the water column rapidly and subsequently will be diluted by both horizontal and vertical diffusion.

The use of chemical dispersants in a spill response will increase the rate of oil removal from the surface, and the effectiveness of a dispersant describes the efficiency of this process. There are many natural processes, such as evaporation, natural dispersion, dissolution and biodegradation which also remove oil from the water surface. These processes compete with the chemical dispersant action. In order to determine the effect of the use of a dispersant, it is necessary to determine the amount of oil removed from the surface, and the time taken for that process to occur.

One method of measuring dispersant effectiveness is to measure the concentration of oil in the water column. Since absolute measurements of this type cannot be readily interpreted, most experimental designs have used data from a control and a treated slick in order to determine dispersant effectiveness. There are many problems, both theoretical and practical, associated with such measurements.

Kuiper, J., Van den Brink, W.J. (eds), Fate and Effects of Oil in Marine Ecosystems. ISBN 90-247-3489-4

It is not possible, at a reasonable cost, to obtain a synoptic measurement of the water column hydrocarbon concentration over the several square kilometers associated with most experimental slicks. Smaller slicks cannot be used since the spreading and edge effects would dominate and thus not be representative of an actual spill.

Experimental studies, (Brown et al, 1986) in large wave basins, have demonstrated that the shape of the dispersed oil cloud is not symmetrical. It has been shown that oil is initially moved into the water column as a highly concentrated dispersed plume, which moves as the result of waves and ocean currents and the coupling between them. This interaction is highly asymmetrical, since the current and wave fields are rarely in the same direction. A slower process, driven by the concentration gradient, creates diffusion plumes of lower oil concentration, but of a much more homogeneous nature than the initial plume structure. The final result consists of a number of small plumes of high oil concentration surrounded by a lower concentration diffusion generated plume.

Acoustic sounding measurements (Hay and Davidson, 1984) have shown that the oil occurs as a series of horizontal layers, with a vertical dimension of a few centimeters. The existence of these layers further complicates the sampling problem for water column measurements.

On the practical side of a field experiment, measurements must be made from small boats to reduce the influence of the sampling on the motion of the oil in the water column. In order for oil to be moved into the water column, wave energy is needed to create turbulence. The wave energy that is required, especially in the case of a heavier oil, may be quite large. The wave heights needed to produce this energy may be larger than those in which small boats can be operated safely. Hence, when ocean conditions would permit effective dispersant action, sampling may not be possible from small vessels.

An alternative to the measurement of hydrocarbon concentrations in the water column is the measurement of oil remaining on the surface. Such measurements can be made using aircraft equipped with remote sensing equipment. This method can cover a large area very quickly, and give a near synoptic view of the slick. However such a methodology is not without its problems. If the dispersant was 100% effective, then there would be no oil on the surface to detect. Thus the signal-to-noise ratio for remote sensing systems decreases with increasing dispersant effectiveness, which is not a desirable experimental situation.

There is no reliable remote sensing technique to measure oil thickness, and hence oil volumes cannot be determined. However, it is possible to estimate oil thickness, and to accurately measure slick areas, and thus determine dispersant effectiveness if a control slick is used as a calibration standard.

2. SYSTEM DESIGN

There are two components to the design of the system: the selection of a suitable sensing system and the integration of these sensors with a data analysis and display system.

2.1. Selection Sensors

Oil can be detected using any of its thermal, acoustic, mechanical, electrical or electromagnetic properties. For a remote sensing system, it is necessary that the atmosphere be transparent to the sampling radiation and that a good coupling can be achieved at the air-liquid interface. This effectively limits the system to the use of electromagnetic radiation, since acoustical energy cannot be readily coupled across the air-liquid boundary. There are limitations in the wavelengths that can be used in the electromagnetic spectrum due to the absorption in the atmosphere due to water, oxygen and nitrogen. The transmission characteristics of the atmosphere are shown in Figure 1.

Bands in the microwave, far and near infrared, visual and near ultraviolet are available to detect oil on water.

FIGURE I SIMPLE REMOTE SENSING SYSTEM

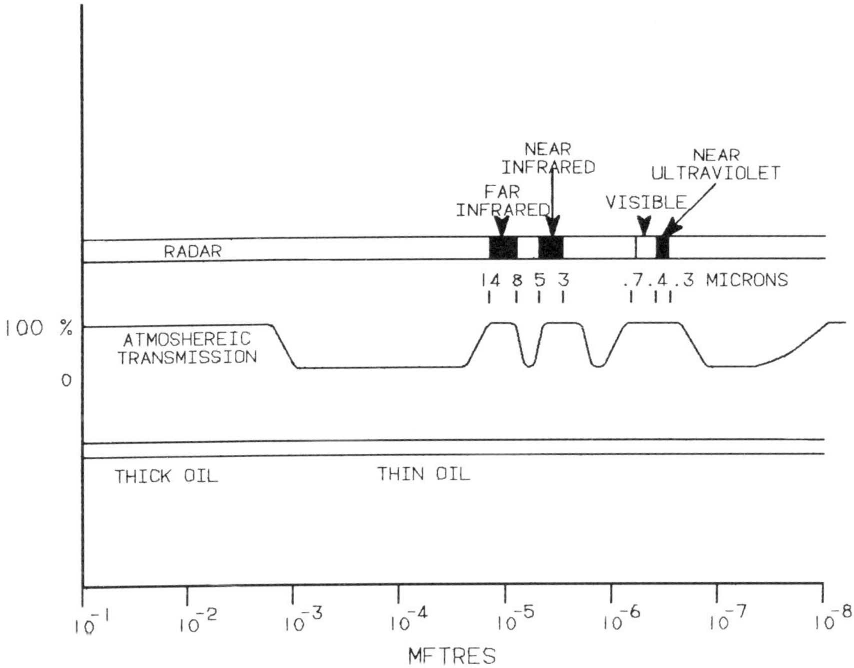

In the microwave region, it is possible to determine the surface roughness by measuring the scattering of electromagnetic radiation (Catoe and MacLean, 1979). Since the viscosity of oil is much greater than water, the high frequency wind generated waves are attenuated due to the presence of oil and thus the surface roughness spectrum has fewer high frequency components. Visually the surface seems smoother, and this can be detected by microwave measurements. If the detection system is to be able to measure small field and wave basin scale slicks, a minimum resolution of a few meters is required. This is not achievable for present day microwave systems. The all-weather capability of microwave systems and their ability to cover large areas is a great advantage for the routine surveillance and detection of oil spills at sea, however.

The far-infrared (8-14 microns) (O'Neil et al, 1979) band of the electromagnetic spectrum is the region of thermal emission, and is thus sensitive to temperature differences of the water surface. Since the oil is a better thermal insulator than water and absorbs rather than reflects sunlight, it will have a different temperature than the surrounding water. This can be detected in the far-infrared with a suitably sensitive detector. As the oil becomes thinner, it will reach thermal equilibrium with the water and thus not be detected in the infrared. A number of commercially available sensors are available with adequate spatial and temperature resolution to detect oil of thickness greater than about 50 microns. This type of sensor was chosen to detect the "thick" oil in a slick.

The use of a sensor operating in the visible region was considered, but no combination of frequencies was found to be sufficiently unique for oil to be useful. There are many other natural features such as seaweed that can be confused with oil in the visible region.

Hydrocarbons are known to reflect strongly in the ultraviolet, and since the atmosphere is transparent in the near ultraviolet, the 400 nm band (.4 μ) was chosen to detect the presence of hydrocarbon on the water surface. A layer as thin as 1 to 10 microns gives a strong signal, so this sensor is used to detect "thin" oil.

2.2. Integration of Units

The total sensing package consists of two sensors and their associated support equipment, an image analysis and recording system and a real-time display. In addition to input data from the sensor package, the image analysis system requires navigational and operating information in order to compute the characteristics of the slick in real-time.

To reduce the cost and improve the reliability of the total package, commercially available units were used where possible. There are three types of sensor outputs that could be used:

1. analogue output
2. digital output
3. video output.

Since the main requirements of the processing were the manipulation of images, it was decided to use a video signal processing system combined with a computer based analysis system. Such systems have been developed for video 'special effects' and could be readily modified for the requirements of the oil spill detection system. A block diagram of the system is shown in Figure 2.

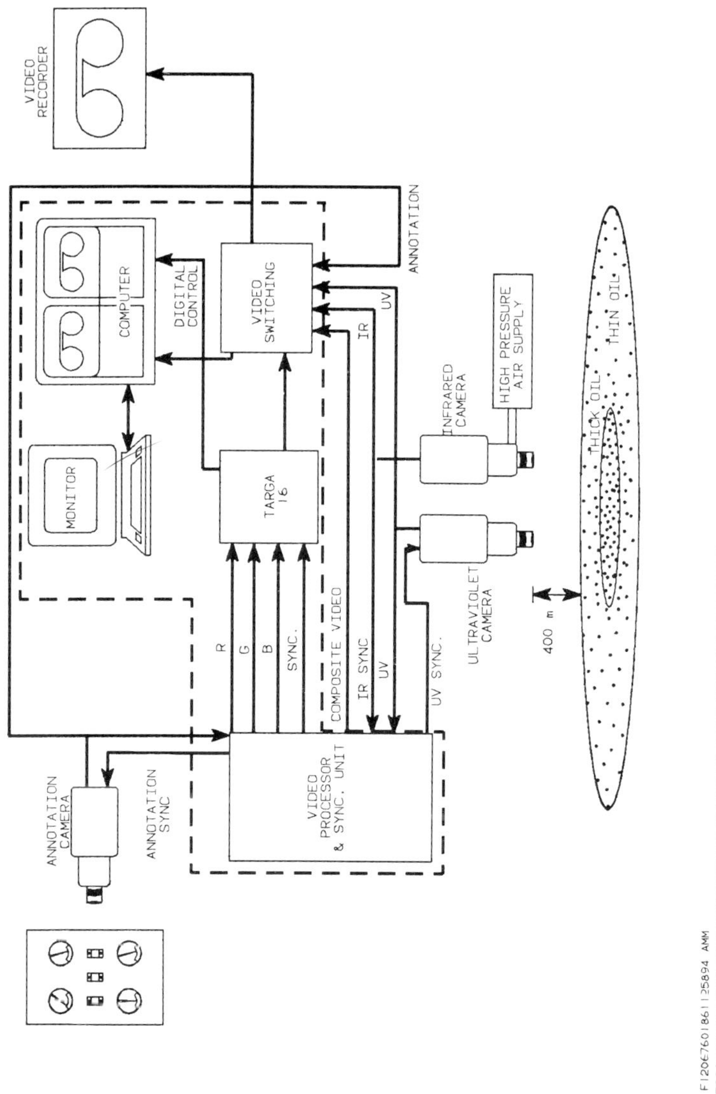

FIGURE 2 BLOCK DIAGRAM OF SIMPLE REMOTE SENSING SYSTEM

The infrared sensor chosen was a Barr and Stroud Model IR18 Imager, which uses a Mullard Sprite detector cooled by a high pressure air refrigeration system. This unit was selected since it could be supplied with a self-contained electrically powered air source, which would not limit the mission time of the IR detector.

The ultraviolet system uses a conventional low-light level video camera, with a filter which only passes the 400 mm band in the near ultraviolet. The low-light level camera is required since a filter which is thick enough to suppress the visual radiation will also attenuate the desired signal. This filter is combined with a polarizing filter which reduces the effects of sun glitter. The direction of polarization can be changed by the operator in order to optimize the image contrast.

The complete system was designed to be used with an "aircraft of opportunity". The aircraft navigation data should be readily integrated into the data collection system. The method chosen was to provide for an annotation camera which points at the aircraft instrumentation, with no electrical connection required to the aircraft navigational aids.

All the sensors chosen produce NTSC video signals, which can be processed by standard video image analysis equipment. The images from the three cameras can be combined in a number of ways, in order to optimize the presentation of the data to the operator. In order to integrate the output of the three cameras they must be synchronized. Since the infrared camera video is generated by a rotating mirror, this signal can be used to control the timing of the other two cameras. This is the function of the processor and video sync unit in Figure 2. The output of this unit is a composite colour video image that contains the infrared image as red, the ultraviolet as blue and the navigation as green.

This composite colour video is the input signal to the computer controlled image processing system. The video is digitized using an AT&T TARGA16 board which is controlled by a Tandy Model 3000 microcomputer. The output is reconverted to video using the TARGA16 board, and displayed on a high-resolution monitor. The data is recorded on an industrial-grade 19mm video recorder.

The data analysis system allows the computation of areas of the slick. Geometric corrections are made before the areas are computed. The slick identification is achieved using density slicing techniques.

3. USE IN EXPERIMENTAL PROGRAM

To study the effectiveness of dispersants in two experimental spills the remote sensing system has been used. The first of these occurred in July 1985 on three freshwater bogs located about 100 km northeast of Edmonton, Alberta, Canada (Quaife et al, 1986) (Figure 3). The experiment used 3 cubic meters of Norman Wells oil on each of two bogs. One slick was untreated, while the other was sprayed with Corexit 9550 applied by an agricultural spray helicopter. The ultraviolet and infrared cameras were mounted in a Cessna 411 and flown at regular intervals over the spills (Goodman et al, 1986). These images were analyzed for spill area as a function of time. This data was used to compute dispersant effectiveness.

FIGURE 3 LOCATION MAP ATHABASCA

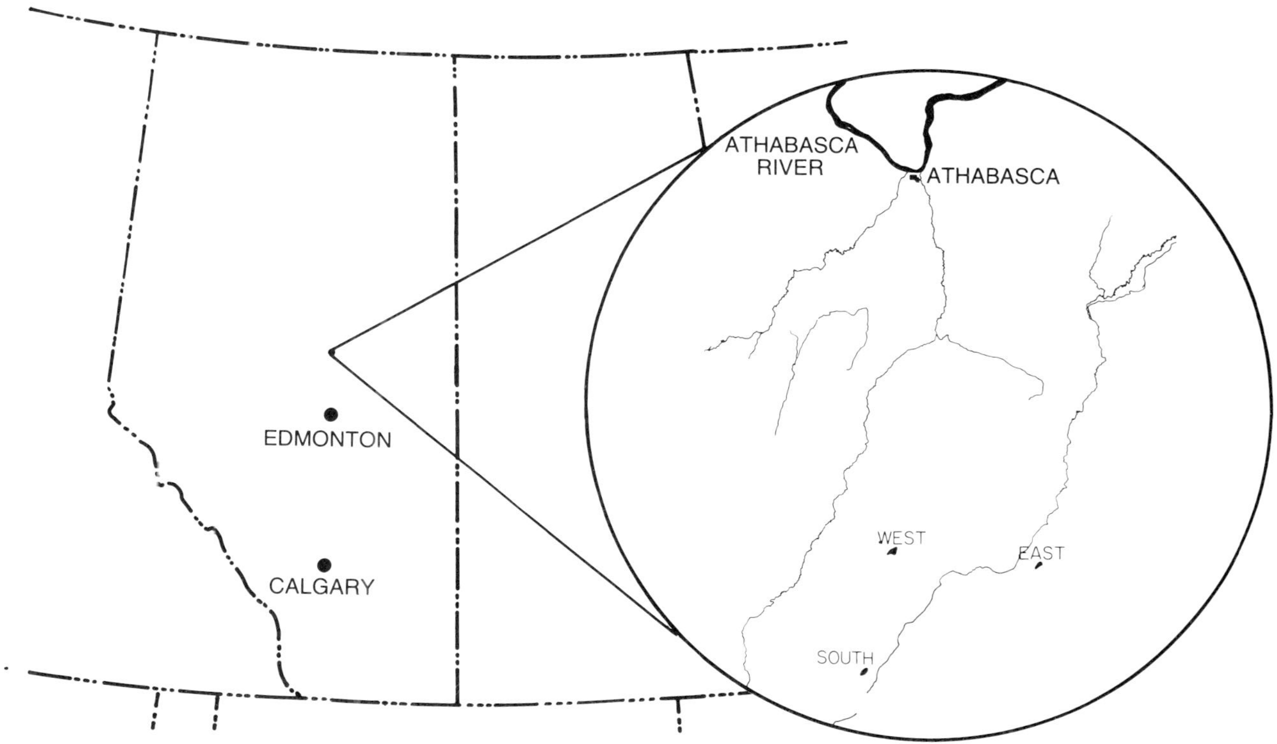

The second experimental spill took place in August 1986 in the Canadian Beaufort Sea near the Arnak drilling site (Figure 4). This experimental program involved an aged sweet mixed blend oil, and two dispersants--CRX-8, an experimental Exxon product, and Enersperse 700 from British Petroleum. Four slicks, each initially containing 2.5 cubic meters of oil were laid down in sequence. The spreading of the slicks was monitored by the remote sensing system. After about half an hour, the dispersants were applied to three of the slicks. Two slicks were treated with Enersperse 700, the third with CRX-8. The slick areas were continuously monitored over the next four hours. One hour after the initial spraying, the CRX-8 slick and one of the Enersperse 700 slicks were resprayed where thick oil was detected. This process was repeated an hour later (Swiss et al, 1987).

4. USE DURING A SPILL-OF-OPPORTUNITY

There was an accidental release of about 488 cubic meters of Arctic diesel from an artificial island (Minuk I-53) in the Canadian Beaufort Sea in September 1985 (Birchard and Nancarrow, 1986) (Figure 4). The remote sensing equipment was used to track this oil, all of which moved offshore. Evaporation and natural dispersion dissipated the slick over a period of five days. The infrared camera was used to monitor for oil on the shore of the outer barrier islands and none was detected.

5. USE TO MEASURE DISPERSANT EFFECTIVENESS

It has been shown that the measurement of the effectiveness of a dispersant in the field must consider not only the removal of oil from the surface, but also the time scale associated with the removal process (Brown and Goodman 1987). This removal process can be best measured by considering the time rate of change of oil concentration in the water column. However, an alternate is to use the amount of oil remaining on the surface, and by making some assumptions regarding evaporation, it is possible to calculate the amount of oil in the water due to dispersant action. While the area of the slick gradually increases due to spreading, the thickness is decreasing due to evaporation, natural dispersion and dissolution, so that the actual volume of oil is decreasing with time. When a dispersant is added, there is a rapid removal of oil into the water column. Some of this oil resurfaces, and the remaining oil on the surface continues to be reduced by evaporation and natural dispersion (Goodman et al, 1984).

The simple remote sensing system data provides an accurate and synoptic measurement of oil surface area, and some discrimination between thick and thin oil. At present, there is no reliable method of determining oil thickness from the air, so some assumptions must be made. If one uses the relative areas of a control and treated slick, then the dispersant effectiveness can be determined without a knowledge of the thickness of the slick. This was the procedure used during the Beaufort trials discussed above.

FIGURE 4 BEAUFORT

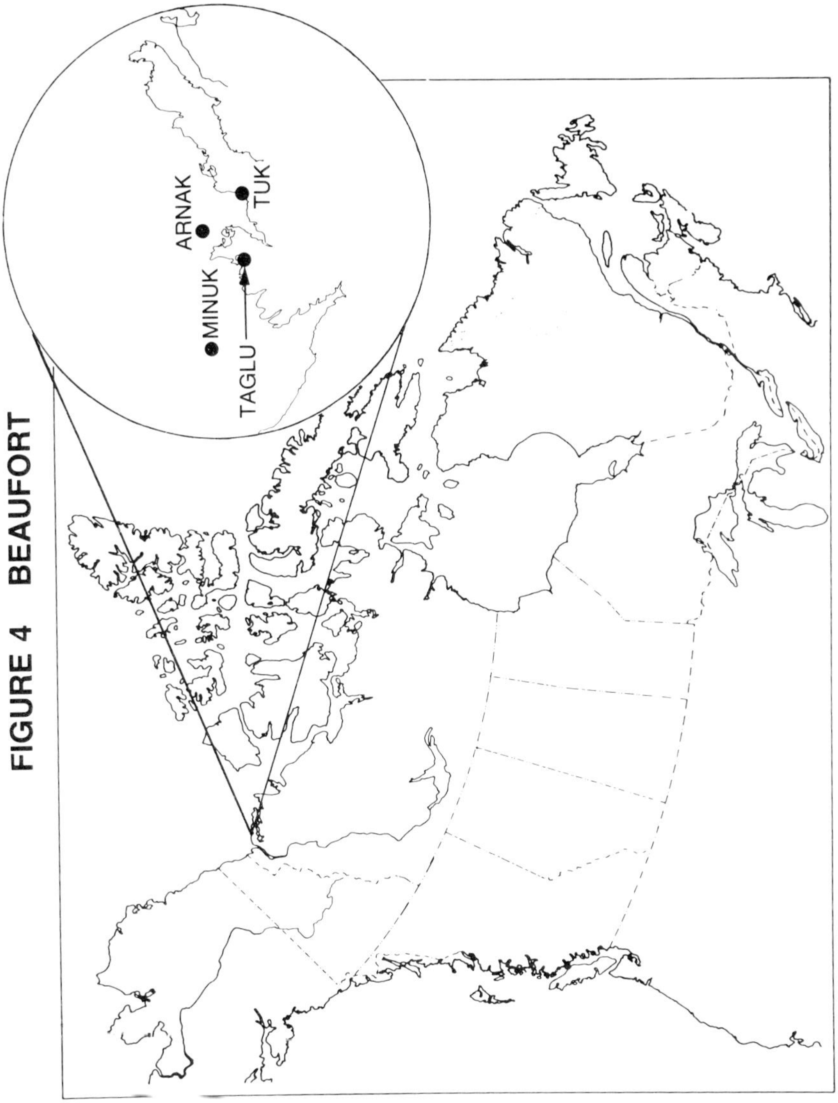

6. CONCLUSIONS

As the result of the use of this equipment for both experimental and actual spills, the simple remote sensing system has been proven to be a useful adjunct to an oil spill response. It is capable of providing synoptic scale data of the total spill at frequent time intervals. No other sampling system has such a capability.

7. ACKNOWLEDGEMENTS

The authors wish to thank Esso Resources Canada Limited for permission to publish this work and for their financial support of some of the experimental programs. Financial support has also been received from the ESRF (Environmental Studies Revolving Funds) and PACE (The Petroleum Association for the Conservation of the Canadian Environment). Many of the concepts in the papers have been discussed with Jim Swiss of Dome Petroleum and Merv Fingas of Environment Canada and their contribution is gratefully acknowledged.

8. REFERENCES

1. Birchard, E.C. and R. Nancarrow (1986). "The Minuk I-53 Artificial Island and Oil Spill." Proceedings of Ninth AMOP Symposium, Edmonton, 1986, pp. 375-378.
2. Brown, H.M., M.N. To and R.H. Goodman (1986). "Experimental and Theoretical Basin Studies of Dispersant effectiveness in Cold Water." Ninth AMOP Symposium, Edmonton, 1986, pp. 635-643.
3. Brown, H.M., R.H. Goodman and G.P. Canevari (1987). "Where Has all the Oil Gone? Dispersed Oil Detection in a Wave Basin and at Sea." 1987 Oil Spill Conference, Baltimore, Md. U.S.A.
4. Brown, H.M. and R.H. Goodman (1987). "The Use of Tests in a Wave Basin to Define Dispersant Effectiveness." Oil Pollution: Fate and Effects of Oil in Marine Ecosystems, Amsterdam, February, 1987.
5. Catoe, C.E. and J.T. MacLean (1979). "A Multispectral Look at Oil Pollution, Detection, Monitoring and Law Enforcement." Canada Centre for Remote Sensing #1019756.
6. Goodman, R.H., M.R. MacNeill and H. Ripley (1984). "Remote Sensing Program for the 1983 Offshore Dispersant Trials." Proceedings of Seventh AMOP Symposium, Edmonton, 1984, pp. 488-505.
7. Goodman, R.H., G. Peterson and R. Fitch (1986). "Use of Remote Sensing During the Freshwater Dispersant Experiment." Proceedings, Ninth AMOP Symposium, Edmonton, 1986, pp. 601-621.
8. Hay, A.E. and L.W. Davidson (1984). "Remote Acoustic Monitoring of the Dispersion of Oil from a Controlled Spill." An experimental study Seventh AMOP Symposium, Edmonton, 1984, pp. 506-525.
9. O'Neil, R.A., R. Vollmers, H. Parker, F. Gunniborg and R. Spanff (1979). "Airborne Remote Sensing of Oil Spills in Coastal Waters." Proceedings Workshop of the NATO-CCMS Pilot Study on the use of Remote Sensing for the Control of Marine Pollution, Washington, D.C., April 1979.
10. Quaife, L.R., C.H. Peabody, H.M. Brown and R.H. Goodman (1986). "Freshwater Oil Spill Research Program - Field Trial." Proceedings, Ninth AMOP Symposium, Edmonton, 1986, pp. 601-621.
11. Swiss, J.R., N. Vanderkooy, S.P. Gill and R.H. Goodman "Beaufort Sea Dispersant Trials." 1987 Oil Spill Conference, Baltimore, Md. U.S.A.

BIODEGRADATION AND PHOTO-OXIDATION OF CRUDE OILS IN A TROPICAL MARINE ENVIRONMENT

R J WATKINSON and D GRIFFITHS

Fermentation and Microbiology Division, Shell Research Ltd., Sittingbourne Research Centre, Sittingbourne, Kent. UK.

1. INTRODUCTION

Several attempts have been made to carry out fate of oil studies directly in the environment. Such studies are necessary to validate the measurements and observations made in laboratory simulations. The problems associated with studies carried out directly in the environment are that there are so many interactions taking place the interpretation of data is extremely difficult, particularly when the environment itself is varying with time. The converse is true for laboratory simulations, since one controls many environmental parameters and limits the number of interactions in order to be able to interpret the data.

There have been relatively few studies of crude oil degradation at sea or simulated field conditions. An interesting way of maintaining oil samples at sea in a convenient form was developed by Gilbert and Higgins (1973). They incorporated the oil in a cellulose filter and deployed these in some protective container. This approach has also been taken by Gibbs and Bavor (1981) by deploying oiled Millipore filters held in nylon net pockets.

The present studies have employed the oiled membrane filter techniques developed by Gilbert and Higgins as the most practical method for following the rate of oil degradation in the marine environment.

2. MATERIAL AND METHODS

2.1 Preparation and deployment of filters

Three oils were applied to cellulose acetate/cellulose nitrate Millipore filters (8 µm, 47 mm diameters). Two different Eastern light crude oils, topped to 200°C, were used directly by immersing the filters in the crudes, avoiding excess oil, and then allowing the filters to dry to constant weight at 30°C. An Eastern heavy crude was used as a 75% solution in dichloromethane and this solution used exactly as the other crudes. Each filter contained 70 - 100 mg of crude oil.

For the preparation of standard curves the oils were diluted in carbon tetrachloride in steps of 20% over the range 0 to 100% oil. The filters were then dried and weighed as above. 15 mm discs were removed from each filter for oil analysis by IR spectroscopy and standard curves of absorbance versus oil mass were constructed.

For deployment in the sea, 34 mm diameter discs were cut from each of the filters and placed into perforated plastic practice golf balls such that they formed an equatorial diaphragm. The balls were sealed and placed in cages with floats such that the golf balls floated at the

Kuiper, J., Van den Brink, W.J. (eds), Fate and Effects of Oil in Marine Ecosystems. ISBN 90-247-3489-4

surface but were constantly immersed. The floating equipment was then tethered by ropes to the end of the pier at Muka Head, Penang.

2.2 Microbiological counts

15 mm discs were cut with a sterilised cork borer from the main filter. The discs were placed in 10 ml sterile Ringers solution along with five 2 mm Ballotini beads. The bottles were vigorously shaken for 5 minutes to release the organisms from the membrane. The resulting solutions were serially diluted in Ringers solution and 0.1 ml of the solutions spread on Marine Agar plates and on to Marine Bushnell-Haas-Oil plates. The total heterotrophs growing on the Marine Agar plates were counted after 40 hours incubation. The oil utilising organisms growing on the oil plates were counted after 7 days.

2.3 Oil analysis

15 mm diameter discs were cut from the recovered filters, extracted twice with 1 ml of carbon tetrachloride and the extracts pooled and made to a final volume of 2 ml. The extracted oil was determined by IR spectroscopy by relating the sum of the transmissions at 2950, 2920 and 2870 cm^{-1} to values obtained from the individual oil standard curves. Photo-oxidation products were determined by observing the IR spectrum in the region of 1850 to 1600 cm^{-1}.

Liquid chromatography of the oil residues was effected by the elution of fractions from dual phase alumina/silica gel columns. Analysis of the individual hydrocarbon fractions was carried out using capillary gas chromatography.

3. EXPERIMENTAL AND RESULTS

The protected oiled filters were deployed in floating cages at the end of the pier at Muka Head. Individual filters were recovered over a total of 20 days. Control filters holding no oil were deployed to measure any extraneous oil that may have contaminated the filters. 15 mm discs were immediately removed for the microbiological counts. The remaining parts of the samples were oven dried at 60°C for later analysis of hydrocarbons.

Leaching controls for the filters were carried out in the laboratory by placing autoclaved oiled filters into sterile sea water medium in 21 conical flasks. The medium also contained 1% mercuric chloride as a means of maintaining sterility. Filters were removed at intervals and the quantity of oil remaining determined by IR spectroscopy of carbon tetrachloride extracts. The data showed that appreciable quantities of oil was washed off the filters accounting for 75-82% of the original oil. Since this occurred within two days and then reached a plateau this has been assumed to occur in the field samples and thus these values have been subtracted from the starting mass of the oil. The loss of this oil mainly due to attrition and dissolution, showed no chemical discrimination in the loss of material (i.e. aromatic and aliphatic hydrocarbons suffered similar % losses) and the assumption is that this dispersed oil is degraded in the environment by the same processes affecting the oil remaining on the filters.

3.1 Biodegradation of oil

Filters were recovered at intervals over a period of twenty days. 15 mm diameter discs were removed from air dried samples and extracted with carbon tetrachloride for IR analysis. The mass of oil obtained from the

Table 1 - Determination of oil on 15 mm discs from deployed filters

Filter	Days exposure	No. of replicates	Oil mass at time zero (mg)	Oil det'd at time T	Oil loss
L1	3	2	9.60	3.60	6.00
L2	3	3	9.31	3.34	5.97
L3	6	2	8.73	2.28	6.45
L4	6	2	9.26	2.25	7.01
L5	10	1	9.81	1.65	8.16
L6	10	1	9.22	1.25	7.97
L7	lost	-	-	-	-
L8	lost	-	-	-	-
L9	lost	-	-	-	-
L10	lost	-	-	-	-
H1	3	3	8.75	3.36	5.39
H2	3	3	9.15	3.68	5.43
H3	6	2	8.98	1.83	7.15
H4	6	2	9.79	1.95	7.84
H5	10	2	10.09	1.98	8.11
H6	10	2	8.18	1.55	6.63
H7	14	1	8.52	1.37	7.15
H8	14	1	8.45	1.33	7.12
H9	20	1	9.26	1.10	8.16
H10	20	1	8.05	1.75	6.30
T1	3	3	7.73	4.53	3.23
T2	6	2	8.38	3.68	4.70
T3	10	2	7.57	1.61	5.94
T4	14	lost	7.42	-	-
T5	20	1	8.27	0.50	7.77

filters is presented in Table 1 and as Figure 1. The rapid disappearance of oil over the first three days parallels the leaching control experiments. This loss was thus taken to arise from attrition and degradation to be responsible for the further losses from this time. The colonisation of the filters by micro-organisms also occurred rapidly in the first three days. The difference between the amounts of attrition between the field and laboratory leaching controls may be due to some protection of the oil slick by the biofilm.

The percentage biodegradation of the remaining oil from day 3 to day 20 is shown in Figure 2 for each of the three crudes. In spite of the faster initial rate of degradation of the light crude oils the heavier crude displayed more extensive degradation by day 20. Unfortunately, within the timescale of the experiments many filters had been colonised by barnacles. In the case of a series of light crude filters and non-oiled control filters the barnacles had completely digested the filters by day 20. The barnacles had presumably been attracted to the rich

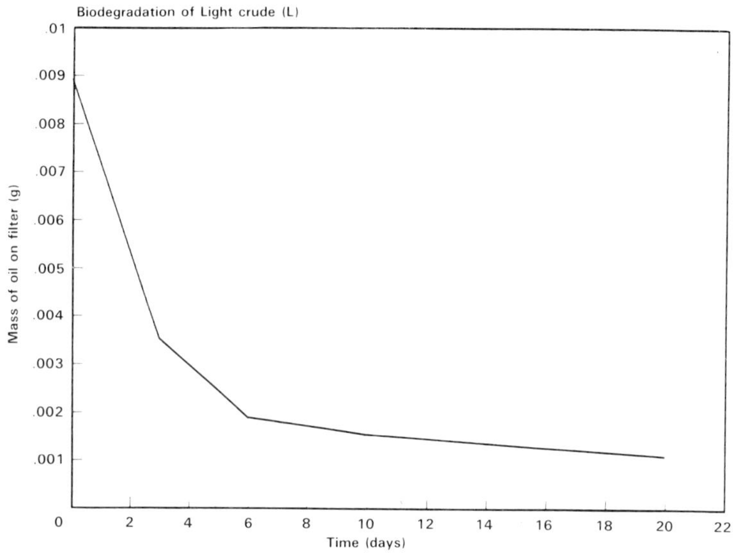

Fig 1 Mass of oil on each filter was determined by extracting 15 mm discs and IR spectroscopy. The IR transmission was related to the standard curves

biofilm on the filters since they had not colonised the golf balls or the metal baskets.

The extensive degradation, shown in the above analysis, of 70-90% is confirmed in the gas chromatograms of the light crudes (Figure 3). The chromatogram of the original crude is dominated by the alkane fraction with pristane showing as a major peak. After three days incubation most of the oil had disappeared with the light ends almost totally absent (see Figure 3b), on a weight basis the pristane had decreased and the alkane:pristane or alkane:phytane ratios indicate some biodegradation had taken place. By day 20 there are no recognisable alkane peaks (Figure 3c), even the pristane peak is absent and we are left with an "unresolved envelope" that is typical of heavily weathered and degraded crudes.

3.2 Photo-oxidation products

It has been demonstrated that photo-oxidation of crude oils gives rise to oxygenated hydrocarbon species which may be detected by the C-O bond frequencies in the IR spectrum in the region of 1700 cm^{-1}. Figure 4 shows the appearance of such a peak, which increases with time of exposure, in the oils recovered from filters. All extracts from each filter for oil analysis were combined with the residual

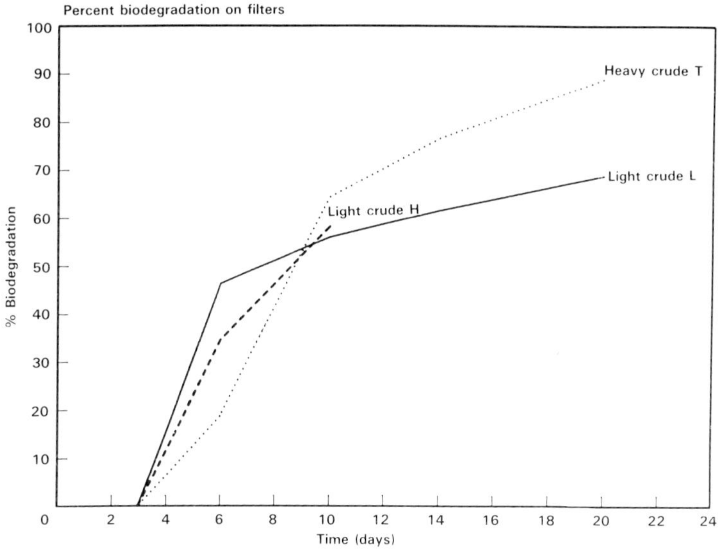

Fig 2 Biodegradation of Heavy and Light crude oils in a marine environment

filter remnants remaining and the total extracted with carbon tetrachloride to provide a minimum volume for IR analysis of the oxidised products. These data are for the series of filters holding a light crude but all the crudes gave similar IR spectra. The solutions were subsequently diluted to allow the determination of the C-H frequencies and thus estimate the total oil present. The data has been normalised to a standard quantity of oil in Figure 5. This graph indicates that all the oils undergo rapid photo-oxidation with the lighter oils showing a faster initial rate of formation of oxidised products.

3.3 Microbial colonisation of the oiled filters

The oiled filters were essentially sterile at the moment they were deployed. Thus colonisation would be dependent upon the indigenous population and their preference for the substrate. Microbial suspensions were released from 15 mm discs as described in Materials and Methods before dilution and plating on Marine Agar and Bushnell-Haas-oil plates. Figure 6 shows the rapid colonisation of the filters by micro-organisms and also reaching a some maximum population by day 6. The numbers represent the organisms recovered from the 15 mm disc. It can be seen that the hydrocarbon utilisers, as determined by

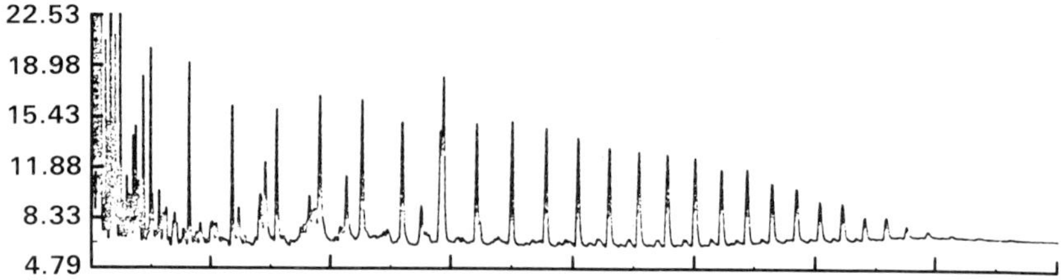

Fig 3a Heavy crude at time = 0

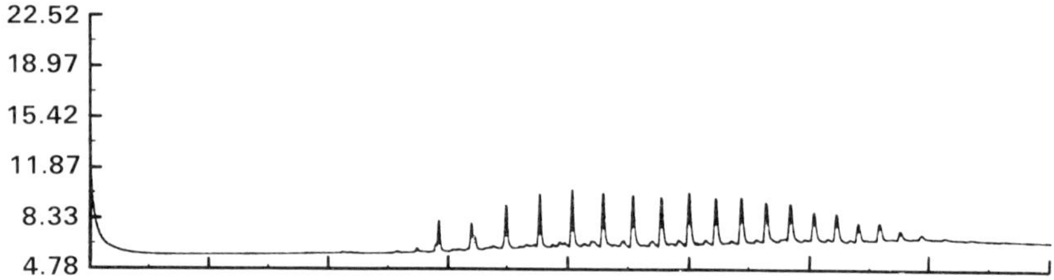

Fig 3b Heavy crude from filters recovered after 3 days

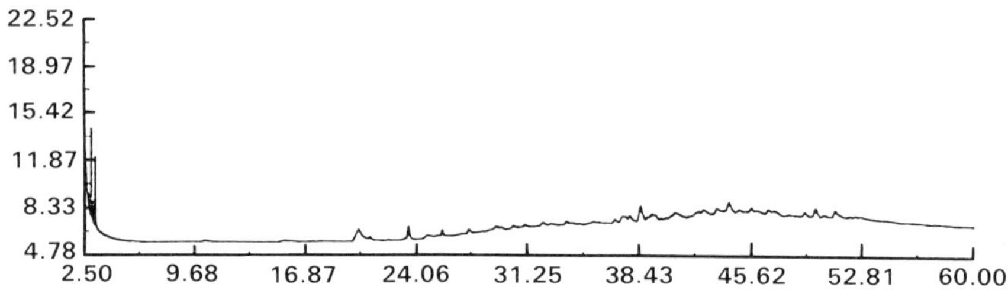

Fig 3c Heavy crude from filters recovered after 20 days

Chromatograms were produced on a Carlo Erba Fractovap 4160
with on column injection
Column: CP Sil 5 CB (26 m x 0.33 mm id)
Temp. Programme: 45 – 300°C at 5°C/min

Fig 3 Chromatograms of original and weathered Temana crude

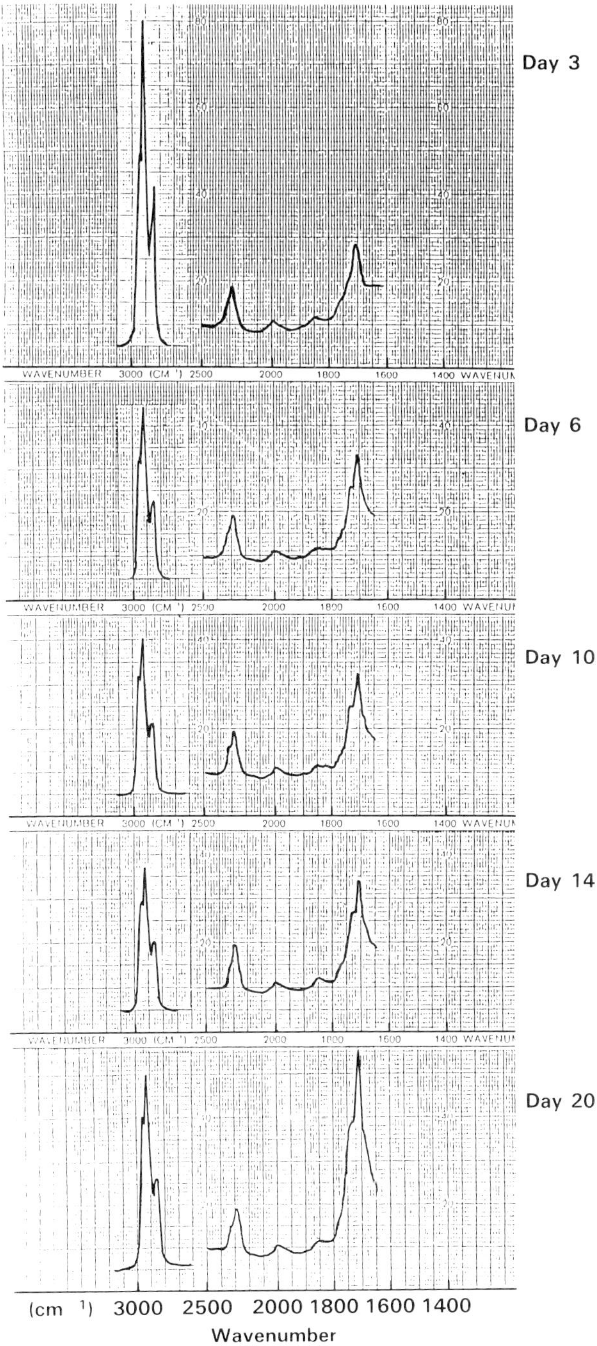

Fig 4 Formation of photo-oxidation products in crude L. All carbon tetrachloride extracts from recovered filters were combined to provide sufficient material for oxidised product determination. The concentration of the oxidised product is related to the absorbance at 1700 cm^{-1}. The samples were diluted ten-fold for the determination of total oil from the spectra between 2600 – 3200 cm^{-1}

selection on Bushnell-Haas medium, represent 5 - 10% of the total heterotroph population. The blank filters of 3 x 10^6 at day 20. The oil degraders on the control samples only represented approximately 1% of these values. The data for the microbial counts are provided in Table 2.

Table 2 - Microbial populations of the filters

Crude	Incubation time (days)	Microbial counts: Heterotrophs	Microbial counts: Oil utilisers
Baram	6	3.5 x 10^7	2.1 x 10^6
	10	2.5 x 10^7	7.3 x 10^5
	20	2.5 x 10^7	7.3 x 10^5
Brunei	6	8 x 10^6	7.9 x 10^5
	10	1.5 x 10^7	2.9 x 10^5
	20	1.4 x 10^7	3.3 x 10^5
Controls (no oil)	6	-	-
	10	2.7 x 10^6	3 x 10^4
	20	3.0 x 10^6	3 x 10^4

4. DISCUSSION

It has been possible with the relatively simple techniques employed in this study to carry out in the marine environment a field study that validates the general findings of other researchers on the fate of oil in these environments.

The fate of oil in the marine environment has been extensively reviewed by van der Linden (1978) and more recently by Floodgate (1984). When crude oil is released into a marine environment, immediate changes occur as a result of several processes. These include evaporation, dissolution, dispersion into the water column, photo-oxidation, biodegradation and emulsification. The oil eventually disappears as a result of the degradative processes and by adsorption to particles and by sinking. The impact of photo-oxidation on the processes has been reviewed by Payne and Phillips (1985). They reported that oxidised products could be seen in the IR spectra of photo-oxidised oils, that some resin or polymer formation may take place and that some water soluble products would be produced and enhance the emulsification of water in the oil. In the experiments reported here one can clearly see that the generation of oxidised materials having the same IR spectra as products generated by the UV oxidation of the same crudes in laboratory photochemical experiments.

Biodegradation causes the major portion of the oil to be completely oxidised to carbon dioxide and water, since this would be a consequence of increasing biomass, although one would expect to produce some oxidised products. One of the first events to take place on exposure of oil films to the environment would be the increase in the number of

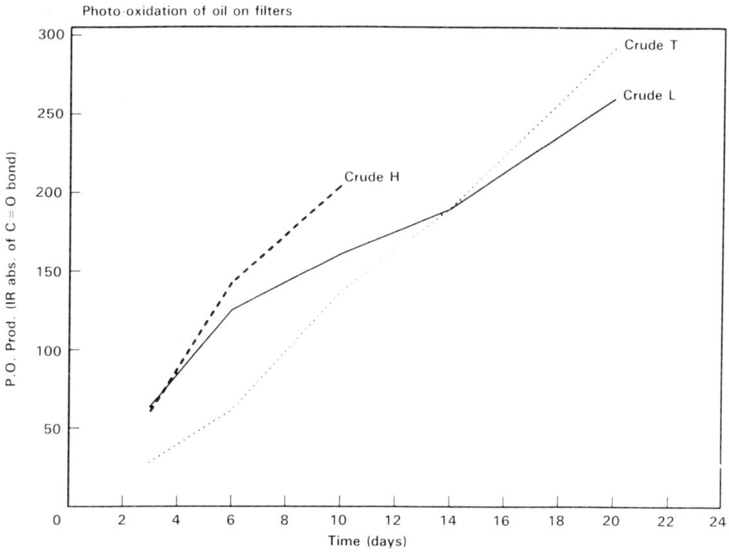

Fig 5 Formation of photo-oxidised products in crude oils. Extracts were subjected to IR analysis as for Fig 15. The absorbance at 1700 cm^{-1} was normalised to the % transmission of the C – H peaks at 2800 – 2950 cm^{-1} such that each concentration of photo-oxidised product relates to the same total oil concentration.

organisms in the microbial population that could utilise the oil components. Thus some selection of the natural population would be expected to occur and this has been clearly seen in these simple experiments. The activity of these hydrocarbon utilising organisms will generates products which provide nutrients for the general heterotrophic population such that this total population will also increase. These interactions have been confirmed by these studies.

The processes contributing to the degradation of the oils in these experiments have been demonstrated to be dissolution, photodegradation and biodegradation. The quantitative contribution of each mechanism is not discernable in these experiments as in the natural environment these processes act in concert leading to the mineralisation of hydrocarbon products.

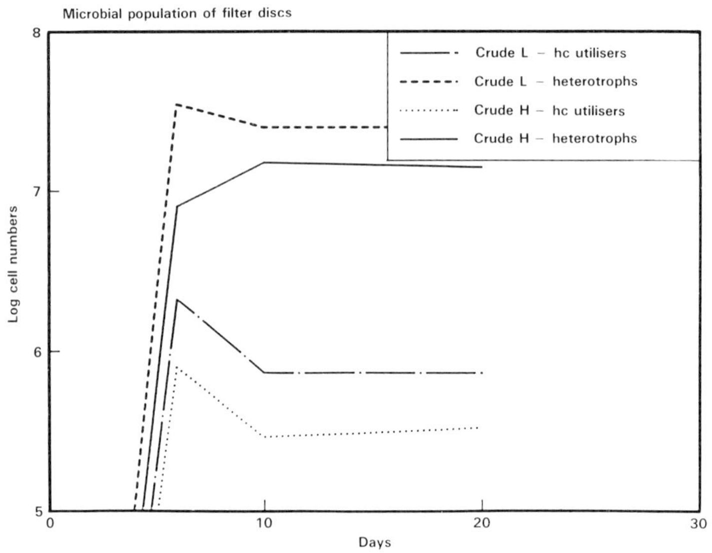

Fig 6 Microbial population of filter discs. Microbial counts for heterotrophs were carried out on Marine Agar plates after incubation for 40 hours. Hydrocarbon utilising colonies were counted on Marine Bushnell-Haas oil medium after incubation for seven days

These studies have demonstrated the relatively fast degradation of three crude oils representing light and heavy crudes appropriate to the South China Seas region. The extent of degradation is reflected in the removal of relatively slowly degradable branched alkanes. These findings are similar to previous findings but indicate the importance of photo-oxidation in the disappearance of oil films in the marine environment.

The authors would like to acknowledge the helpful discussions of Dr. P. Gilbert of SRC, Dr. Nor Mahadi of University Kebangsaan, Malaysia, Kuala Lumpar and the assistance of the staff at University Sain's Malaysia and the field station at Muka Head, Penang.

5. REFERENCES:

Van der Linden, A. C., (1978). Developments in biodegradation of hydrocarbons-I Watkinson, R. J. Applied Science. p 165.

Floodgate, G., (1984). Petroleum Microbiology. Ed. Atlas, R. M. Macmillan p 355.

Gilbert, P. D., and Higgins, I. J., (1978). J. Gen. Microbiol. 108, 63 - 70.

Gibbs, C. F., and Bavor, H. J., (1981). Tech. Report 29. Ministry for Conservation Fisheries and Wildlife Division (Australia). "Degradation rates of crude oils in low nutrient marine environment".

Payne, J. R., and Phillips, C. R., (1985). Environ. Sci. Technol. 19, 569-579.

OIL IN SEDIMENTS OF THE HUMBER ESTUARY FOLLOWING THE 'SIVAND' OILSPILL INCIDENT

D.I. LITTLE

Oil Pollution Research Unit, Field Studies Council,
Orielton Field Centre, Pembroke, Dyfed, U.K.

ABSTRACT: The 'Sivand' oil spill of 28th September 1983 was unusual in that few comparable quantities (6,000 tonnes) of crude oil have been spilled in a turbid estuary. This study aimed to contribute to knowledge of the spatial and temporal distribution of the oil, and to examine its degree of degradation in two sedimentologically contrasted shorelines where considerable ecological and conservation interest are focussed (Humberston Fitties, a sandflat, and Blacktoft Sands, a muddy shore in the inner estuary). Methods included replicate sediment core sampling, gas liquid chromatography and various sedimentological measurements.

Soon after the spill, the sands at Humberston Fitties had oil concentrations of > 50,000 ppm. Over the 12 month study period this was reduced to *ca*. 3,000 ppm. This was expected since coarser sediments generally promote oil weathering. The mud sediments showed lower levels of contamination initially (*ca*. 2,000 ppm) followed by only slight reduction over the 12 months, although there is evidence that 'Sivand' oil was re-deposited on the study site after natural cleanup of the original oil was almost complete.

Good drainage at both sites permitted initial oil infiltration. The presence of root macropores at the muddy site permitted penetration, weathering and, perhaps, release of the oil more easily than if the oil had been bound in the sediment matrix. Sediment mobility at the muddy site contrasted with the high shear strength and cohesiveness of the oiled sandflat. The oiled, muddy surface layers of the sandflat were very stable throughout the study period, showing a highly significant correlation between % mud and total hydrocarbon concentration (Figure 1 and 2). Interstitial water movements were a major process in degradation of the oil at the sandy site, but no dramatic sediment turnover or erosion occurred which might have removed the oil. The steep gradients in oil concentration and composition with depth contrasted with the muddy site, where 'free' oil penetrated as a coherent mass *via* root macropores.

The study illustrates the importance of sedimentological factors in determining the fate of spilled oil and thus in the prediction of ecological damage and calculation of mass balances of oil spills. Work so far published on the 'Sivand' spill has emphasised the effectiveness of the cleanup operation and the lack of biological effects, but may have under-estimated the amount of oil trapped and transported by sediments. Of the 6,000 t spilled, 1,500 t were estimated to have evaporated, 450 t were physically recovered and 1-2,000 t were chemically dispersed (Nichols and Parker, 1985). From this it may be estimated by difference that over 2,000 t were either naturally dispersed or stranded on shorelines. Extrapolating from the intertidal sediments in the present study, only

Kuiper, J., Van den Brink, W.J. (eds), Fate and Effects of Oil in Marine Ecosystems. ISBN 90-247-3489-4

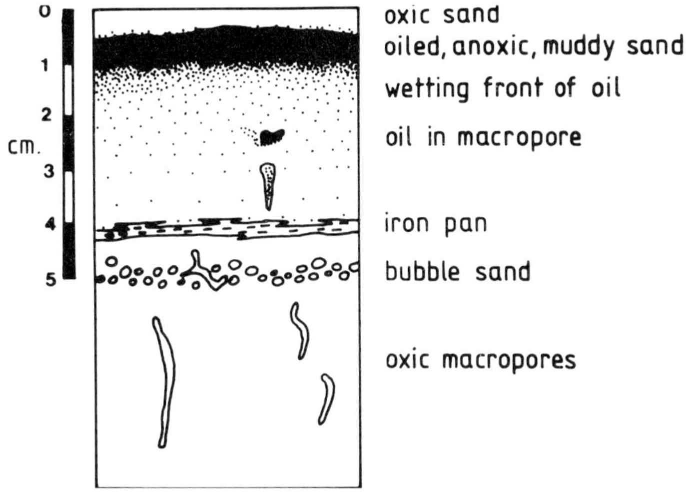

FIGURE 1. Sketch of sediment structures and redox profile of an oiled sand core at Humberston Fitties.

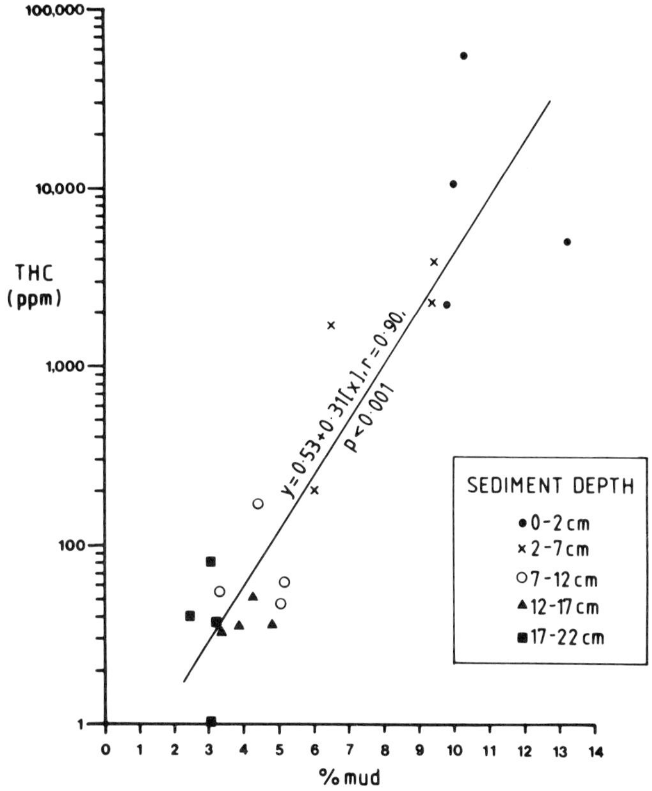

FIGURE 2. Sediment % mud plotted against total hydrocarbon (THC) concentrations on all four sampling dates in 1984 at Humberston Fitties.

about 100-250 t of the stranded oil would have been incorporated into shorelines, so obviously the vast majority of the oil spill was dispersed into the water column. It is not known whether most of this has left the estuary or whether it is trapped in fluid mud layers and on suspended particulate matter (SPM). Data on the oil content of Humber SPM suggest that 4×10^6 t of silt would be sufficient to account for the spilled oil.

REFERENCE

1. Nichols JA and Parker HD: Dispersants: Comparison of laboratory tests and field trials with practical experience at spills. Proceedings of the 1985 Oil Spill Conference (Prevention, Behaviour, Control, Cleanup). American Petroleum Institue, Washington, DC. 421-427, 1985.

LABORATORY AND FIELD INVESTIGATION OF THE ACCUMULATION AND DEPURATION OF PETROLEUM HYDROCARBONS BY BLACK MUSSELS.

ROBERT PETER MASON
SEA FISHERIES RESEARCH INSTITUTE
PRIVATE BAG X2
ROGGEBAAI 8012
SOUTH AFRICA

Black mussels, choromytilus meridionalis, were exposed for both short-term periods of less than one week and for longer-term periods of one to three months to water-soluble fractions of Qatar light crude oil under laboratory conditions. After exposure, the mussels were transferred to clean seawater and depurated for periods of two weeks to three months. In addition, mussels from chronically polluted field sites were depurated in the laboratory. Mussels were digested in alkali overnight and extracted with ether before being passed through a silica-alumina column (Mason (in prep.)). The column effluent was collected for flourescence analysis at an excitation wavelength of 280nm and an emission wavelength of 374nm.

Results showed that the accumulation of hydrocarbons was rapid. Levels substantially higher than background levels were recorded within the first twenty-four hours of exposure to water concentrations of less than 1ppm. The initial rate of accumulation was found to decrease as the concentration in the water was decreased, as found by other authors (Stegeman and Teal 1973). After long term exposure a maximum level of accumulation was found. Mussels exposed to 0,1ppm accumulated 52ug/g into their tissue within the first ten days. In the following 78 days tissue levels varied by less than twenty percent with no overall increase in concentration. Evidence suggests that the amount of petroleum hydrocarbons accumulated at this maximum accumulation level depends on the concentration in the water column.

After short-term exposure, the mussels lost their accumulated burden rapidly with the majority of the accumulated hydrocarbons being lost within two weeks (Figure 1). Rates of depuration were similar for mussels exposed, in different experiments, to 0,3; 0,7 and 1,5ppm for periods of less than a week. Half-lives for depuration varied between five and six days and were similar to those reported by Lee (1977) for laboratory exposures of two to seven days. In contrast, mussels exposed for extended periods lost their accumulated hydrocarbons much more slowly with half-lives for depuration of 16 and 29 days after exposure to 0,15ppm for 46 and 95 days respectively. A linear relationship was found between half-life and exposure period. The results indicate that exposure time rather than exposure period is the major factor in determining the rate of loss of accumulated hydrocarbons.

Kuiper, J., Van den Brink, W.J. (eds), Fate and Effects of Oil in Marine Ecosystems. ISBN 90-247-3489-4

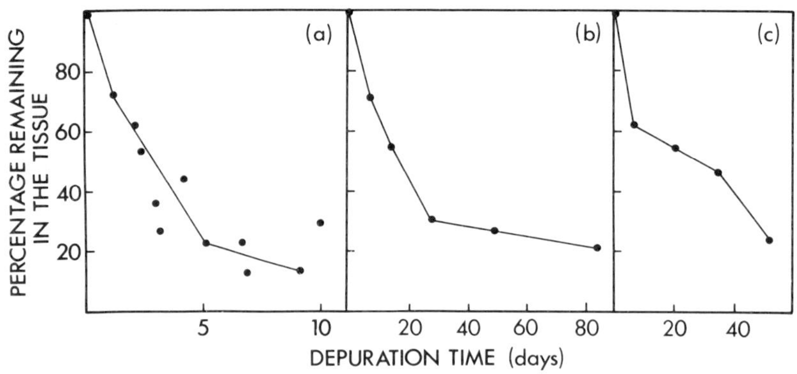

Figure 1: Rate of loss of accumulated hydrocarbons by black mussels after exposure under laboratory conditions: (a) to 1,5 ppm for 3,8 days; (b) to 0,15 ppm for 46 days and (c) to 0,15 ppm for 95 days.

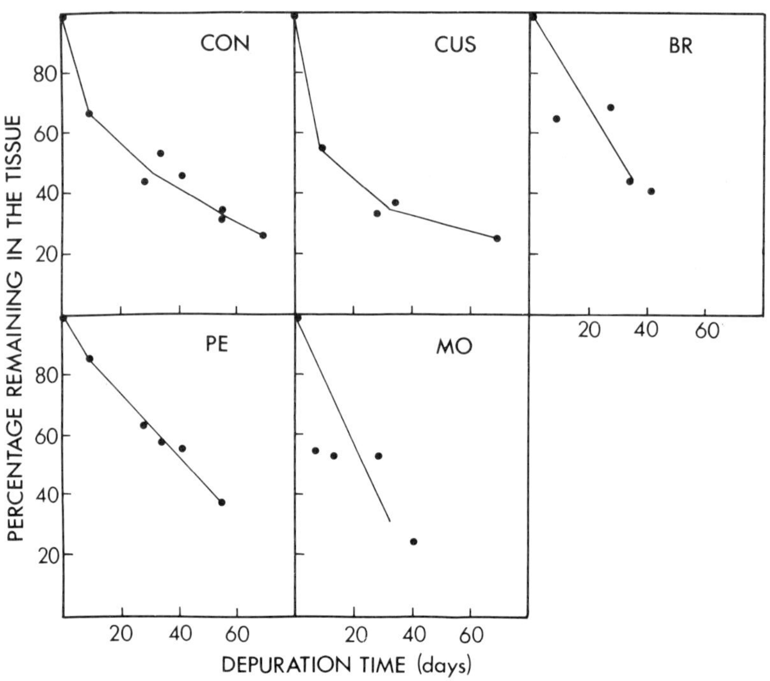

Figure 2: Rate of loss of cronically accumulated hydrocarbons by black mussels collected from various sites around the Cape Peninsula.

Mussels collected from chronically polluted field sites lost their accumulated hydrocarbons at rates similar to those found after long-term exposure. For mussels taken from sites close to the laboratory, and therefore from a similar environment to that created in the laboratory, depuration half-lives of between 37 and 41 days were found. Water concentrations in the vicinity of the sampling sites were measured over a week and were generally found to be less than 0,1ppm. The tissue levels measured prior to depuration were between 15 and 40ppm. The chronically polluted mussels from different stations had similar rates of loss even though their tissue concentrations when sampled differed by more than a factor of two. These results correlate well with the results of the long-term laboratory studies. Finally, mussels were collected from a field site after a fuel oil spill and depurated in the laboratory. A depuration rate similar to that found after short-term laboratory exposures was found (half-life of 3,5 days).

The results of the laboratory depuration trials of mussels exposed both in the laboratory and in the field indicate that the exposure regime has a marked effect on the rate at which the mussels lose their accumulated burden.

REFERENCES

1. Lee RF: Accumulation and turnover of petroleum hydrocarbons in marine organisms. In Fates and effects of petroleum hydrocarbons in the marine environment. Wolfe DA(ed). New York: Pergamon Press, 1977.

2. Mason RP: Laboratory investigations of the accumulation and depuration of petroleum hydrocarbons by black mussels *Choromytilus meridionalis*. Submitted to S. Afr. J.mar.Sci., 1986.

3. Stegeman JJ, Teal JM: Accumulation, release and retention of petroleum hydrocarbons by the oyster, *Crassostrea virginica*. Mar.Bio. 22, 37-44, 1977.

SECTION II

EFFECTS OF OIL ON SPECIES, POPULATIONS AND ECOSYSTEMS, BOTH IN LABORATORY TESTS AND IN THE FIELD (INCLUDING MODEL ECOSYSTEMS) AND EFFECTS OF THE TREATMENT OF OIL SPILLS

THE EFFECT OF THE WATER SOLUBLE FRACTION OF NORTH SEA CRUDE OIL ON THE TRANSPORT MECHANISM OF TAURINE IN ERYTHROCYTES FROM FLOUNDER, *PLATICHTHYS FLESUS* (L.).

L.-O. REIERSEN[1] **& K. FUGELLI.**
Department of Biology, University of Oslo, Norway.
[1] Present address: The Norwegian State Pollution Control Authority. P.O. Box 8100, Dep., 0032 Oslo 1, Norway.

ABSTRACT
Flounder erythrocytes has been used as a model to study the effect of petroleum hydrocarbons on amino acid transport systems in cell membranes. The effect of water soluble fraction (WSF) of North Sea crude oil on the transport velocity of taurine into the cells - influx, has been measured using 14-C labled taurine. Two different methods of exposure were performed; flounder were exposed *in vivo* for 1 month in flow-through aquaria to 0.1-0.3 ppm WSF, and erythrocytes were collected from un-contaminated fish and exposed *in vitro* for 20 min - 23 hours, to salt media containing different oil-concentrations (approx. 1 ppb-10 ppm). Taurine influx was inhibited after both exposure methods used, at all concentrations tested and already after 20 min. exposure. Both the sodium dependent and the sodium independent transport mechanisms were affected. The *in vivo* studies showed that fish in low condition were more vulnerable to the oil-exposure than fish in good condition. The basic regulatory mechanism maintaining a stable cell volume during changes in plasma osmolality was intact in oil-exposed fish. The *in vitro* studies showed that the negative effect on the transport mechanisms was reversible after washing the oil-exposed cells in un-contaminated salt media. The cell volume was not affected under the performed tests although the taurine transport mechanism which take part in the cell volume regulation was inhibited.

1. INTRODUCTION

Petroleum hydrocarbons disolved in seawater accumulate rappidly in fish (1,2) and can provoke changes in fish morphology and development (3,4,5, 6). Such hydrocarbons and their metabolites can be found in tissue lipids as in cell membranes (4,7,8,9). On the basis of such observations it has been assumed that petroleum hydrocarbons could affect cell membrane transport systems for inorganic ions and organic molecules. Normal cellular function, the maintenance of a stable cell volume and the composition of the intracellular environment are dependent on these transport systems (10).

However, little is known about the effect of pollutants on such mechanisms in marine animals (11,12). At the Department of Biology, University of Oslo, amino acid transport systems have been studied in flounder erythrocytes and especially the taurine transport which consists of two different mechanisms - one sodium dependent and one sodium independent. The sodium independent mechanism is stimulated by lowered osmolality (13). These systems play an essential role in the maintenance of normal cell volume, (13,14,15,16,17,18).

In the present study we have used flounder erythrocytes as a model to

Kuiper, J., Van den Brink, W.J. (eds), Fate and Effects of Oil in Marine Ecosystems. ISBN 90-247-3489-4

investigate sublethal effects of oil-concentrations on cell membrane transport mechanisms. This study was a part of the Norwegian Marine Pollution Research and Monitoring Program (FoH).

2. MATERIAL AND METHODS

Flounder, Platichthys flesus (L.), were caught by gillnets in the inner Oslofjord, Norway. The fish (30-40 cm) were kept in seawater aquaria (30 $^0/_{00}$, 8-10 ^{0}C) for at least 10 days prior to experiments. The fish were not fed during the captivity. The condition of the fish, expressed as the condition factor C.F., was determined using the equation $C.F.=W/L^3 \cdot 100$, where W is the weight and L the length of the fish. In all in vitro experiments and in two of the in vivo series fish in good condition were used (C.F. > 1.0). In one in vivo serie fish were starved two months prior to the experiments to achieve fish with low condition factor (C.F. <1.0).

2.1. Oil exposure

Two types of oil exposure were performed.

2.1.1. Fish were exposed for 1 month in vivo to seawater containing 0.1-0.3 ppm of water soluble fraction (WSF) of North Sea crude oil. The erythrocytes were then sampled and the tests of taurine transport activity performed. The WSF of crude oil was produced continously as described by Johannessen (19) and pumped through a semi-open system to the test aquaria (20). The exposures were performed in flow-through aquaria (240 l) and the flow was 40l day^{-1}. To maintain adequate oxygen tension for the fish the aquaria were aerated and the oxygen content in the water was monitored every day.

Three separate series were run, each with a test and a control group (5-10 fish in each group). All the experiments were conducted with a 12:12 hours light:dark cycle. Usually, one control group was taken when the oil-exposure started and a second control at the end of the exposure period.

2.1.2. The concentration of oil (WSF) in the water was monitored every day by means of fluorescence spectroscopy, calibrated with gas-liquid chromatography (GLC) analysis. The latter was performed at the Central Institute of Industrial Research, Oslo. The GLC-analysis on the WSF was made by two different extracts; on di-chloro methane extracts analysing hydrocarbons with a boiling point higher than that of toluene (C7), and the second by head space technique analysing the volatile components with a boiling point less than C7.

2.1.3. The in vitro studies were performed on erythrocytes obtained from non-contaminated fish. Prior to the test of taurine transport activity the cells were exposed to oil-containing saline media for 20-1380 min. The saline media containing WSF of crude oil were produced as discribed by Østgaard & Jensen (21); crude oil and media (1:10) were slowly stirred for 10 days in a glass jar in total darkness. The resulting test medium (stock solution) was stored in darkness at 4 ^{0}C for a maximum period of three weeks. Dilutions of the stock solution were made by addition of different volumes of oil-free media. The concentration of oil (WSF) in the stock solution was determined with the GLC method after extracting the hydrocarbons with di-chloro methane.

2.2. Sampling and handling of the erythrocytes

Blood was obtained by puncturing the bulbus arteriosus with a heparinized syringe. Erythrocytes were separated from plasma by centrifugation at (4000 rpm for 20 sec.) 10 ^{0}C. The cells were washed three times in normal medium (330 mOsm, 160 mM Na, 3.2 mM KCl, 1.3 mM $MgSO_4$, 3.75 mM $CaCl_2$, 5.85

mM Choline-Cl, 12.2 mM TRIS-HCl, 0.3 mM Taurine and 0.3 mM GABA. pH = 7.4 at 10 °C).

The equilibration of the cells in different media and the influx measurements were performed in a water bath at 10 °C. To achieve a proper stirring the erythrocytes and media were mixed in plastic syringes (5 ml) which contained an air buble (0.5 ml) and were rotated 30 times $min.^{-1}$.

2.3. Measurements of taurine influx

After a preincubation period the initial transport velocity (influx) of taurine into the cells was measured, by exposing the eyrthocytes for 20 min. to media containing 14-C labelled taurine.

2.3.1. The effect from in vivo exposure of fish to oil-containing water on the sodium and the osmolality dependent taurine transport systems was tested in oil-free media. The erythrocytes from one fish were divided in batches and prior to the influx studies, they were equillibrated (preincubated) to media containing different concentration of sodium (0, 113 or 160 mM) and osmolalities (255 or 330 mOsm). Sodium-free media were made by replacing sodium chloride with choline chloride, and lowered osmolality media (255 mOsm) by reducing the concentration of sodium- or choline-chloride respectively. A more detailed description of the media composition, incubation procedure and the 14-C taurine method used has been presented (13,17,18).

2.3.2. Effects of different oil concentrations on taurine transport was studied in vitro. Erythrocytes from one fish only was used in each experiment. After 4 hours equilibration in normal medium, the cell sample was divided in two. One fraction was exposed to oil-containing media and the other was kept in oil-free media (control). The oil-exposures lasted 20 or 240 min.

2.3.3. The reversibility of the observed effect was tested in vitro in experiments where cells initially were equilibrated to normal media, then half the cells were exposed to oil-containing media for 4 hours. The other half was a control. The cells exposed to oil were then divided in two batches - one was continued exposed to oil, and the other was washed three times in oil-free media and finally equillibrated in oil-free media for a period of 19 hours. During the experiments the different media were replaced at fixed intervals. The taurine influx was tested in cell samples from all groups.

2.4. The cell volume

The cellular water content (% H_2O) was used as a measure of cell volume and the water content was determined as the difference between wet weight and weight after drying the cells at 105 °C for 24 hours.

2.5. Statistics

The Student's t-test was used to test the significance between the exposed and the control groups. Differences with a p value less than 0.05 were considered significant.

3. RESULTS

3.1. The effect of 1 month in vivo exposure to 0.1-0.3 ppm of WSF

3.1.1. Effects on the condition factor (C.F.) of the flounder in the three series used in the experiments is shown in Table 1. No significant differences were observed between the control and the oil-exposed fish in each serie, or between the two control groups in each serie.

3.1.2. Effects on the taurine influx is shown in Table 2. Taurine influx in cells from fish with low condition factor (serie I, Table 1) was reduced significantly ($p<0.01$) both in sodium media and in sodium-free media (serie I, Table 2).

In erythrocytes from fish in good condition (series II & III, Table 1) a reduction in taurine influx was observed in 6 out of 8 groups tested (series II & III, Table 2), the reductions were, however, not significant.

TABLE 1. The condition factor (C.F.) of the flounder used in three experiments. Controls were taken at the beginning (C1, C3, C5) and the end (C2, C4, C6) of each experiment. O1, O2, O3 are fish exposed for 1 month to 0.1-0.3 ppm WSF of crude oil. Mean values±S.D., n the number of fish.

Serie	Group	n	C.F.± S.D.
	C1	8	0.94±0.11
I	O1	8	0.88±0.03
	C2	6	0.90±0.06
	C3	8	1.01±0.13
II	O2	8	0.98±0.14
	C4	8	1.07±0.11
	C5	8	1.18±0.11
III	O3	5	1.17±0.17
	C6	8	1.10±0.15

TABLE 2. Taurine influx in erythrocytes from flounder exposed for 1 month to 0.1-0.3 ppm WSF of crude oil. The groups are the same as in Table 1. Mean values±S.D., n the number of fish. The osmolality of the testmedia was 330 mOsm.

			Taurine influx (μmole·g dry wt^{-1}·min^{-1})·10^3			
			Sodium concentration in the media (mM)			
			0		160	
			Taurine concentration in the media (mM)			
Serie	Group	n	0.3	12	0.3	12
I	O1	8	0.16±0.07*	-	2.8±1.0*	-
	C2	6	0.41±0.09	-	7.6±2.3	-
II	O2	8	0.26±0.13	5.4± 2.7	5.7±1.9	18.7±5.0
	C4	8	0.41±0.27	11.3±10.6	6.4±2.3	23.4±8.9
III	O3	5	0.38±0.13	9.4±5.4	4.5±1.8	19.2±7.8
	C6	8	0.26±0.11	6.4±2.9	5.5±1.6	24.0±5.8

* The difference between control and oil-exposed group is significant ($p<0.01$).

In these experiments there were no significant differences in cell volume, measured as the water content, between the erythrocytes from the controls and the oil-exposed fish. (Preincubation and influx at 330 mOsm,

Table 3).

3.1.3. Effects of oil-exposure on the osmolality dependent transport mechanisms was tested on cells from oil-exposed fish. Before the taurine influx measurements the cells were preincubated for 4 hours in low osmolality media (255 mOsm). The influx was measured either in the same medium or in normal osmolality medium (330 mOsm). Taurine influx in cells from oil-exposed fish was significantly reduced in three of four different tests (Table 4).

TABLE 3. The water content in flounder erythrocytes expressed as percentage water. The groups are the same as in Table 1. Mean values±S.D., n the number of fish.

		Water content (g cellular water·g wet weight^{-1})·100					
		Media osmolality (mOsm)					
	Preincub.:	330		255		255	
	Influx :	330		255		330	
Serie	Group		n		n		n
	C1	65.6±1.2	16	-		-	
I	O1	64.6±1.1	6	-		-	
	C2	65.0±0.6	4	-		-	
	C3	63.4±0.3	4	-		-	
II	O2	64.5±2.0	8	70.4±4.5	7	64.0±1.9	4
	C4	64.5±1.7	5	70.6±2.3	7	64.6±1.6	4
	C5	66.5±1.0	4	71.0±1.8	5	66.3±1.3	3
III	O3	66.1±0.3	2	70.1±2.8	4	63.7±0.3	2
	C6	65.7±1.6	5	69.1±1.1	6	64.5±0.3	4

TABLE 4. Effects of oil-exposure on the osmolality dependent taurine transport in flounder erythrocytes. The groups are the same as in Table 1. The media contained 12 mM taurine and 113 mM sodium. Mean values±S.D., n the number of fish.

			Taurine influx (μmole·g dry wt^{-1}·min^{-1})·10^3	
			Medium osmolality (mOsm)	
		Preincub.:	255	255
Serie	Group	n Influx :	255	330
II	O2	7	32.6±12.2*	15.4±6.7*
	C4	8	52.1±14.4	25.1±8.9
III	O3	5	30.4±9.8*	17.6±7.2
	C6	8	56.0±9.4	24.3±11.1

* The difference between control and oil-exposed group is significant ($p<0.05$).

The normal stimulatory respons on taurine influx from reduced osmolality (13), was present in cells from oil-exposed fish, but reduced in magnitude compaired to the controls (preincubation and influx at 255 mOsm, Table 4).

The reversibility of this stimulatory respons was also present in the cells from the oil-exposed fish (preincubation at 255 and influx at 330 mOsm, Table 4).

In the results shown in Table 4, no effect of oil-exposure was observed on the cell volume (preincubation at 255 and influx at 255 and 330 mOsm, Table 3). Cells from oil-exposed and control fish responded similarly to changes in medium osmolality, and their volumes (e.g. water content) were the same in media with identical osmolality.

3.2. The effect of in vitro exposure

3.2.1. The effect of 4 hours in vitro exposure of erythrocytes to 3-5 ppm WSF of crude oil on taurine influx was tested at four different taurine concentrations. The results of three experiments are shown in Table 5. The reduction in taurine influx in oil-exposed cells is expressed as percentage of the corresponding control. The reduced influx was most pronounced at the highest medium taurine concentrations (4 and 12 mM), and significant at 12 mM taurine. In expereiment I (Table 5) the influx values at 0.3 mM taurine was very low in the control and made it difficult to observe any inhibitory effect.

TABLE 5. Inhibition of taurine influx in flounder erythrocytes, exposed 4 hours *in vitro* to oil-containing media (3-5 ppm). The values are expressed as percentage reduction compared to corresponding controls measured in oil-free media (100%).

	Taurine influx (% of control)			
Taurine in media (mM):	0.3	1.0	4.0	12.0
Experiment nr.				
I	0.0	6.0	21.1	32.5
II	9.0	7.3	52.3	54.6
III	23.7	34.8	48.8	33.3
$\bar{X}\pm$S.D.	10.7±11.6	16.0±16.3	40.7±17.1	40.1±12.5*

* The reduction is significant ($p<0.05$).

Taurine influx at different medium concentrations of taurine in presence and absence of sodium, in oil-containing and oil-free media, is shown in Fig. 1. In oil-exposed cells the taurine influx in sodium media was reduced, but no effect was observed in sodium-free media. This means that it is the sodium dependent transport component that is inhibited (curve b as compaired to the control, curve a, Fig. 1). In four separate experiments the same effect was observed on the sodium dependent componet. The effect on the sodium independent component was, however, not so consistent.

3.2.2. The effect of 20 min. exposure of cells to oil-containing media was tested and the same effect was observed as after 4 hours exposure. This indicates that the inhibition occured immediately when petroleum hydrocarbons were presented to the cells.

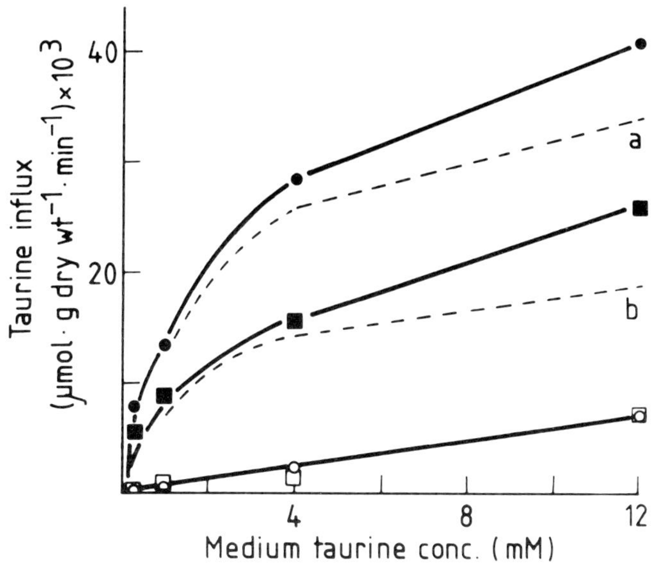

FIGURE 1. Taurine influx at four different medium concentrations of taurine. ● ○ : taurine influx in control cells, ■ □ :taurine influx in oil-containing media (3-5 ppm WSF). Filled symbols are from sodium-media and open symbols from sodium-free media. Curve a and b are produced from a subtraction of the values in sodium-free media from the respective values in sodium-media. These curves rerpresent the sodium dependent influx (a in control, b in oil-exposed cells).

3.2.3. The effect of 4 hours in vitro exposure to different concentrations of WSF of crude oil on the taurine influx is shown in Fig. 2. The influx in cells exposed to oil-containing media are expressed as percentage of the corresponding control (100 %) in each experiment. A redused taurine influx was observed at all oil-concentrations tested. The inhibition of the influx increased with increasing medium concentration of oil (Fig. 2). The lowest concentration tested was approximately 0.001 ppm (0.01 % WSF) which resulted in 10 % inhibition. At the highest concentration tested - 3-5 ppm (100 % stock solution), 40 % inhibition was observed. The cell volume was the same in the oil-exposed as in the control cells.

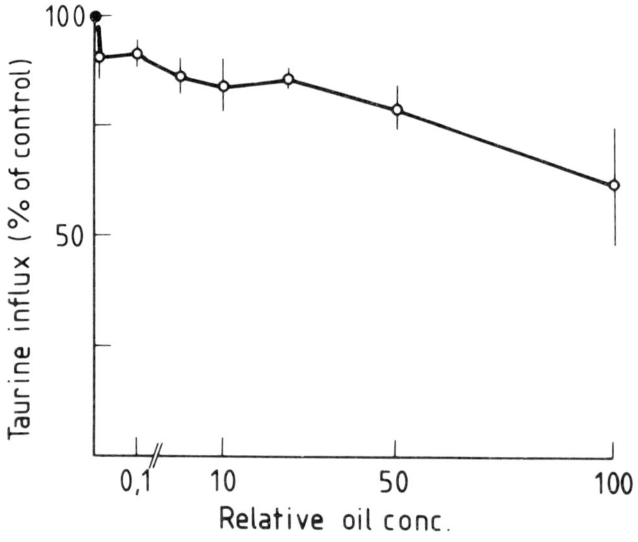

FIGURE 2. Taurine influx at different dilutions of oil-containing medium (100 % WSF). The values are given as percentage of taurine influx in control medium (330 mOsm, 160 mM Na, 4.0 mM taurine). Mean values±S.D. (n = 4).

3.2.4. Reversibility of the observed inhibitory effect of oil on taurine influx was tested in 4 experiments. The influx in oil-exposed cells and cells reequilibrated in oil-free medium are expressed as percentage of the control (100%) (Table 6). After 4,6 and 23 hours exposure to oil-containing media, taurine influx was redused to approximately 80 % of the control. In cells reequilibrated in oil-free media, the influx values increased to 95-98 % of the coresponding control (6 and 23 hours, Table 6).

TABLE 6. Reversibility of the inhibitory effect of oil-containing media on taurine influx in flounder erythrocytes. The cells were first equilibrated 4 hours to oil-containing medium, then washed and incubated in oil-free medium. Taurine influx is expressed as percentage of the corresponding control values (100%). Mean values±S.D. of 4 tests are given.

	Taurine influx (% of the corresponding control)		
Incubation time (hr):	4	6	23
Control	100	100	100
Oil exposed	76.5±4.0	80.4±3.1	82.7±10.1
Reequilibrated	76.5±4.0	94.6±3.1	97.9± 5.1

4. DISCUSSION

4.1. The oil concentrations used in the experiments

In this project two different methods of oil-exposure were performed; in vivo exposure of fish to WSF of crude oil, and in vitro exposure of erythrocytes to media containing WSF of crude oil.

4.1.1 In the in vivo experiments the oil concentration was measured both by fluorescence spectroscopy and GLC. Both the volatile components (with a boiling point less than C7) and the less volatile components (boiling boint higher than C7) were analysed by GLC. The totale concentration varied between 0.1-0.3 ppm, but less than 2 % of this concentration of petroleum hydrocarbons was due to the volatile components with a boiling point higher than touluene (C7) (20). Volatile components evaporate very rapidly from sea water (20,21,22). During the in vivo exposure the seawater was aerated and the dosage system for WSF containing water was semi-open allowing volatile components to evaporate. The concentration used was therefore comparable to that measured after oil spills (23).

4.1.2. In the in vitro studies only the less volatile components were analysed by GLC (extraction by di-chloro methane). The concentration of petroleum hydrocarbons in the stock solution varied between 3-5.3 ppm. According to Østgaard & Jensen (21) the volatile components not analysed by the extraction method used, could in this case give a contribution of approximately 6-10 ppm, leading to a total concentration of 9-15 ppm. However, just opening the bottle containing the stock solution, and mixing/ transfering oil-contaminated media to the incubation syringes, allowed volatile components to escape and thus reduce the concentration. The exact concentration used in the in vitro tests are therefore uncertain. The maximum concentration used might however, have been between 3-15 ppm and the minimum between 0.0003-0.0015 ppm (approximately 0.001 ppm).

4.2. The effect on taurine influx

The results shows that both type of oil-exposure inhibited taurine transport (Table 2,4,5,6 and Fig. 1 and 2). Both the sodium dependent and the sodium independent transport component were inhibited after in vivo exposure (Table 2). The sodium dependent taurine transport is assumed to be a carrier mediated system (13). The observed inhibition indicates therefore an effect on this membrane transport system.

The results on the sodium independent component were not so clear. The independent component might reflect simple diffusion (Fig. 1), but it also seems to contain a component of mediated exchange diffusion - involving special carrier molecules (13). The observed responses on the sodium independent transport components may therefore reflect this complexity.

A significant reduction in taurine influx was observed on both transport mechanisms involved in cells from fish in low condition (serie 1,Table 2). Fish in low condition seems to be more vulnerable to oil-pollution than fish in good condition, and this may be due to a genereal reduced ability to resist infections and pollutants. At present it is not possible to give an exact explenation of the mechanisms involved, but the inhibition of the sodium dependent carrier mediated system might be due to structural transformations of the cell membrane, blocking of the active sites of the carrier molecules, or by inhibition of the energy supply to the transport.

It has previously been assumed (9) that exposure of fish to oil-contaminated water might lead to a reduced energy supply in the fish as a secondary effect due to damages of gill lamellae and thus reduced oxygen uptake. An exposure of fish to oil-contaminated water might therefore lead to a reduced energy supply to active transports in cell membranes and thereby affect both the extracellular anisosmotic and the isosmotic intracellular volume regulation in fish.

However, a reduced influx of the osmotically active cellular solute taurine did not manifest in any effect on the cell volume, either the cells were exposed in vivo (1 month) or in vitro (20 - 1380 min.) to oil-containing water (Table 3). These results indicate that the effect of petroleum hydrocarbons on the cell membrane e.g. bond to lipids (7), is not serious for the cells. The flounder erythrocytes seems to have compensating mechanisms by which the volume is maintained at normal size although the membrane transport of a component involved in regulation of the cell volume is inhibited.

The individual variation between the flounders in biochemical and physiological processes such as osmoregulation and cell membrane transport activity is large. This was the case also for our results both from oil-exposed and control groups. This situation makes it difficult to obtain statistically significant differences between the two groups (Table 2). More precise information on the degree of inhibition might have been achieved if erythrocytes had been collected from the same fish prior to exposure, during oil-exposure and after exposure, but an attempt on this was unfortunatly unsuccessfully.

Our results suggests that the inhibition of taurine influx in erythrocytes seemed to be reversible (Table 6). The results shows that inhibition occured as soon as petroleum hydrocarbons were presented to the cells and persisted until the hydrocarbons were washed out (Table 6). Previous investigations have shown that fish metabolise petroleum hydrocarbons (24,25,26) and that accumulated hydrocarbons are released from tissues when the fish is returned to hydrocarbon-free seawater (1,2,27). These observations and the results presented here, implies that the inhibition of transport mechanisms in cell membranes might be a temporal effect acting until the accumulated petroleum hydrocarbons either are metabolised, or excreated.

4.3. Conclusions

Sublethal concentrations of WSF of oil inhibited the transport mechanism of taurine in flounder erythrocytes both in vivo and in vitro. The inhibition occured imidiately after adding petroleum hydrocarbons to the cells, but the inhibition was reversible if oil-exposed cells were washed and incubated in oil-free media. Although a basic mechanism involved in regulation of the cell volume was inhibited by oil-exposure, the erythrocytes were able to maintain a constant cell volume.

In spite of the large individual variations between flounders in cell membrane transport activity, we belive that this erythrocyte system is suitable to study effects of lipophilic polutants/contaminants on cell membrane mechanisms.

A direct transfer of the results obtained on this transport system to other transport mechanisms is not recomended, but the observed effects may give indications of what can be expected as a concequence of oil-exposure.

5. AKNOWLEDGEMENTS

This project was sponsored by the Norwegian Marine Pollution Research and Monitoring Program (FoH). The studies were performed at The Department of Biology, University of Oslo, and the GLC analyses by R. Berg, R. Lichtenthaler, F. Oreld and T. Øfsti at the Central Institute of Industrial Research, Oslo. The authors wish to thank prof. R. Lange for helpful comments on this paper.

REFERENCES

1. Neff J.M., B.A. Cox, D. Dixit & J.W. Anderson: Accumulation and release of petroleumderived aromatic hydrocarbons by four species of marine animals. Mar. Biol., 38(3), 279-289, 1976.
2. Solbakken J.E., S. Tilseth & K.H. Palmork: Uptake and elimination of aromatic hydrocarbons and a chlorinated biphenyl in eggs and larvae of cod Gadus morhua. Mar. Ecol. Prog. Ser., 16, 297-301, 1984.
3. Kuhnholdt W.W.: Investigation on the toxicity of seawater-extracts of three crude oil on eggs of cod (Gadus morhua L.). Ber. dt. wiss. Kommn. Meeresforsch., 23, 165-180, 1974.
4. Hawkes J.W.: The effects of petroleum hydrocarbon exposure on the structure of fish tissues. In: Wolf D.A.(ed): Fate and Effects of Petroleum Hydrocarbons in Marine Organisms and Ecosystems. pp. 115-128. Oxford, Pergamon Press, 1977.
5. Haensly W.E., J.M. Neff, J.R. Sharp, A.C. Morris, M.F. Bedgood & P.D. Boem: Histopathology of Pleuronectes platessa L. from Aber Wrac'h and Aber Benoit, Brittany, France: long-term effects of the Amoco Cadiz crude oil spill. J. Fish Diseases, 5, 365-391, 1982.
6. Tilseth S., T.S. Solberg & K. Westrheim: Sublethal effects of water-soluble fraction of Ekofisk crude oil on the early larval stages of cod (Gadus morhua L.). Mar. Environ. Res., 11, 1-16, 1984.
7. Stone W.L.: Hydrophobic interaction of alkanes with lipoesoms and lipoproteins. J. biol. Chem., 250, 4360-4370, 1975.
8. Sabo D.J. & J.J. Stegemann: Some metabolic effects of petroleum hydrocarbons in marine fish. In: Vernberg F.J., A. Calabrese, F.P. Thurberg & W.B. Vernberg (eds): Physiological Responses of Marine Biota to Pollutants. pp.279-287. New York, Academic Press, 1977
9. Kiceniuk J.W. & R.A. Khan: Toxicology of chronic crude oil exposure: Sublethal effects on aquatic organisms. In: Nriagu J.O.(ed): Aquatic Toxicology, Advances in Environmental Science and Technology, Vol 13, pp. 425-436, New York, John Wiley and Sons, 1983.
10. Macknight A.D.C. & A. Leaf: Regulation of cellular volume. Physiol. Rev., 57 (3), 510-573, 1977.
11. Vernberg F.J., A. Calabrese, F.P. Thurberg & W.B. Vernberg (eds): Physiological Responses of Marine Biota to Pollutants. 462 pp. New York, Academic Press, 1977.
12. Reiersen L.-O.: The Biology of Common Flatfishes; A Selected Bibliography. 225 pp. ICES. C.M. G:76, 1981.
13. Fugelli K. & L.-O. Reiersen: Volume regulation in flounder erythrocytes. The effect of osmolality on taurine influx. In: Barker Jørgensen C. & E. Skadhauge (eds): Osmotic and Volume Regulation. pp. 418-428, Copenhagen, Munksgaard, 1978.
14. Fugelli K.: Regulation of cell volume in flounder (Pleuronectes flesus) erythrocytes accompanying a decrease in plasma osmolality. Comp. Biochem. Physiol., 22, 253-260, 1967.
15. Fugelli K.: Gamma-aminobutyric acid (GABA) in fish erythrocytes. Experientia, 26, p 361, 1970.
16 Fugelli K.: Amino acid transport coupeled to cell volume regulation in anisosmotic media. In: Gilles R. (ed): Animals and Environmental Fitness. pp. 27-41, New York, Pergamon Press, 1980.
17. Fugelli K. & K.E. Zachariassen: The distribution of taurine, gamma-aminobutyric acid and inorganic ions between plasma and erythrocytes in flounder (Platichthys flesus) at different plasma osmolalities. Comp. Biochem. Physiol., 55A, 173-177, 1976.

18. Fugelli K. & H. Rohrs: The effect of Na and osmolality on the influx and steady state distribution of taurine and gamma-aminobutyric acid in flounder (Platichthys flesus) erythrocytes. Comp. Biochem. Physiol., 67A, 545-551, 1980.
19. Johannessen K.I.: A flow-through extraction and dosage system for long-term bioassays with water-soluble oil hydrocarbons. Mar. Environ. Res., 9, 123-129, 1983.
20. Reiersen L.-O. & J.A. Berge: Increased swimming acrtivity of sand goby, Pomatoschistus minutus, exposed to low concentrations of water soluble fraction of North Sea crude oil. In: Gray J.S. & M.E. Christiansen (eds): Marine Biology of Polar Regions and Effects of Stress on Marine Organisms. pp. 379-387, Chichester, John Wiley and Sons, 1985.
21. Østgaard K. & A. Jensen: Preparation of aqueous petroleum solutions for toxicity testing. Environ. Sci. Technol., 17(9), 548-553. 1983.
22. Johnson J.C., C.D. McAuliffe & R.A. Brown: Physical and chemical behavior of small crude oil slicks on the ocean. In: McCarthy Jr. L.T., G.P. Lindblom & H.F. Walter (eds): Chemical Dispersants for the Control of Oil Spills. ASTM STP 659, pp. 141-158, Philadelphia, American Society for Testing and Materials, 1978.
23. Berge J.A., K.I. Johannessen & L.-O. Reiersen: Effects of the water soluble fraction of North Sea crude oil on swimming activity of the sand goby, Pomatoschistus minutus (Pallas). J. Exp. Mar. Biol. Ecol., 68, 159-167, 1983.
24. Lee R.F. & G.H. Dobbs: Upptake, metabolism and distribution of polycyclic aromatic hydrocarbons by marine fish. Mar. Biol., 17, 201-208, 1972.
25. Pedersen M.G., W.K. Herschberger & M.R. Juchau: Metabolism of 3,4-benzopyrene in rainbow trout (Salmo gairdneri). Bull. environ. Contam. Toxicol., 12, 481-486, 1974.
26. Corner E.D.S.: The fate of fossil fuel hydrocarbons in marine animals. Proc. R. Soc. (Ser. B), 189, 391-413, 1975.
27. Anderson J.W.: An assesment of knowledge concerning the fate and effects of petroleum hydrocarbons in the marine environment. In: Vernberg W.B., F.P. Thurberg, A. Calabrese & F.J. Vernberg (eds): Marine Pollution: Functional Responses. pp. 3-21, New York, Academic Press, 1979.

ASSESSMENT OF THE VULNERABILITY OF MARINE MAMMALS TO OIL POLLUTION

F.R. ENGELHARDT

Canada Oil and Gas Lands Administration,
Ottawa, Ontario, K1A 0E4, Canada

1. INTRODUCTION

A great deal of attention is paid to marine mammals in an evaluation of possible impacts from oil spills. The general topic has been reviewed previously with a management and fisheries orientation (1) and with specific reference to the Arctic environment (2). This paper will assess risk from an oil spill to marine mammals from the perspective of overall vulnerability of an individual, population or species.

A number of reports exist which link accidental oil spills with effects on marine mammals (Table 1), ranging from no visible effects, to coating by oil, clinical effects and the deaths of individuals. These reports are not definitive in that exposure conditions are not known, nor is the type of petroleum always the same, ranging from Bunker C (eg. Arrow, Kurdistan spills) to gas condensate (eg. Scotian Shelf blowout).

It is logical to assume that marine mammals would be as sensitive to petroleum hydrocarbons as are the land mammals (3). The expression of this sensitivity in terms of potential for damage is a function of the sum of the biological and physical factors. Establishing this sum is not simple, however, because the data base for marine mammal-oil interaction remains poor. Inferences have to be drawn from reports of oiling at sea, as well as the patchy data set from experimental oil exposure studies.

2. EXPOSURE PARAMETERS

A survey of physical and chemical characteristics, as well as the environmental fate, of spilled oil is useful to assess the probability for and consequence of exposure in marine mammals. The various species of marine mammals differ in their habits and habitats and the likelihood of exposure to oil varies similarly among the species. Further, the weathering of oil is an important influence on the nature of the toxic effect.

Table 2 presents an overview of modes of exposure of marine mammal groups to oil, where such exposure would significantly affect the individual animal. The table does not account for the young of a species, which may behave differently than the adults. For example, young seals are born on land or ice and may thus not be exposed to oil in or on the water.

2.1 Hydrocarbon composition

The composition of spilled oil can be expected to be an important factor in its effects on marine mammals. The proportion of low molecular weight hydrocarbons, for instance, relates directly to the acute toxicity of the oil. Low molecular weight hydrocarbons volatilize readily at the water surface and are known to cause nervous system effects when

Kuiper, J., Van den Brink, W.J. (eds), Fate and Effects of Oil in Marine Ecosystems. ISBN 90-247-3489-4

TABLE 1: Effects of oil spill incidents reported for marine mammals

Source of oil[1)]	Species[2)]	Location	Type of effect	Reference
-	Seal	Irish Sea	Coating	4
-	Seal	Antarctic	Coating	5
Torrey Canyon	Seal	English Channel	Coating, death	6
-	Harp seal	Gulf of St. Lawrence	Coating, death	7
-	Seal	Arctic	Coating	8-10
-	Sea otter Killer whale	Alaska	Coating, death Illness, death	9
Arrow	Grey seal pups	Nova Scotia	Death	11
Santa Barbara channel blowout	California sea lion, northern fur seal	California	Coating	12-16
-	Seal	Netherlands	Coating	17
-	Grey seal	Wales	Coating, death	18
-	Seal	Baltic Sea	Coating, death	19
Kurdistan	Grey seal, harbour seal	Cabot Strait	Coating, death	20
Sullom Voe oil terminal	European otter	Shetland Islands	Coating, clinical changes, death	21
Regal Sword	Humpback, fin and right? whales, white-sided dolphins	Cape Cod	None apparent	22
Nowruz oil field	Dolphins, dugong	Arabian Gulf	Death	23
blowout	Pilot whale, grey seal	Scotian Shelf	None apparent	24

1) unidentified sources are frequently from ship discharges

2) harp seal - _Phoca groenlandica_; sea otter - _Enhydra lutris_; killer whale - _Orcinus orca_; grey seal - _Halichoerus grypus_; California sea lion - _Zalophus californianus_; northern fur seal - _Callorhinus ursinus_; harbour seal - _Phoca vitulina_; European otter - _Lutra lutra_; humpback whale - _Megaptera novaeangliae_; fin whale - _Balaenoptera physalus_; right whale - _Balaena glacialis_; white-sided dolphin - _Lagenorhynchus acutus_; dugong - _Dugong dugon_; pilot whale - _Globicephala melaena_.

Table 2: Modes of exposure of marine mammals to oil as related to fate of a 'typical' spill of medium weight crude oil[1].

Fate parameter for spilled oil	Marine mammal group[2]							
	Baleen whales	Toothed whales	Fur seals	Phocid seals	Walrus	Sea otter[3]	Sirenians	Polar bear
Surface, open water - fresh	Igf	If	IC	I	IG	ICF	IF	-
- weathered	icgf	-	C	C	C	C	F	-
Ice - fresh	icgf	if	ic	ig	I	-	-	IF
- weathered	cgf	-	C	C	C	-	-	CGF
Dispersed	gf	-	-	-	F	-	-	-
Sedimented	gf	-	-	f	F	-	-	-
Beached - fresh	-	-	IC	I	I	IC	-	ICGF
- weathered	-	-	C	C	C	C	-	CGF

Notes:

1) Code: I,i - exposure by inhalation
C,c - exposure by coating
G,g - exposure by direct ingestion
F,f - exposure through contaminated food
Lower case letters indicate that only some of the species in a given group are susceptible to the exposure mode

2) This overview does not account for differences attributable to the behaviour of juveniles of a species.

3) Includes both the sea otter (*Enhydra lutris*) and the European otter (*Lutra lutra*).

inhaled. Skin and eye lesions may result from contact. In this instance, marine mammals are at risk when surfacing through the air-water interface to breathe.

Conversely, the medium and heavy weight petroleums contain and volatilize proportionately smaller amounts of the acutely toxic, low molecular weight hydrocarbons. Toxicity is more likely to be due to coating, whereas this more viscous oil tends to be more sticky and thus less easily washed off by seawater - it is noteworthy that many of the reports of oil coated seals are linked with spills of Bunker C.

2.2 Weathering of Oil

Spilled oil undergoes a number of physical and chemical changes which influence its composition and distribution at sea. The details of these processes have been extensively investigated (26, 27), but are outside the scope of this paper. It is relevant to note that as oil weathers at the sea surface, in the water column or on the beaches, there is a rapid liberation of the volatile/soluble component, with major losses occurring in less than one day. This suggests that acute inhalation toxicity of oil spilled at sea is a consideration for a short time period only. This is true in particular for single spills, while in the case of a continuously releasing blowout fresh oil will be present at the blowout site for the duration of the event and weathered oil at distance from the site of origin.

Dissolution and dispersion also occur during the weathering of oil. The lower molecular weight hydrocarbons can dissolve into the water under an oil slick, particularly the aromatic hydrocarbons. Whole oil may be dispersed as droplets in the water, either by wave action or by application of chemical dispersants. These aspects are likely to be of little direct effect on marine mammals since levels of oil in the water column tend to be low because of dilution. For example, concentrations of oil under slicks have been measured to be only a few to tens of ppm in the top metre, decreasing exponentially with depth and time. On a strictly theoretical basis, and based on the reasonable prediction that oil slicks have an average thickness of 0.1 mm, the highest concentration in the top 1 metre of water would be 100 ppm if the oil were thoroughly dispersed into it (28). In fact, however, the concentrations tend to be less because much of the volatile component of the oil evaporates, and also, dispersion is not complete.

2.3 Sediments

Oil entrained in the water column may enter marine sediments by association with inorganic materials, such as adsorption to silt particles, or by biological processes, including incorporation into zooplankton fecal pellets (26). In either case, significant exposure of marine mammals to oil in sediments would occur only in very restricted circumstances, requiring that the oil spill would influence the specific area which forms the restricted feeding ground of a population using benthic organisms as food. Walrus (*Odobenus marinus*) may be a good case in point, in that they use traditional haul-out areas and feed on benthic infauna. Even in these instances, direct toxicity effects would be unlikely. Exposure would more probably occur through the food vector.

2.4 Shorelines

Contamination of shorelines by spilled oil may affect marine mammals if the oil is beached in haul-out areas. Seals, sea lions, walrus and sea otters may be vulnerable to this type of exposure by oil. In the Santa Barbara Channel spill, for instance, California sea lions and northern fur seals were coated with beached oil, although no significant mortality could be attributed to this coating (13-16, 29). It has been suggested that coating of adult pelage may secondarily lead to the ingestion of oil by nursing young.

Oil on shorelines tends to be more weathered and thus more viscous. The physical effects of coating may thus be more serious in this setting than for exposure to a fresh spill at sea.

2.5 Oil and Ice

The behaviour of oil in ice is important in evaluating exposure of marine mammals in the Arctic or Antarctic. The topic has recently seen detailed review (30, 31). Ice modifies the spreading and fate of an oil spill since oil tends to concentrate in ice leads and in breathing holes, or by holding free oil on the sea surface up against an ice edge. The thickness of the oil layer tends to be greater than in an open sea spill, and low temperatures decrease the evaporation rate to slow down weathering. Viscosity also increases to promote a tendency for external coating.

Certain marine mammal species are particularly vulnerable to this problem. Narwhal (*Monodon monocerus*) and bowhead whales (*Balaena mysticetus*) make preferred use of ice edges and leads in their migration into the Arctic (32). Ringed seals depend on the ice surface to locate their birth lairs (33); these subnivean hollow structures include a breathing hole to connect to the water. Therefore, an under-ice oil spill in the transition zone region of the Arctic ice pack, where the birth lairs are found, could put ringed seals at risk. Oil under ice may also be a hazard to ringed seals because they may be using under-ice air pockets on extended sub-ice travels (34, 35).

2.6 Contamination of Food

Oil on the sea surface is unlikely to cause contamination of food species used by marine mammals, except in specialized circumstances. It is probable that skim-feeding baleen whales would ingest oil at the sea surface along with near-surface zooplankton (assuming of course that these species would feed in an oil slick, an assumption which is supported by only one report (22), on the *Regal Sword* spill). An additional possibility is the case of polar bears (*Ursus maritimus*), which may feed on their prime food source, ringed seals, even if these have been contaminated through surface coating. Experimental evidence and field reports suggest that polar bears do not have an aversion to ingesting petroleum (36), and may even seek it out.

Oil which has moved into the water column or into sediments may also contaminate marine mammal food sources. Invertebrates are well known for their potential to accumulate hydrocarbons, as demonstrated in both bivalves and crustaceans, including zooplankton (37,38). Fish, on the other hand, have less of a tendency to bioaccumulate oil fractions, perhaps because of their more effective ability to metabolize hydrocarbons (39). Examples of marine mammals at potential risk are walrus, bearded seals (*Erignathus barbatus*), crabeater seals (*Lobodon carcinophagus*), sea otters and a number of species of baleen whales, all of which rely on invertebrates for a large part or all of their diet.

3. EFFECTS OF EXPOSURE TO OIL

Based on experimental evidence, and to a lesser degree on reports of oiling at sea, one can segregate the major effects of oil in accordance with the mode of exposure. Coating, inhalation and ingestion of oil appear to differ from each other in their major effects because of differences in the fractions of oil to which the animal is exposed (ie. volatiles vs. non-volatiles) and differences in target sites for toxicity.

The effects of petroleum on marine mammals have been recently reviewed by the author in other publications (2,40). The material presented in this section is drawn from these reviews.

3.1 Coating

Adherence of oil to the skin or fur of marine mammals is a problem mainly in those species with a well-developed pelage, that is, fur seals, sea otters and the polar bear. There is little evidence of coating with oil in species with poorly developed or no pelage, which include the phocid or hair seals, walrus, cetaceans and sirenians. While the skin of most cetaceans is usually quite smooth over the entire body surface, there are exceptions. Some cetaceans have a roughened body surface, including the gray whale (*Eschrichtius robustus*) with large numbers of attached barnacles, the right whale with prominent rostral callosities (42) and the bowhead whale with dozens to hundreds of eroded areas particularly involving the skin of the head (43). Such roughened areas may increase the likelihood of oil adherence. An additional problem in baleen whales may be that baleen plates and fibres can be coated with oil.

A survey of the recorded effects of external coating is presented in Table 3. Thermoregulatory problems appear to be the most significant initial effect in the fur bearing marine mammals (44-47).

3.1.1 Physical/behavioural. Studies on oiled baleen have shown that oil will adhere, heavy oil in particular, to decrease water flow and filtration capacity (48,49). The significance to the animal is probably minor, since a few hours loss of food input to a great whale should be of little consequence (49). In the case of seals, coating with a heavy oil has been suggested to interfere with swimming ability, perhaps causing death by drowning (7,18). Seals appear to be able to rid themselves of oil coating if the oil is not of heavy weight, occurring rapidly in the case of ringed seals (25) and more slowly in elephant seal pups (29).

A significant direct behavioural effect in sea otters (21, 45, 50, 51) and polar bears is that these species are able to and will groom oil fouling their pelts.

3.1.2 External lesions. The skin of whales is known to be a complex tissue with high metabolic activity. In situ experiments, in which small skin areas of live dolphins and a sperm whale were exposed to gasoline or crude oils, showed that the skin can be damaged, although recovery was possible (49). Ringed seals showed no apparent skin damage, but did experience eye lesions when exposed to a simulated oil slick for 24 hours (25, 52).

This limited experimental data base suggested that lesions of external surfaces may be expected, particularly if the oil is light and aromatic, therefore more reactive, or the duration of coating is prolonged such was the case in polar bears (36).

3.1.3 Uptake of hydrocarbons. It is possible that across-skin transfer of hydrocarbons can occur, especially if the oil coating is fairly fresh. Both seal and cetacean skin are well vascularized, and uptake may occur when this tissue is in a vaso-dilated/non-diving state. It is generally difficult to segregate this route of uptake from other pathways, such as inhalation or ingestion.

3.1.4 Thermoregulatory. The predominant direct effect of oil coating is on body temperature balance, especially in those marine mammal species which rely on fur for their thermoinsulation. Disturbance of the boundary layers of trapped air or water in the fur can have potentially serious effects on the individual animal. Studies have been carried out on sea otters and polar bears to demonstrate this effect. Some comparative information is available for pinnipeds.

In general, the physical consequence of increased conductance across an insulating fur layer is a tendency for decreased body core temperature. The threat of hypothermia is compensated by an increase in metabolism. Sea otters showed little change in core temperature although subcutaneous temperatures below the oiled area dropped markedly. Presumably, core temperature was maintained by a near doubling of body metabolism. Such an increase occurs at the expense of body energy supplies, which are unlikely to be available indefinitely. It is probable that hypothermia would be the ultimate result of exposure in a natural setting. Polar bears and sea otters are particularly sensitive, although it may be inferred that the newborns of phocid seal species which rely on lanugo (a pre-natal fur) and metabolic activity for thermal balance (eg. harp seals) would also be at risk from an oil spill.

3.2 Inhalation

Inhalation would be a problem particularly with unweathered oils, and especially if they were of light weight. Volatile hydrocarbons will be

Table 3 - Effects of external coating with oil reported for marine mammals

Type of effect	Marine mammal group	Exposure parameters	Description of effect	Reference
Physical/ behavioural	Baleen whales	*In vitro* study; light and heavy oils	Water flow through baleen decreased, especially with heavy oil; filtration effectiveness decreased	48,49
	Toothed whales	-	Oil clogging blowhole of dolphin (cause of death?)	14
	Fur seals	Medium oil spill	Coating up to 75% of body area, clean within 1 mo.	29
	Phocid seals	Bunker C spill	Heavy coating and possible swimming problems; oil plugged nares (cause of death?)	7,9,11,18
		Light oil; experimental slick	Complete coating; clean within 1 d. in seawater	25
	Sea otters	Medium oil; experimental exposure to range of coverage	Grooming of fur; poor cleaning by seawater alone	45,50,51
		Bunker C spill	Grooming of fur	21
	Polar bear	Medium oil; experimental slick	Grooming of fur; poor cleaning by water alone	36
External lesions	Toothed whales	*In situ* exposure of small areas of skin to gasoline and crude oils	Epidermal and dermal damage; circulatory perfusion decreased; skin enzymes affected; recovery from damage possible	49
	Phocid seals	Light oil; experimental slick	Corneal and conjunctival damage to eyes; recovery possible	25,52
	Polar bear	Medium oil; experimental slick	Hyperkeratinitis and alopecia of pelage	36
Uptake of hydrocarbons	Phocid seals	Light oil; experimental slick	Low levels of hydrocarbons in blubber and other tissues	41
Thermo- regulatory	Phocid seals	Medium oil; *in vitro* studies	No change in heat conductance of pelt of adult; increase in young	44
		Light oil; experimental slick	No change in body core temperature	52
	Fur seals	Medium oil; *in vitro*	Heat conductance across pelt increased	44
	Sea otter	Medium oil; exposure to range of coverage; also *in vitro* studies	Heat conductance across pelt increased; metabolism increased; subcutaneous temperature decreased in oiled areas	44,45,46
	Polar bear	Medium oil; experimental slick; also *in vitro* studies	Heat conductance across pelt increased; metabolism increased; wind enhanced effect; subcutaneous temperature increased to enhance heat loss.	36,47

released to their greatest degree during the early stages of weathering. The acute effects of volatile petroleum fractions have been well documented in humans and laboratory animals (3), but there is a lack of definitive information for marine mammal species. None the less, the dependency of marine mammals on the air-water interface, where the concentration of evaporated volatiles would be the greatest, creates a potential vulnerability to toxicity effects from inhalation. Speculatively, aerosols of oil and seawater formed through wave action may also be inhaled.

The effects of exposure to volatile petroleum hydrocarbons in land mammals range from upper respiratory tract irritation, abnormal nervous system functions, noted as quivering, loss of coordination and convulsions, to anaesthesia, respiratory failure and death. The evidence for inhalation toxicity in marine mammals is limited to a study on ringed seals (52). Within minutes of exposure to a light crude oil slick, the seals showed signs of nervous system disturbance, such as quivering and some uncontrolled whole body movement, which persisted for the duration of the 24 hours of exposure. This exposure did not cause any respiratory tissue damage, and the animals recovered from the nervous system effects soon after removal from the slick. Uptake of hydrocarbons by inhalation, as well as simultaneous transfer across the skin, occurred into various body tissues, although the levels were only a few ul/kg. Clearance of absorbed hydrocarbons was rapid and apparently effective, using both renal and biliary routes of excretion (41).

3.3 Ingestion

The effects of ingestion of petroleum have been examined in four groups of marine mammals: toothed whales, phocid seals, sea otters and polar bears (Table 4). Additionally, baleen whales are vulnerable to ingesting oil when their baleen structures are coated.

3.3.1 Uptake of hydrocarbons. There is little question that petroleum hydrocarbons can be absorbed from the gastrointestinal tract to be distributed to various body tissues, as has been demonstrated in ringed seals and polar bears (36, 41, 53, 54). Storage in adipose tissues is of special interest in that long term effects may result from a later rapid release of hydrocarbons when the energy store is utilized. A case in point is the blubber of seals, used during seasonal migrations and also during the lactation period. Presumably, contaminating hydrocarbons would follow the same route as blubber fats which are transferred into milk, thereby exposing nursing young. The experimental findings that polar bears had no detectable hydrocarbons in their adipose tissue may be explained in that these animals were not eating after exposure to oil, and probably utilized their fat for energy in the post-exposure period (36).

A survey study for petroleum residues was carried out on thirteen species of stranded baleen and toothed whales. Petroleum hydrocarbon residues, using naphthalene as an indicator, were found in many tissues but especially in blubber (49). Similarly, the presence of hydrocarbons of possible petroleum origin was detected in a number of tissues of four species of seals from the Bering Strait, and especially in blubber, (56).

3.3.2 Metabolism and excretion. Both renal and hepatic/biliary mechanisms appear to be effective in clearing accumulated hydrocarbons from tissues. In both ringed seals and polar bears, hydrocarbon levels declined with time after the exposure period (36, 53-55). This is presumed to involve an active metabolism of aromatic hydrocarbons, promoted by an induction of mixed function oxygenases (MFO) in liver and

Table 4 - Effects of ingestion of oil reported for marine mammals

Type of effect	Marine mammal group	Exposure parameters	Description of effect	Reference
Uptake of hydrocarbons	Phocid seals	5ml light crude/ day for 5 days; fed in fish diet	Hydrocarbons in all ringed seal tissues examined; predominantly in blubberand liver	41,53
	Polar bear	medium crude; ingested through grooming of fur	Hydrocarbons in wide range of tissues, mainly in kidney, brain and bone marrow; none in adipose tissue	36,54
	Toothed whales	*in vitro*, phenanthrene	MFO system found in liver tissue of minke whale; induced by phenanthrene;	58
Metabolism and excretion	Phocid seals	5 ml light crude /day for 5 days; fed in fish diet	Urine of ringed seals showed very high levels of metabolized hydrocarbons; small amounts in bile; aryl hydrocarbon hydroxylase activity increased	53,55
	Polar bear	Medium crude; ingested through grooming of fur	Hydrocarbons in urine and bile	36,54
Clinical effects	Toothed whales	3.4-5 ml hydraulic oil/day; 5 days per week for three months; mineral oil control	No changes in plasma chemistry of dolphins; no histopathological changes	49
	Phocid seals	5 ml light crude /day for 5 days; fed in fish diet	No significant changes in clinical parameters of ringed seals for blood or plasma; suggestion of kidney histopathology; elevated cortisol levels, depressed thyroid hormones; cortisol breakdown rate doubled	25,52,55
		Up to 75 ml light crude; single dose	No sign of pathology in harp seal pups	52
	Sea otters	Medium crude; ingested through grooming of fur	No clinical abnormalities reported	45,50
		Bunker C; ingested through grooming of fur	Fatal haemorrhagic gastroenteropathy	21
	Polar bear	Medium crude; ingested through grooming of fur	Peripheral haemolysis, erythropoietic disfunction, renal lesions; death 4 to 5 weeks after exposure, with anaemia, elevated clinical enzymes, low blood sodium; elevated cortisol, depressed thyroid hormones	36,54

other tissues. This metabolic system, common to all mammals, has been demonstrated as being inducible in ringed seals, especially in kidney tissue (55). The MFO system has also been shown to exist in grey seal liver, to a greater degree in adults than in the young (57). This suggests that young seals are less able to rid themselves of accumulated hydrocarbons, thus increasing their vulnerability relative to adults seals.

The MFO system occurs also in whales, and has been identified by the presence of cytochrome P-450 in liver tissue of a number of cetacean species (49). The presence of this marker enzyme of the MFO system suggests that cetaceans should be able to detoxify petroleum hydrocarbons. A recent report characterizes the MFO system in the minke whale, *Balanoptera acutorostrata*, in some detail (58). Four MFO enzymes were identified in liver tissue of adult whales, but were incompletely represented in foetal liver tissue. *In vitro* tests in this species showed the MFO system to be inducible by phenanthrene, substantiating its ability to metabolize aromatic hydrocarbons.

3.3.3 *Clinical effects*. As shown in Table 4, seals and whales appear to be little affected by ingestion of petroleum oils, at least in small quantities of a few millilitres per day. Neither bottlenosed dolphins, *Tursiops truncatus* (49), nor ringed seals (25, 52), showed significant changes in routine clinical parameters. Histopathological changes were also not prevalent. There were, however, changes measured in the levels of two hormones, cortisol and thyroxine (55), showing a probable stress response to oil ingestion. Further, the fact that the cortisol breakdown rate doubled in the same animals which showed elevated MFO enzyne activities has led to speculation that a long term exposure to oil may become a physiological problem (55). The inference is that prolonged elevation of MFO activities could lead to eventual exhaustion of systemic reserves of cortisol, as well as of other steroid hormones, since it is the MFO system which breaks down steroids as a normal physiological function, even when not induced by aromatic hydrocarbons. The biological consequences of steroid abnormalities could be far-reaching if they occurred during hormonally dependent periods such as migration or reproduction phases.

In contrast to the lack of pathology noted for seals and dolphins, European otters showed damage to the gastrointestinal tract from ingestion of oil (21). Other studies on sea otters concentrated on thermoregulation effects, which were more immediate than any clinical changes. The polar bear studies, however, demonstrated that oil ingestion could have major clinical consequences. Pathological changes were observed in plasma, in whole blood and in tissues. The effects took over a month to develop, when hematopoietic and renal damage became acutely life-threatening (36, 54). Differences between the experimental results from seals and dolphins, as compared to those from sea otters and polar bears, may be related to the quantity of oil ingested (assumed to be large in the case of the polar bears), or perhaps more indirectly to gastrointestinal functions. Gastrointestinal clearance in seals, for instance, is known to be more rapid than for many mammals, which would decrease the amount of ingested oil which could be absorbed. It is difficult to verify such speculation through the sparse experimental data base for marine mammals; none the less, assessment of risk from oil pollution to marine mammals requires that such qualified inferences be made.

4. DETECTION AND AVOIDANCE

Undoubtedly the question of risk of an oil spill to marine mammals is related to whether or not they are able to detect oil at sea, and if they would avoid it given the opportunity. There is no definitive statement possible at present on this issue, considering the differences between experimental and field observations.

Reports on the ability of bottlenosed dolphins to detect oil have demonstrated that captive trained individuals were able to avoid an oil slick by visual, tactile and/or echolocation means, and once detected, would avoid the slick covered area (49, 59, 60). Detection or avoidance was less effective for a sheen of oil under low light conditions. The tactile sense seemed to be the most important of the several detection mechanisms.

On the basis of field observations, however, marine mammals may not necessarily avoid oil-covered waters. Gray whales migrating off the coast of California continue to swim and to surface through slicks formed from natural oil seeps (49). Several species of whales and dolphins were observed to swim and to feed in the slick area formed from the Regal Sword spill of Bunker C and No. 2 fuel oil (22). Feeding activity was noted both at the surface and below surface, and behaviour was apparently normal. Further, sea otters would not avoid waters covered with oil from a terminal spill of Bunker C in Shetland, but continued to dive through the oil layer during feeding excursions (21). Otters in test tanks also did not consistently avoid oil covered waters (51). At a condensate blowout from an exploratory well on the Scotian Shelf, a pilot whale, as well as a grey seal, were seen swimming in the condensate slick (24). Although these observations do not answer the question of detection ability, they show that these species at least do not always avoid oil on the sea surface. While there have been no written reports documenting detection and avoidance of oil by polar bears, it may be that this species is not adverse to ingesting oil. A number of anecedotes have stated that polar bears will drink petroleum oils found in camps or caches raided by them and captive polar bears will apparently lick lubricants from door hinges (36).

5. VULNERABILITY FACTORS

The various pieces of information describing effects of oil on marine mammals can be brought together with the oceanographic and biological variables for an assessment of vulnerability. This assessment may be accomplished through a series of interlinked questions, where a positive answer to one or several of them translates into a statement of vulnerability for a given marine mammal species. Appropriate questions which may be asked are as follows:

1. Is oil from a spill going to occupy an area of the sea, or of a shoreline, which is a habitat for the marine mammal species at a time when oil will be present?
2. Is the area in question 1 an exclusive or critical habitat for a population of the species?
3. Is the species threatened or endangered so that damage to a given population could be critical to species survival?
4. Are the individuals expected to not avoid oil exposure, either voluntarily, or by physical circumstance?
5. If the individuals avoid oil, would this displace them from a critical habitat?

6. Are the conditions of exposure such that a direct effect due to the toxicity of oil may be expected?
7. Is oil going to contaminate or cause the loss of an important food source for the population?
8. If food species are contaminated, are individuals of the marine mammal population likely to ingest enough oil to cause direct toxicity effects?
9. ...to cause latent physiological toxicity effects?
10. ...to cause a transfer of contamination to fetus or nursing young?

This evaluation of risk of oil pollution can be developed further by more detailed questions and qualifications, but, as they stand, they serve to integrate a disparate information base in a critical manner for impact assessment purposes. It can be concluded, however, that population-significant impact of oil on marine mammals is likely only in special circumstances, restricted to localities which may, at a certain time of year, host a large proportion of a sensitive population. Species which are considered as threatened or endangered are additionally vulnerable to oil impact.

REFERENCES

1. Geraci JR & St Aubin DJ: Offshore Petroleum Resource Development and Marine Mammals: A Review and Research Recommendations. Mar. Fish. Res., 42, 1-12, 1980.
2. Engelhardt FR: Effects of Petroleum on Marine Mammals. Chapter 7, in: ed. FR Engelhardt, pp. 217-43: Elsevier Applied Science Publishers, London, 1985.
3. Engelhardt FR: Environmental Effects of Petroleum on Mammals. Rev. Environ Toxicol. 1, 319-37, 1984.
4. Davis JL: Observations on the Grey Seal (Halichoerus grypus) at Ramsey Island, Pembrokeshire. Proc. Zool. Soc. London, 119, 673-32, 1949.
5. Lillie H: Comments in Discussion. Proc. Int. Conf. on Oil Pollution of the Sea, London, pp. 31-3, 1954.
6. Spooner MF: Biological Effects of the Torrey Canyon Disaster. J. Devon Trust Nat. Conser., 1967 Suppl., 12-19, 1967.
7. Warner RF: Environmental Effects of Oil Pollution in Canada: an Evaluation of Problems and Research Needs. Can. Wild. Serv. Ms. Rep. No. 645, 16-17, 1969.
8. Hess R & Trobaugh L: Kodiak Islands Oil Pollution. Event No. 26-70, Smithsonian Inst. Cent. Shortlived Phenomena, Washington, DC, Annu. Rep., pp. 150-3, 1970.
9. Morris R: Alaska Peninsula Oil Spill, Event No. 36-70, Smithsonian Inst. Cent. Shortlived Phenomena, Washington, DC, Annu. Rep., pp. 154-7, 1970.
10. Muller-Willie L: How Effective is Oil Pollution Legislation in Arctic Waters? Musk-Ox 14, 56-7, 1970.
11. Anon: Report of the Task Force - Operation Oil (Clean Up of the Arrow Oil Spill in Chedabucto Bay) to the Minister of Transport, Information Canada, Ottawa. No. 2, 46-7, 1970.
12. Nelson-Smith A: The Problem of Oil Pollution of the Sea. Adv. Mar. Biol., 8, 215-306m 1970.
13. Simpson JG & Gilmartin WG: An Investigation of Elephant Seal and Sea Lion Mortality on San Miguel Island. Bioscience, 20, 289, 1970.

14. Brownell R: Whales, Dolphins and Oil Pollution. In: Biological and Oceanographic Survey of the Santa Barbara Oil Spill, 1969-1970, Vol. 1, ed. D. Straughan: Allan Hancock Found. Sea Grant Publ., No. 2, University of Southern California, pp. 255-76, 1971.
15. Brownell R & Le Boeuf B: California Sea Lion Mortality; Natural or Artifact. In: Biological and Oceanographic Survey of the Santa Barbara Channel Oil Spill, 1969-1970, Vol. 1., ed. D Straughan: Allan Hancock Found. Sea Grant Publ., No. 2, University of Southern California, pp. 287-306, 1971
16. Straughan D: Oil Pollution and Wildlife and Fisheries in the Santa Barbara Channel. In: Trans. 36th North American Wildlife and Natural Resources Conference Symposium VII, ed JD Trefethen: Wildlife Institute, Washington, DC, pp. 219-29, 1971.
17. Van Haaften JL: Die Bewirtschaftung von Seehunden in den Niederlanden. Beitr. Jagd and Wildforsch., 8, 345-9, 1973.
18. Davis JE & Anderson SS: Effects of Oil Pollution on Breeding Grey Seals. Mar. Poll. Bull., 7, 115-18, 1976.
19. Pfister K: The 1979 Baltic Oil Spill. Ministry of Interior, Dept. Environmental Protection, Helsinki, Finland, 53 pp, 1980.
20. Parsons J, Spry J & Austin T: Preliminary Observations on the Effect of Bunker C Fuel Oil on Seals on the Scotian Shelf. In: Scientific Studies During the Kurdistan Tanker Indicent: Proceedings of a Workshop, ed. JH Vandermeulen: Bedford Inst. Oceanography, Dartmouth, Nova Scotia, Rep. BI-R-80-3, pp. 193-202, 1980.
21. Baker JR, Jones AM, Jones TP & Watson HC: Otter *Lutra lutra* L. Mortality and Marine Oil Pollution. Biol. Conserv., 20, 311-21, 1981.
22. Goodale DR, Hyman MAM & Winn HE: Cetacean Responses in Association with Regal Sword Oil Spill. Chapter CI, in: A Characterization of Marine Mammals and Turtles in the Mid-and North Atlantic Areas of the US Outer Continental Shelf, ed. RK Edel, MA Hyman & MF Tyrell: Cetacean and Turtle Assessment Program, Annual Report 1979, University of Rhode Island, pp. XI-1-XI-15, 1981.
23. Anon.: Death Toll Mounts in Gulf. Mar. Poll. Bull. 14, 322.
24. Martec Limited: Report on the Environmental Program Associated with the Blowout at Shell et al., Uniacke G-72, September 1984. Shell Canada Resources Limited, Dartmouth, Nova Scotia, Canada, 1984.
25. Geraci JR & Smith TG: Direct and Indirect Effects of Oil on Ringed Seals (*Phoca hispida*) of the Beaufort Sea. J Fish. Res. Board Canada, 33, 1976-84, 1976.
26. Mackay D: The Physical and Chemical Fate of Spilled Oil. Chapter 2, in: Petroleum effects in the Arctic Environment, ed. FR Engelhardt, pp. 37-61: Elsevier Applied Science Publishers, London, 1985.
27. National Research Council: Oil in the Sea, Inputs, Fates and Effects: National Academy Press, Washington, D.C., 1985.
28. McAuliffe CD: Organism Exposure to Volatile Hydrocarbons from Untreated and Chemically Dispersed Crude Oils in Field and Laboratory. Proc. Ninth Arctic Marine Oilspill Program Tech. Seminar, Edmonton, Alberta, P8-497-524, 1986.
29. Le Boeuf B: Oil Contamination and Elephant Seal Mortality: a 'Negative' Finding. In: Biological and Oceanographical Survey of the Santa Barbara Channel Oil Spill, 1969-1970, Vol. 1, ed. D Straughan: Allan Hancock Found. Sea Grant Publ., No. 2, University of Southern California, pp. 277-81, 1971.
30. Clark RC Jr. & Finley JS: Occurrence and Impact of Petroleum on Arctic Environments. In: The Arctic Ocean, ed. L Rey: Macmillan Press, London, pp. 295-341, 1982.

31. Dome Petroleum Ltd, Esso Resources Canada Ltd & Gulf Canada Resources Inc.: Hydrocarbon Development in the Beaufort Sea - Mackenzie Delta Region, Environmental Impact Statement, Calgary, Alberta, 7 Vols, 1982.
32. Braham H Fraker M & Krogman B: Spring Migration of the Western Arctic Population of Bowhead Whales. Mar. Fish. Rev. 42, 36-46, 1980.
33. Smith TG & Stirling I: The Breeding Habitat of the Ringed Seal (Phoca hispida), The Birth Lair and Associated Structures. Can. J. Zool., 53, 1297-1305, 1975.
34. Bertram GCL: The Biology of the Weddell and Crabeater Seals, with a Study of the Comarative Behaviour of the Pinnipedia. Br. Mus. (Nat. Hist.): British Graham Land Expedition 1934-1937, London, Scientific Reports, 1, 1-139, 1940.
35. Milne AR: Use of Artificial Subice Pockets by Wild Ringed Seals (Phoca hispica). Can. J. Zool., 52, 1092-3, 1974.
36. Øritsland NA, Engelhardt FR, Juck FA, Hurst RJ & Watts PD: Effect of Crude Oil on Polar Bears. Dept. of Indian Affairs and Northern Development Canada, Environmental Studies No. 24, 268 pp., 1981.
37. Varanasi U & Malins DC: Metabolism of Petroleum Hydrocarbons: Accumulation and Biotransformation in Marine Organisms. In: Effects of Petroleum on Arctic and Subarctic Marine Environments and Organisms, Vol. II, Biological Effects, ed. DC Malins, Academic Press, New York. pp. 175-270, 1977.
38. Neff JM: Polycyclic Aromatic Hydrocarbons in the Aquatic Environment - Sources, Fates and Biological Effects: Elsevier Applied Science Publishers Ltd, London, 262 pp., 1979.
39. Payne JF: Mixed-function Oxygenases in Biological Monitoring Programs, Review of Potential Usage in Different Phyla of Aquatic Animals. In: Ecotoxicological Testing for the Marine Environment, eds. G Personne et al., pp. 625 - 55: State Univ. Ghent and Inst. Mar. Scient. Res., Bredene, Belgium, Vol. 1, 1984.
40. Engelhardt FR: Petroleum Effects on Marine Mammals. Aquat. Toxicol., 4, 199-217, 1983.
41. Engelhardt FR, Geraci JR, & Smith TG: Uptake and Clearance of Petroleum Hydrocarbons in the Ringed Seal, Phoca hispida. J. Fish. Res. Board Canada, 34, 1143-7, 1977.
42. Slijper EJ: Whales, 2nd Edn, Cornell University Press, New York, 511 pp., 1979.
43. Albert TF: Some Thoughts Regarding the Effect of Oil Contamination on Bowhead Whales, Balaena mysticetus. In: Tissue Structural Studies and Other Investigations on the Biology of Endangered Whales in the Beaufort Sea, ed. T Albert, Appendix 9, Report to Bureau of Land Management, US Dept. of Interior, Anchorage, Alaska, pp. 945-53, 1981.
44. Kooyman GL, Davis RW & Castellini MA: Thermal Conductance of Immersed Pinniped and Sea Otter Pelts Before and After Oiling with Prudhoe Bay crude, in: Fate and Effects of Petroleum Hydrocarbons in Marine Ecosystems and Organisms, ed. DA Wolfe, Pergamon Press, New York, pp. 151-7, 1977.
45. Costa DP & Kooyman GL: Effects of Oil Contamination in the Sea Otter, Enhydra lutris. Report Outer Continental Shelf Environmental Assessment Program: NOAA, Alaska, 42 pp., 1980.
46. Costa DP & Kooyman GL: Oxygen Consumption, Thermoregulation, and the Effect of Fur Oiling and Washing on the Sea Otter, Enhydra lutris. Can J. Zool., 60, 2761-7, 1982.

47. Hurst RJ & Øritsland NA: Polar Bear Thermoregulation: Effect of Oil on the Insulative Properties of Fur. J. Therm. Biol., 7, 201-8, 1982.

48. Braithwaite LF, Aley MG & Slater DL: The Effects of Oil on the Feeding Mechanism of the Bowhead Whale. Final Report to Bureau of Land Management, US Dept. of Interior, Washington, DC, 45pp., 1983.

49. Geraci JR & St Aubin DJ: Study of the Effects of Oil on Cetaceans. Report for US Department of the Interior, Bureau of Land Management, Washington, DC, Contract No. AA-551-CT9-29, 274 pp., 1982.

50. Williams TD: Chemical Immobilization, Baseline Parameters and Oil Contamination in the Sea Otter. Marine Mammal Commission Rep. No. MMC-77/06, Washington, DC, 27 pp., 1978.

51. Siniff DB, Williams TD, Johnson AM & Garshelis DL: Experiments on the Response of Sea Otters _Enhydra lutris_ to Oil Contamination. Biol. Conser., 23, 261-72, 1982.

52. Smith TG & Geraci JR: The effect of contact and Ingestion of Crude Oil on Ringed Seals of the Beaufort Sea. Beaufort Sea Project Tech. Rep. 5, 67 pp., 1975.

53. Engelhardt FR: Petroleum Hydrocarbons in Arctic Ringed Seals, _Phoca hispida_, Following Experimental Oil Exposure. Proc. Conf. on Assessment of Ecological Impacts of Oil Spills, Keystone, Colorado, American Institute of Biological Sciences, pp. 614-28, 1978.

54. Engelhardt FR: Oil Pollution in Polar Bears: Exposure and Clinical Effects. Proc. Fourth Arctic Marine Oilspill Program Tech. Seminar, Edmonton, Alberta, pp. 139-79, 1981.

55. Engelhardt, FR: Hydrocarbon Metabolism and Cortisol Balance in Oil-Exposed Ringed Seals, _Phoca hispida_. Comp. Biochem. Physiol. 72C, 133-6, 1982.

56. Shaw DG, Else P, Gritz R, Malinky G, Mapes G, McIntosh D, Schwartz J, Smith E & Wiggs J: Hydrocarbons: Natural Distribution and Dynamics on the Alaskan Outer Continental Shelf. Section VIII: Hydrocarbons of Bering Sea Biota. In: Environmental Assessment of the Alaskan Continental Shelf: Annual Reports of Principal Investigators for the Year Ending March 1978, Vol. VIII: Contaminant Baselines: NOAA (EAL), US Department of Commerce, Boulder, Colorado, pp. 559-67, 1978.

57. Addison RF & Brodie PF: Characterization of Ethoxyresorufin O-de Ethylase in Grey Seals _Halichoerus grypus_. Comp. Biochem. Physiol., 79C, 261-3, 1984.

58. Goksøyr A: Initial Characterization of the Hepatic Microsomal Cytochrome P-450 - System of the Piked Whale (Minke) _Balaenoptera acutorostrata_. Mar. Environ. Res. 19, 185-203, 1986.

59. Smith TG, Geraci JH & St Aubin DJ: Reaction of Bottlenose Dolphins, _Tursiops truncatus_, to a Controlled Oil Spill. Can. J. Fish. Aquat. Sci. 40, 1522-1525, 1983.

60. St Aubin DJ, Geraci JR, Smith TG & Friesen TG: How do Bottlenose Dolphins, _Tursiops truncatus_, React to Oil Films Under Different Light Conditions? Can. J. Fish. Aquat. Sci. 42, 430-436, 1985.

EFFECTS OF DIFFERENT OIL DOSES, DISPERSANT AND DISPERSED OIL ON MACROFAUNA IN MODEL TIDAL FLAT ECOSYSTEMS

ROB DEKKER & GODFRIED W.N.M. VAN MOORSEL

Netherlands Institute for Sea Research
P.O. Box 59, 1790 AB Den Burg, Texel, Netherlands

1. ABSTRACT

Artificial intertidal mudflats were used to study effects of 'Forties' oil, a dispersant (Finasol OSR5) and dispersed oil. Oil was added in a 0.5 and a 0.1 mm layer. In the latter treatment, two exposure times were used. The effects on the development of stocked macrobenthic infauna species are presented here. As the experiments lasted 10 months, both short and long-term effects and interactions could be investigated.

The most important effects of oil were instant mortality in Cerastoderma edule (at the high dose), and a dose dependent mortality in Macoma balthica. Dispersant-alone had limited effects, but the temporary appearance at the sediment surface of a high proportion of the bivalves is of potential importance in natural situations. Dispersion of oil had more severe short-term effects (high mortality in C. edule, M. balthica and Arenicola marina) than mere presence of oil. Increased mortality of C. edule, due to frost at the end of the first experiment, was only observed in oil and dispersed-oil treatments. This is one of the ecotoxicological effects observed due to the large-scale and the long-term nature of the experimental set-up. Population development of the amphipod Corophium volutator was retarded, depending on oil or dispersed-oil dosage or exposure time. Negative effects of oil and dispersed oil on parameters like length growth and individual dry weight were compensated for by better growth of surviving specimens. Interspecific food competition may have been involved.

2. INTRODUCTION

Effects of toxic substances on marine macrobenthic animals are generally obtained from laboratory experiments. It is doubtful, however, if the results of such typical short-term dose-effect studies can be extrapolated safely to field conditions. Field research includes a natural situation, but the complexity of interactions between the many species present, the drawback of large year to year variations and logistic problems may constitute limitations.

Model ecosystems bridge the gap between the two above mentioned approaches. It is shown how different levels of oil pollution, combating oil spills with dispersants and dispersants alone effect the development of macrofauna in model ecosystems simulating an intertidal soft-bottom environment. Initial macrofauna densities can be replicated.

The first 2 y of the 3-y-OPEX project are described. OPEX, an acronym for Oil Pollution EXperiment, is a multidisciplinary project in which the Netherlands organisations TNO, RIN and NIOZ participate. More information (e.g. on the fate of oil, primary production, development of bacteria and meiobenthos) will be reported elsewhere. A preliminary report with all raw data obtained during the first two years of OPEX has been published (1).

The objective of OPEX is to collect information on effects of oil spills

Kuiper, J., Van den Brink, W.J. (eds), Fate and Effects of Oil in Marine Ecosystems. ISBN 90-247-3489-4

on tidal flat ecosystems and to provide basic knowledge for optimal treatment of such spills.

3. MATERIAL AND METHODS

3.1. Technical set-up

3.1.1. Facility. Eight out-door MOTIFs (MOdel TIdal Flats), experimental facilities of the Research Institute of Nature Management at the island of Texel (Netherlands), were used. The set-up of the MOTIFs is based on the design of indoor tidal flat ecosystems developed by de Wilde and Kuipers (2). Each MOTIF consists of a concrete and/or brick basin of 6 x 3.5 m^2 and 1.2 m deep (Figs 1 and 2). A .45 m high brick wall separates the 'tidal flat' from the 'tidal channel'. The tidal-flat area (18 m^2) is filled with natural sediment (see below), to the level of the wall with a slope of about 2°. Ropes over the MOTIFs prevent bird predation. A movable gantry is used to sample .1 m above the tidal flat to prevent excess disturbance of the sediment (Figs 1 and 2).

3.1.2. Tidal regime. MOTIFs are coupled in pairs. Each six hours, via time-switch operated valves, water is pumped from one MOTIF into the other, to create a diurnal tide. The sea water volume of each pair of MOTIFs is 12.3 m^3. The total duration of flood or ebb is 1¾ h. Once a week, the tidal

FIGURE 1. Model Tidal Flats as used in OPEX.

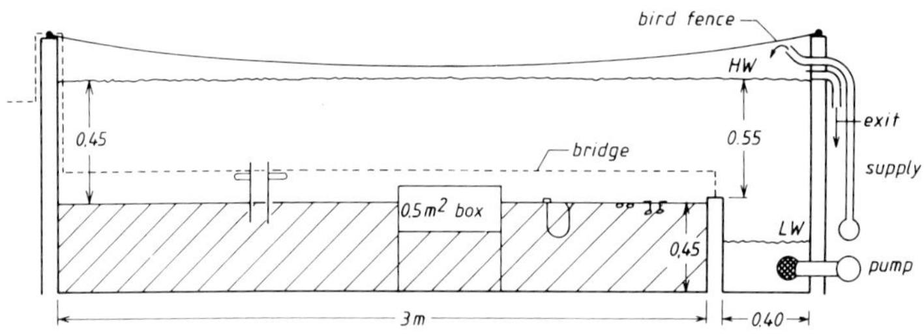

FIGURE 2. Cross section through MOTIF.

regime is reversed to prevent the same MOTIFs from always falling dry at the same time of the day. In each MOTIF, a wave board is deployed to mix the water column.

3.1.3. Sea water supply. All MOTIFs are permanently supplied with Wadden Sea water: 120 l h^{-1} $MOTIF^{-1}$ in 1984 and 50 l h^{-1} $MOTIF^{-1}$ in 1985. The average residence time of the water was 2 and 5 days, respectively. Five days is similar to the average residence time of water in the western part of the Wadden Sea. An overflow discharges surplus water from each basin.

3.1.4. Sediment. The sediment used was collected from a natural tidal flat near the southern tip of Texel. Median grain size was 230 µm. It was subjected to several climatological conditions for two weeks prior to its transfer into the MOTIFs in 1984. This was long enough to get rid of most of the associated fauna, only *Hydrobia ulvae* survived the two-week storage period. In March 1985 all sediments were removed from the MOTIFs. A mixture of sediment from the 1984 controls and new sedemnt was used for the 1985 experiment. Details about technical set-up may be found in (1).

3.2. Experimental set-up

3.2.1. Macrofauna introduction. Experiments started in March 1984 and 1985. Large macrofauna species, *Cerastoderma edule*, *Macoma balthica* and *Arenicola marina*, were stocked in the MOTIFs in densities similar to average densities on intertidal flats in the nearby Wadden Sea (3). Together, these three species account for about 75% of the total infaunal biomass in the Wadden Sea. *Nereis diversicolor*, *Corophium volutator* and *Hydrobia ulvae* were introduced in low densities.

3.2.2. Oil and Finasol introduction. In the 1984 experiment a relatively large spill was simulated in the MOTIFs. Apart from the effects of untreated and dispersant treated oil, the effects of the dispersant alone were also studied. In 1985 relatively small spills were simulated. Effects of dispersant-treated oil and untreated oil with different residence times were studied. Table 1 provides a synopsis of both experiments. Oil doses were added during falling water, which simulates a "worst case" oil spill. Moreover, the oil was distributed evenly over the tidal flats during stranding.

In both experiments, a North Sea crude oil from the Forties field was used. The lighter fraction (boiling point < 200°C) was removed before application to simulate a certain weathering period, in the order of one day. Finasol OSR5 was chosen as a dispersant, mainly because this product is held in stock by the Dutch authorities responsible for combat of oil spills in the Dutch sector of the North Sea.

On May 7, 1984, a layer of topped Forties with an average thickness of

TABLE 1. Summary of MOTIF treatments and fate of oil and dispersant in OPEX 1984 and 1985 experiments

Treatment		Year	MOTIFs	Oil in sediment	Oil in water	Dispersant in water
Control	C	1984	7 8	-	-	-
Control	C	1985	5 6	-	-	-
Finasol	Fi	1984	5 6	-	-	++
Forties .1 mm 1 day	Fo1.1	1985	1 2	+	+	-
Forties .1 mm 3 days	Fo1.3	1985	7 8	++	+	-
Forties .5 mm 5 days	Fo5.5	1984	3 4	+++	+	-
Forties .1 mm + Finasol	Fo1+Fi	1985	3 4	±	+++	+
Forties .5 mm + Finasol	Fo5+Fi	1984	1 2	+++	++	++

.5 mm (Fo5), was introduced in four MOTIFs. In two of them, the oil was treated with 450 ml spray of Finasol OSR5 (Fo5+Fi). Application of Finasol was only partially successful. Highest oil concentrations in water directly after addition were about 10% of calculated values. Consequently, the amount of undispersed oil in Fo5+Fi was not much less than in Fo5.5. Five days after oil treatment, floating oil was removed from the MOTIFs (Fo5.5, Fo5+Fi). During the year, oil concentrations in Fo5+Fi MOTIFs were as high as in Fo5.5 MOTIFs, probably due to a constant supply of undispersed oil from the sediment. The same amount of Finasol alone was added to two other MOTIFs (Fi). Nothing was added to the two remaining MOTIFs, which served as controls (C).

On May 6, 1985, an oil layer with an average thickness of .1 mm was introduced in four MOTIFs. From two MOTIFs all floating oil was removed one day after addition (Fo1.1). In the other two, residence time was three days (Fo1.3). Another two MOTIFs received the same amount of oil, premixed with 90 ml of the dispersant (Fo1+Fi). This method of addition was more effective than in 1984. Highest oil concentrations in the water were close to the calculated maximum (94%). These concentrations were twice as high as in 1984, and the amount of undispersed oil was much less than Fo5+Fi. Two remaining MOTIFs served as controls (C). The distribution of controls and contaminated systems over the MOTIFs in 1985 was different from that during the preceding experiment. This was done to correct for possible MOTIF effects. The water had an average residence time of three days during the first week after addition.

3.3. Sampling methods

3.3.1. Visual surface censuses. During the first 2 months after treatment, casts of *Arenicola marina* were counted regularly to obtain an impression of lugworm activity. Other visual particularities such as crawl tracks and appearance at the sediment surface were noted.

Empty shells ("doublets") of bivalves found at the sediment surface were removed and counted once a fortnight. Not each case of bivalve mortality results in the appearance of a doublet. Yet this method adequately signals sudden changes in mortality.

Visual censuses were carried out more frequently directly after oil addition.

3.3.2. Macrofauna samples. During each of the monthly Small Macrofauna Samplings (SMS), benthic macrofauna was extracted from 32 cores with a diameter of 66 mm and a length of .15 m. Total surface area per SMS was .109 $m^2.MOTIF^{-1}$. Fair estimates of the development of small macrofauna species (Nereis diversicolor, Corophium volutator and Hydrobia ulvae) were obtained. Specimens of Macoma balthica and Cerastoderma edule were present in too low densities to be of any value and were restocked.

During Large Macrofauna Samplings (LMS), four square .25 m^2 box cores were sampled to the bottom of the MOTIFs. LMSs gave larger numbers of M. balthica, C. edule and A. marina, but were taken less frequently than SMSs (four times per experiment, viz. in June, August, October and January/February of the next year). Numbers and biomass values are compared with populations that were introduced.

Samples were sieved over 1 mm square gauze. The material retained, including the living organisms, was collected in plastic bags and stored for maximally 2 days at 3°C until further analysis. After sieving, the remaining sediment was returned to the open core holes.

3.4. Analyses

Samples were sorted in shallow white dishes to the species level. Shell length of bivalves was measured with calipers to the nearest .1 mm. In Cerastoderma edule the "winter ring" was also measured. Because cockles laid down this ring at the start of the experiment, individual growth since this time could be calculated. Bivalves were boiled briefly and soft parts were removed from the shells. The reproductive stage was determined with a stereo loupe.

Arenicola marina and Cerastoderma edule were further treated individually and Macoma balthica in .5 mm size groups. Other species were pooled per MOTIF per sampling date only. All samples were placed in porcelain cups and dried to constant weight in a ventilated stove (at least 3 days) at 60°C. After determination of the dry weight, samples were combusted in a furnace for two hours at 560°C. This resulted in a weight loss: the ash-free dry weight (AFDW), which is used as a biomass value in this paper. Individual AFDW and total biomass m^{-2} were calculated. In the gastropod Hydrobia ulvae whole animals including shells were dried and ashed. Soft tissue AFDW of these animals was calculated as 90% of the total AFDW (4).

4. RESULTS

4.1. Direct effects

4.1.1. Cerastoderma edule. In 1984, the highest numbers that appeared at the sediment surfaces were found in the Fi treatment. The majority of the animals crept back into the sediment and only low mortality was seen (Fig. 3). In the field, this behaviour would probably result in higher mortality due to bird predation. Mortality in Fo1.1 and Fo1.3 was comparable to controls. In Fo5.5, 15% of the population appeared at the sediment surface and short term mortality was 25%. Fo5+Fi resulted in an emergence of 60% and a mortality of 35%. The effects of Fo1+Fi were much more pronounced. After emergence almost 100% died. The most probable explanation of this phenomenon is the higher concentration of oil in the water in 1985. But there was also a difference in the populations that were introduced. In 1984, the MOTIFs were stocked with first year cockles. The next year, this same year class (then second-year cockles) was used, due to limited availability of first year animals. Shells of the 1985 animals showed low growth during the preceding year, which suggests a relatively bad condition.

Shell growth in Fo treatments and in Fo5+Fi was initially inhibited compared with the controls (Fig. 4).

A long-term effect was seen in the winter of 1985: cockles from Fo5.5 and

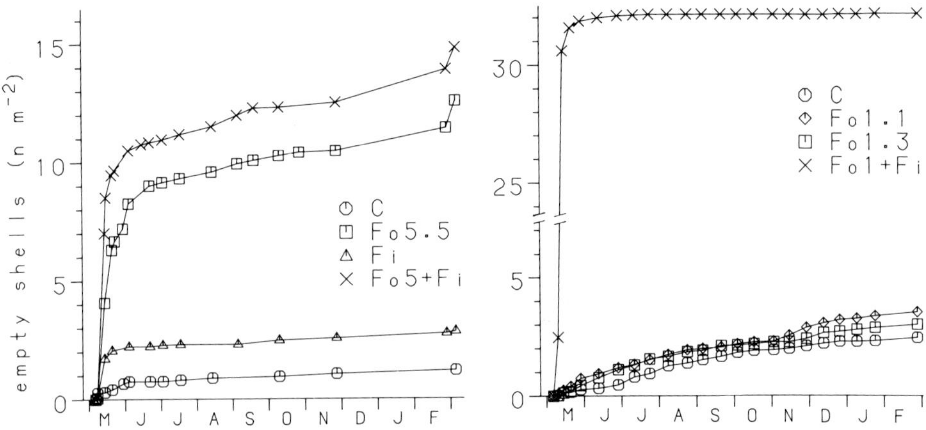

FIGURE 3. Cerastoderma edule. Mortality expressed as cumulative number of empty shells that appeared at the sediment surface. Note different ordinates. Left OPEX 1984, right OPEX 1985.

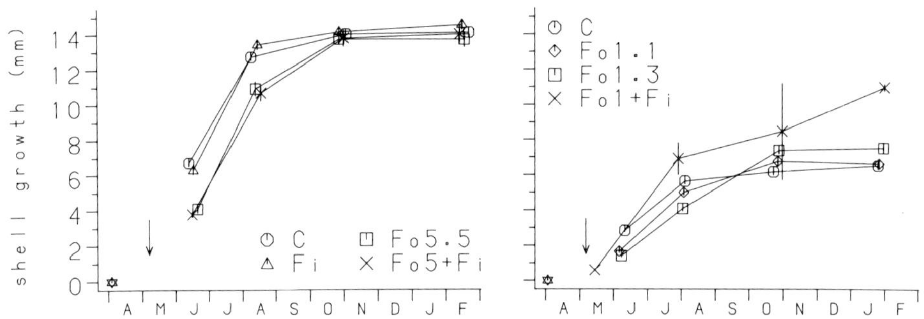

FIGURE 4. Cerastoderma edule. Shell-length increase since introduction. Increase in 1985 during dosage (arrow) is indicated by cockles that died in the Fo1+Fi treatment. Asterisks denote time and value of introduction. Bars denote standard error. Left OPEX 1984, right OPEX 1985.

Fo5+Fi suffered increased mortality during a severe frost period (Fig. 3). This was not found in the C and Fi MOTIFs and suggests a reduced condition as a result of chronical oil exposure. In this phenomenon it is interesting to realize that the weakest individuals probably had already died during the previous period of the experiment.

4.1.2. Macoma balthica. In comparison to Cerastoderma edule, only few M. balthica emerged at the sediment surface. They live deeper in the sediment and seem to be more resistant to stress than shallow living C. edule. Highest numbers appeared in Fi, but after a few days these specimens crept back into the sediment and mortality was similar to controls. In the three Fo treatments there was a clear dose-dependent mortality (Fig. 5). In Fo5.5 this was about 60% at the end of the experiment. Emergence of empty shells suggested a stronger short term effect of Fo1+Fi in comparison to Fo1.3.

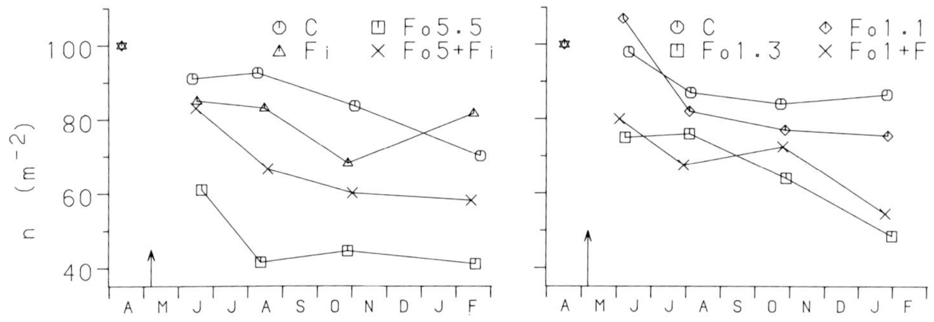

FIGURE 5. Macoma balthica. Density. Further explanation cf. Fig. 4.

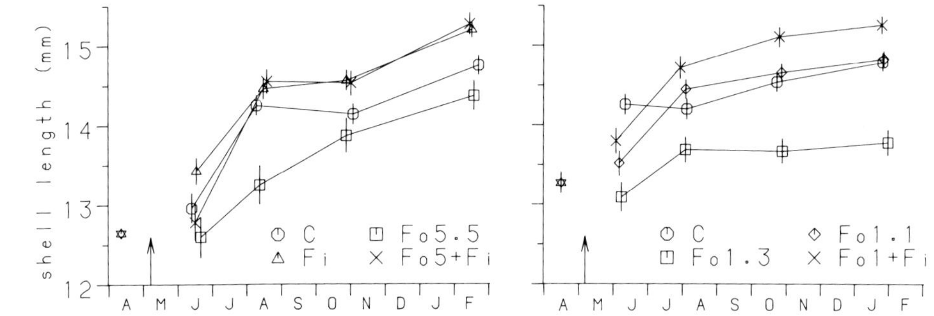

FIGURE 6. Macoma balthica. Shell length. Further explanation cf. Fig. 4.

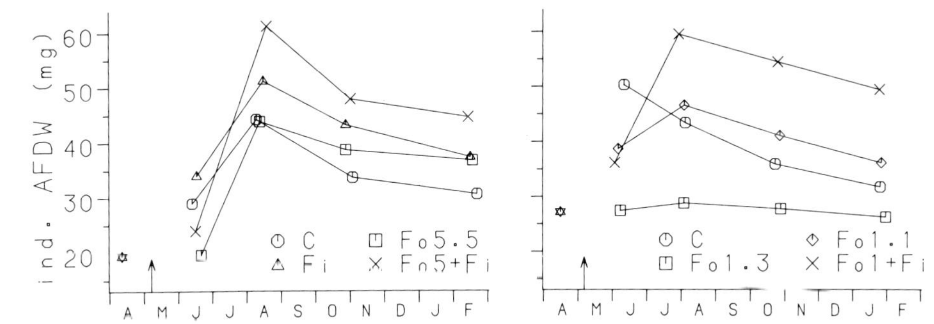

FIGURE 7. Macoma balthica. ind. AFDW. Further explanation cf. Fig. 4.

Both Fo+Fi treatments resulted in a mortality of 40%.

Increases in shell length and ind. AFDW were lowest in Fo5.5 and Fo1.3 treatments (Fig. 6, 7).

4.1.3. Arenicola marina. In both Fo+Fi treatments adults stopped bioturbation activity soon after contamination. A few days later about 4 ind. m^{-2} left their burrows and died soon afterwards. After two weeks, surviving specimens resumed activity. A decrease in bioturbation was also observed in Fo5.5. Surface counts during LMS 1 in 1984 revealed reduced adult densities in Fo5.5 and Fo5+Fi (11-12 ind. m^{-2}) compared to Fi and C (16 and 19 ind. m^{-2}, respectively). In 1985, only Fo1+Fi treatment resulted in reduced adult densities (Fig. 8). Ind. AFDW decreases during autumn, as a result of spawning. The change was highest in Fi and both controls, suggesting a reduced spawning capability of A. marina in all oil-treated MOTIFs. Due to large variation in ind. AFDW, however, differences were not significant.

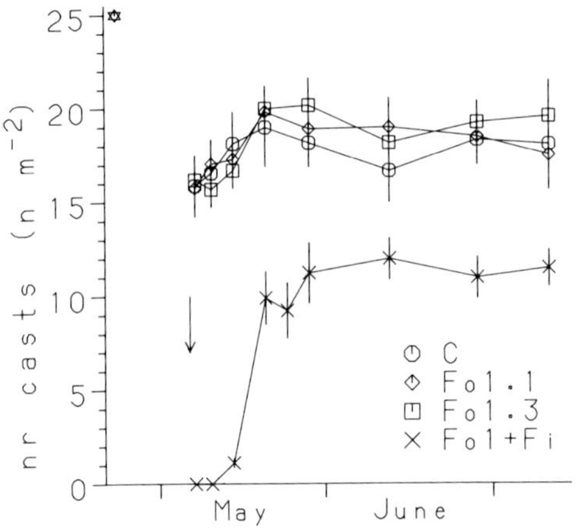

FIGURE 8. Arenicola marina. Bioturbation activity OPEX 1985 expressed as fresh casts observed at the surface. Further explanation cf. Fig. 4.

4.1.4. Corophium volutator. Densities in C. volutator developed in a dose-dependent way. Highest contaminations (Fo5.5 and Fo5+Fi) caused rapid elimination of the populations without recovery afterwards (Fig. 9). In C and Fi, probably because bird predation lacked, populations reached maximum densities of about 10,000 ind. m^{-2}. Less severe contaminations in 1985 showed recovery of the populations. This took place first in Fo1.1, then in Fo1.3 and last in Fo1+Fi. The delays resulted in concomitant reductions in

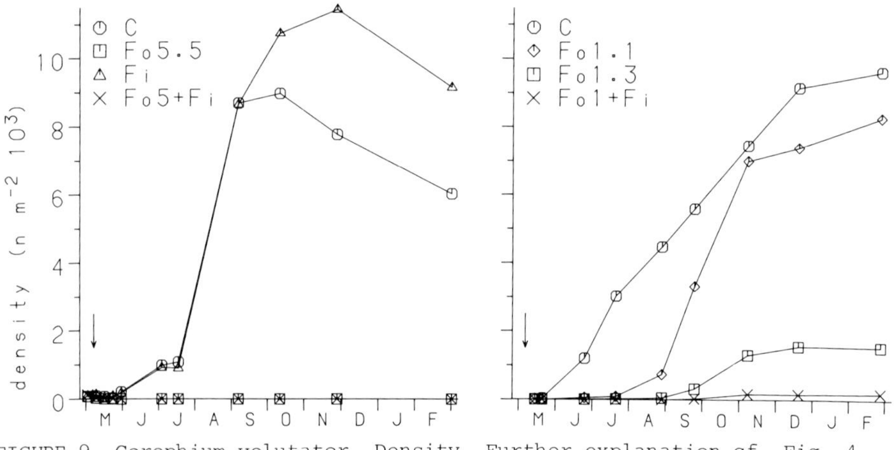

FIGURE 9. Corophium volutator. Density. Further explanation cf. Fig. 4.

abundances at the end of the experiment. The recovered populations probably originated from the reproduction of some surviving specimens. Introduction via the supply water proved to be possible only to a very limited extent (5). Under field conditions recolonization will probably occur faster.

4.2. Indirect effects

4.2.1. *Cerastoderma edule* and *Macoma balthica*. In oil treatments shell growth as well as ind. AFDW were equal or higher (except for *M. balthica* in Fo1.3) than in controls at the end of the experiments (Figs. 4, 6, 7, 10). In dispersed oil treatments these parameters were always among the highest values. High mortality in *C. edule* resulted in reduced filtering of phytoplankton. This improved food condition for the surviving *C. edule* and even for the primarily deposit feeding *M. balthica*.

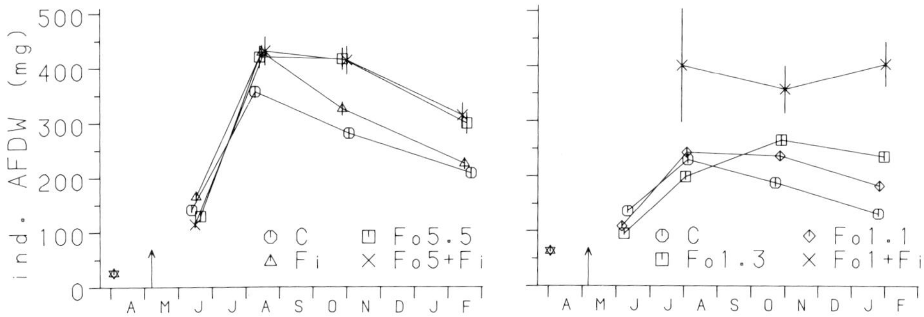

FIGURE 10. *Cerastoderma edule*. ind. AFDW. Further explanation cf. Fig. 4.

4.2.2. *Hydrobia ulvae*. Juvenile settlement was highest in the most contaminated treatments, viz. Fo5.5 and both Fo+Fi, and less or absent in other treatments (Fig. 11). After juvenile settlement, numbers rapidly fell to low levels. Despite recruitment in several MOTIFs, *H. ulvae* nowhere developed high densities. Obviously, the set up (e.g. sediment type) does not allow such population developments. Remarkably, in those treatments where

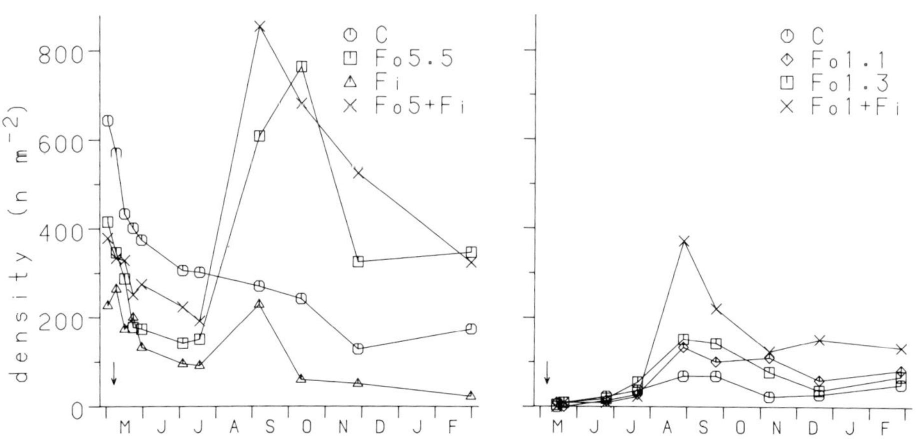

FIGURE 11. *Hydrobia ulvae*. Density. Further explanation cf. Fig. 4.

Corophium volutator had no or bad recovery, H. ulvae settlement was relatively successful (Fo5.5, Fo5+Fi and Fo1+Fi). In natural situations, an inverse relation between densities of H. ulvae and C. volutator has also been found (6). As both species are small, mobile surface deposit feeders, competition for food is probably responsible for this phenomenon.

4.3. Biomass

4.3.1. Arenicola marina. In 1984, before the addition of oil and dispersant, considerable settlement of juvenile A. marina took place. Settlement was not equal in all MOTIFs. Later on, juveniles were absent in MOTIFs 1 and 2 (Fo5+Fi), and very abundant, up to 371 ind. m^{-2}, in MOTIF 5 (Fi). Intermediate densities were found in other MOTIFs. The inadequate sea water supply system was probably responsible for these differences. These juveniles contributed considerably to the total biomass, especially in MOTIF 5. In 1985 juvenile settlement hardly occurred.

4.3.2. Nereis diversicolor. Reproduction took place during summer which resulted in increasing abundance in all MOTIFs. In 1984, this increase was very rapid, but in 1985 it lasted for at least 4 months, possibly as a result of a prolonged reproductive season and/or reduced growth of juveniles. Within the MOTIF pairs, density differences were so large that no clear differences between treatments could be observed. Differences between MOTIFs in 1984 were less prominent than those in 1985 (Fig. 12). Ind. AFDW, however, were very different in 1984, with highest values in Fo5.5 and Fo5+Fi. This resulted in higher total biomass values in these treatments than in C and Fi. In 1985, ind. AFDW values were rather variable, but nevertheless led to fairly equal total biomass values of N. diversicolor in all 1985 treatments.

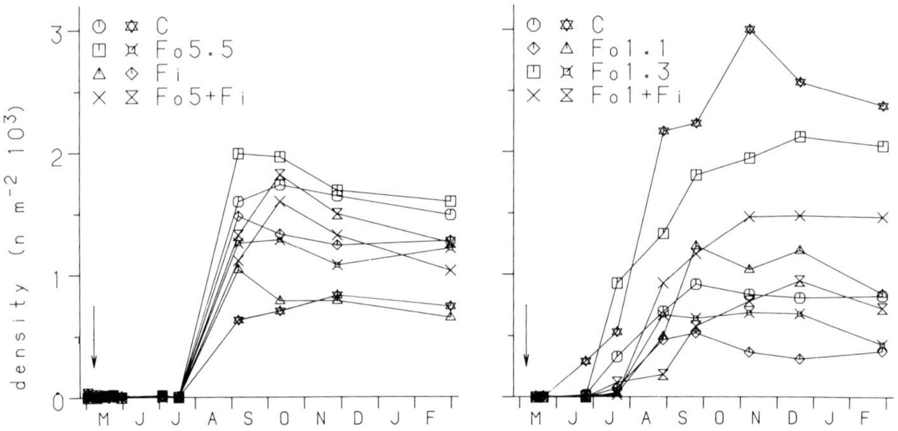

FIGURE 12. Nereis diversicolor. Density in different MOTIFs. Further explanation cf. Fig. 4.

4.3.3. Total Biomass. The change in development of total biomass in MOTIF pairs is remarkably similar (Fig. 13). Lowest similarity is seen in both Fo+Fi. In the short term (based on comparison of biomass introduced with first LMS), an increase is seen in C, Fi and Fo1.1, but not in Fo5.5, Fo1.3 and Fo5+Fi. A decrease even occurred in Fo1+Fi. Short-term effects are predominantly reflected by adult effects: Cerastoderma and Arenicola.

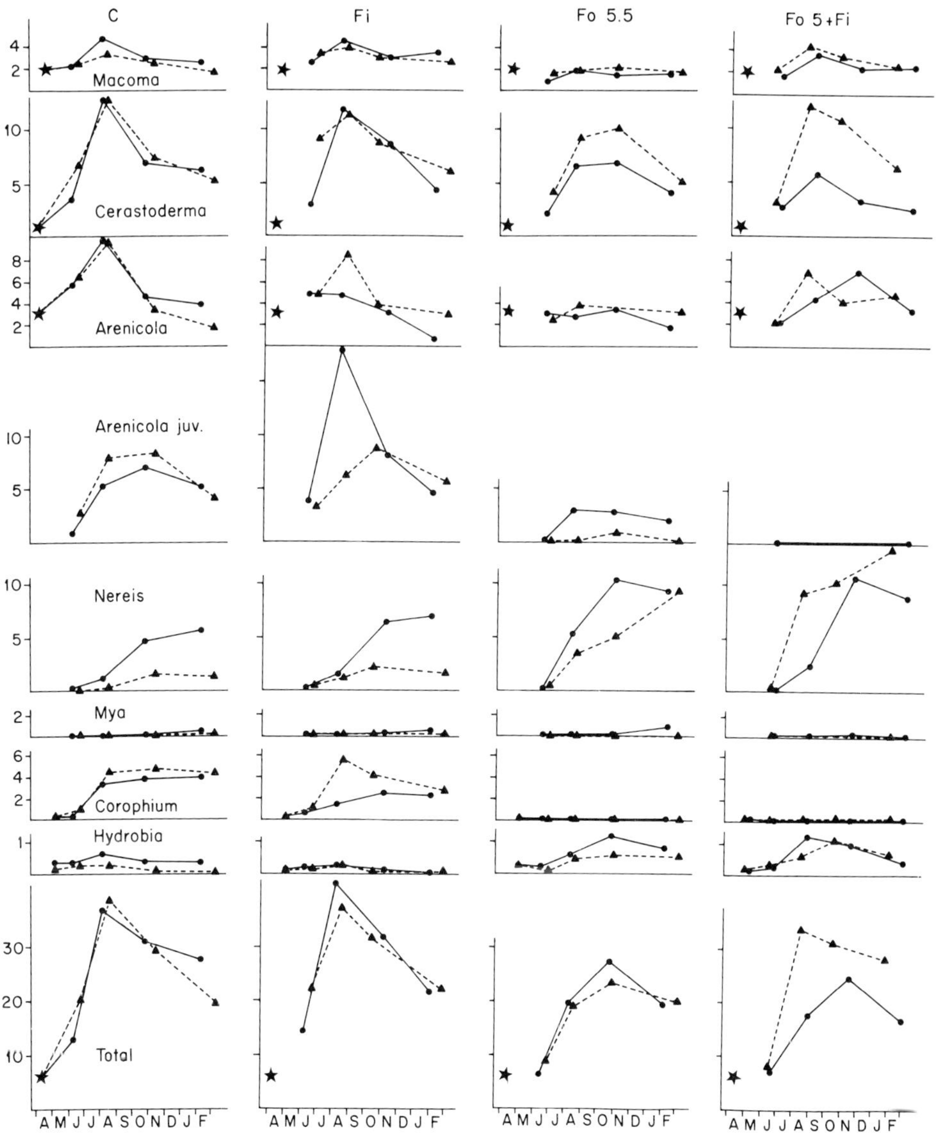

FIGURE 13. Total biomass and species biomass expressed as $AFDW.m^{-2}$. Note different scales in Hydrobia and Total. Circles and continuous lines: even MOTIF numbers; triangles and dashed lines: odd MOTIF numbers. Further details as in legend to Fig. 4.

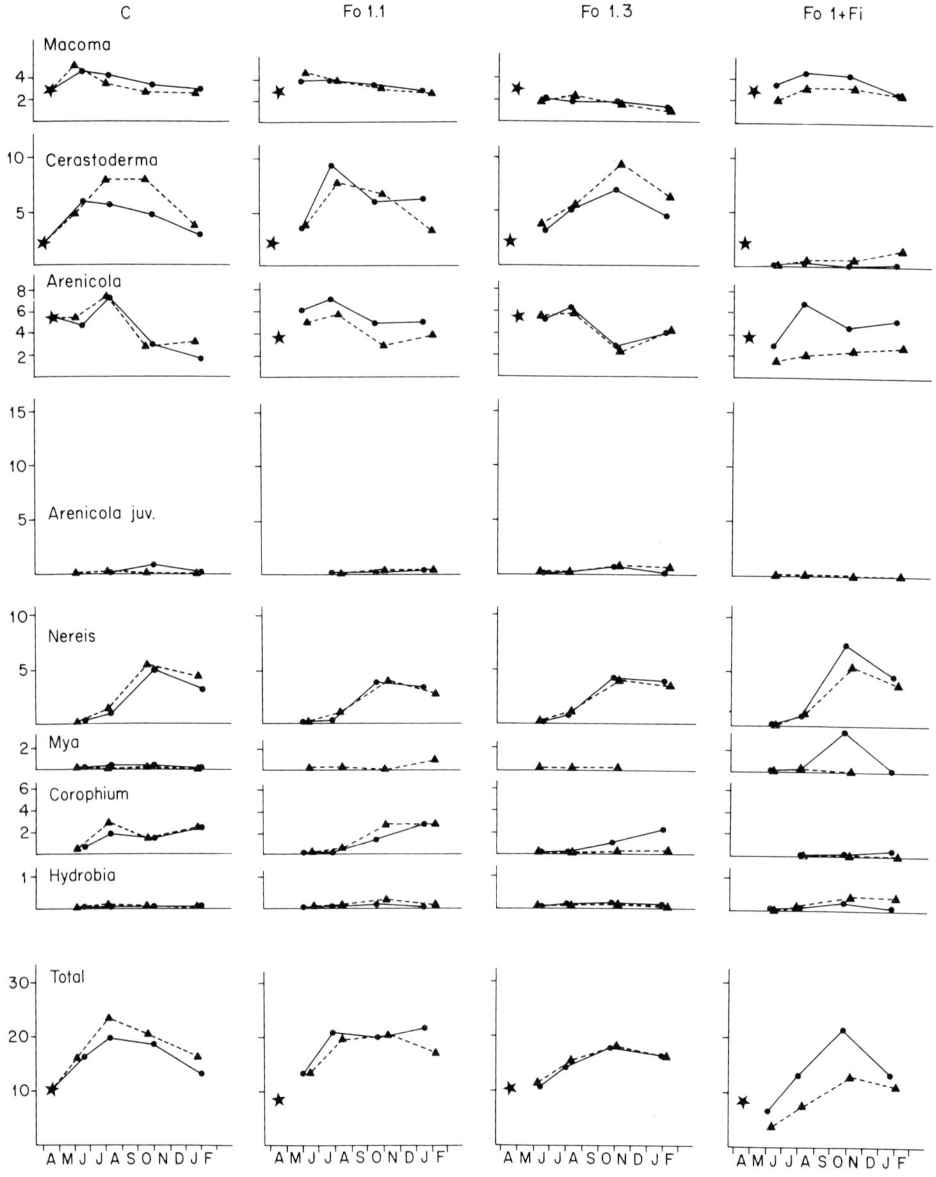

Short-term increase is linked with a maximum in late summer (LMS 2). When short-term increase did not occur, a maximum was reached during autumn (LMS 3). An exception is one of the Fo5+Fi treatments. In Fo1.1 an intermediate result is seen.

Largest contributions to changes in total biomass values originated in Arenicola juveniles and adults, Cerastoderma, Nereis and to some extent Macoma and Corophium. High settlement of Arenicola juveniles in 1984 in C and Fi shows that the MOTIFs have a potential carrying capacity of

36-42 g m^{-2}. But other MOTIFs in 1984 also had high maxima and end values in comparison to 1985 (max 13-25 g). Lower biomass values in 1985 are reflected by lower chlorophyll-a values, than in 1984. During the summer of 1984, chlorophyll-a concentrations in the water were about twice the level of 1985 (1). This may have been caused by a higher water supply in 1984. Another possibility is a lower competition for food in 1984 because mussels (*Mytilus edulis*) were less abundant that year. In both cases, food seemed to be a limiting facotr. *Nereis* biomass was responsible for differences between MOTIF pairs, as was *Mya* in 1985.

Although in both years various treatments resulted in differently structured macrofauna assemblages, the final total macrobenthic biomass values in each MOTIF, within each experiment, were similar. Different macrofauna assemblages converted primary produced matter into a total biomass with the same efficiency.

5. DISCUSSION

Compared with the field situation, especially the control MOTIFs show a similar behaviour. Growth and growth period in *Cerastoderma edule* and *Macoma balthica* are within the range of data observed in the field. The same holds true for growth and spawning period of *Arenicola marina*.

Comparable field experiments have been carried out with Arabian Light crude oil and the dispersant Finasol OSR5 in the German Wadden Sea. High doses (0.5 l oil m^{-2}) were used on emerged intertidal sediments (7, 8) or lower doses injected in sea water during rising tide in so called 'Bremerhaven Caissons' (9, 10). These experiments were carried out with mature benthic ecosystems, in contrast to the clearly immature systems in the MOTIFs. Moreover, field experiments enabled the study of long-term effects on juvenile settlement in many macrobenthic species. As a result of the long sea water supply of the MOTIFs, input of pelagic larvae was unpredictable and settlement much reduced compared to natural systems.

Entrainment of oil in the sediments, in both Fo and Fo+Fi treatments, was also observed by van Bernem (7). In the OPEX experiments it was facilitated by the well drained character of the sediments, and by bioturbation of *Arenicola marina*. Farke *et al*. (9, 10) found low infiltration of oil into deeper sediment layers. This probably was due to their very different experimental design.

In the Bremerhaven Caissons (10), the recovery of *Macoma balthica* in terms of repopulation of the contaminated area by juveniles was better in dispersed oil than in oil treatments, but Rachor (8) found more severe effects, both in the short and long term.

Especially in Fo+Fi, the feeding activity by *Arenicola marina* stopped soon after the addition of oil and dispersant. This was also observed by Rachor (8) and Farke *et al*. (10). As an effect of this reduced bioturbation, lower amounts of oil were buried in the sediment. Thus oxidation and removal of oil by the next tide was promoted.

Nereis diversicolor does not seem to be directly influenced by oil or dispersant, which was also found by Rachor (8). No direct mortality or reduced numbers were observed. In some treatments, *N. diversicolor* populations even seemed to benefit from the treatments. This myst be considered as an indirect effect: improved food conditions as a result of reduced competition. *Nereis*, an opportunistic species under field conditions, behaved in the same way in the MOTIFs.

6. CONCLUSIONS

1. Dispersant sprayed on an oil film resulted in dispersion of only 10% of the floating oil. It proved to be less successful than premixing the dispersant with the oil before application in the MOTIFs.

2. Oil stranded on the tidal flats partly remained in the sediment due to sticking to the sediment particles and burial by bioturbation activity of Arenicola. This trapped oil delivered components to the interstitial and overlying water for a prolonged period.

3. Addition of Forties alone caused the disappearance of the crustacean Corophium and a dose-dependent mortality in the molluscs Cerastoderma and Macoma.

4. Dispersant alone caused Cerastoderma and Macoma to emerge from the sediment, exposing them to bird predation under field conditions.

5. Dispersion of oil caused high mortalities in Corophium, Cerastoderma and Arenicola, and moderate mortality in Macoma. In general, dispersion of oil aggravated the effects of oil addition. After 10 months, the effects on the remaining fauna were not less than after treatment with oil only.

6. After a 10-month experimental period, total macrobenthic biomass was similar in all treatments. However, heavy oil pollution and dispersion of oil resulted in differently structured communities in comparison with the controls.

7. ACKNOWLEDGEMENTS

We want to thank P.A.W.J. de Wilde and J.J. Beukema for their valuable suggestions, M. Grashof, F. Ellenbroek, L. van der Laag, E. Berghuis, A. Kok for their practical assistance and F. Eijgenraam for providing computer programs for data storage and statistical analyses. The reprographic department of NIOZ helped in preparing the figures. B. Bak improved the English and J. Schröder typed the manuscript.

8. REFERENCES

1. Kuiper J, H van het Groenewoud, N Admiraal, M Scholten, PAWJ de Wilde, GWNM van Moorsel, R Dekker, WJ Wolff & C Brouwer: The influence of dispersants on the fate and effects of oil in model tidal flat ecosystems. Report MT-TNO nr R 86/182: 1-94, 1986.
2. Wilde PAWJ de & BR Kuipers: A large indoor tidal mud-flat ecosystem. Helg. wiss. Meeresunters. 30: 334-342, 1977.
3. Beukema JJ: Biomass and species richness of the macro-benthic animals living on the tidal flats of the Dutch Wadden Sea. Neth. J. Sea Res. 10: 236-261, 1976.
4. Dekker R: Numbers, growth, biomass and production of organic and calcareous matter of Hydrobia ulvae (Gastropoda: Prosobranchia) in the western Dutch Wadden Sea. NIOZ Int. Rep. 1979-15: 1-27, 1979.
5. Kuiper J, R Dekker, PAWJ de Wilde & WJ Wolff: De ontwikkeling van een modelecosysteem ten behoeve van ecotoxicologisch onderzoek. Report MT-TNO nr R 84/141: 1-95, 1984.
6. Barnes RSK & RN Hughes: An introduction to marine biology. Blackwell Sci. Publ., Oxford, London, 1982.

7. Bernem K-H van: Experimentelle Untersuchungen zur Wirkung von Rohöl und Rohöl/Tensid-Gemischen im Ökosystem Wattenmeer. I. Eindringverhalten und Persistenz von Rohölkohlenwasserstoffen in Sedimenten nach experimenteller Kontamination. Senckenbergiana marit. 16: 13-30, 1984.
8. Rachor E: Experimentelle Untersuchungen zur Wirkung von Rohöl und Rohöl/Tensid-Gemischen im Ökosystem Wattenmeer. XIII. Makrofauna. Senckenbergiana marit. 16: 225-234, 1984.
9. Farke H, K Wonneberger, W Gunkel & G Dahlmann: Effects of oil and a dispersant on intertidal organisms in field experiments with a mesocosm, the Bremerhaven Caisson. Marine Environ. Res. 15: 97-114, 1985.
10. Farke H, D Blome, N Theobald & K Wonneberger: Field experiments with dispersed oil and a dispersant in an intertidal ecosystem: fate and biological effects. Proc. 1985 Oil Spill Conf., Washington: 515-520, 1985.

EXPERIMENTAL APPLICATION OF OIL-BASED MUDS AND CUTTINGS TO SEABED SEDIMENTS

I.M.T. DIXON

Oil Pollution Research Unit, Field Studies Council,
Orielton Field Centre, Pembroke, Dyfed, U.K.

ABSTRACT: Between September 1984 and July 1985, a field experiment was carried out in Milford Haven to follow the macrofaunal effects and subsequent recovery from a single application of used diesel and 'low-tox' oil based muds (OBM). Six treatments, including two levels of cuttings addition, were investigated and each was replicated three times in a randomised block experimental design. The cuttings' treatments were designed to give surficial sediment hydrocarbon concentrations of about 5000 ppm (high dose) and 400 ppm (low dose). Treatments were applied by divers to marked seabed plots (2 m x 2 m). Core samples were taken for hydrocarbon, sedimentological and macrofaunal analysis prior to treatment and then subsequently at 2 weeks, 1 month and then 2 monthly intervals for a total period of 10 months.

Prior to treatment no hydrocarbon or biological gradients across the experimental site were discernible. Following treatment, sediment hydrocarbon concentrations tended to fall rapidly within the first month, followed by a period of slower removal. Evidence of OBM contamination had disappeared from the low-dose plots after 4-6 months. In high dose plots, hydrocarbon levels had fallen almost to within the background range of values by 10 months, but slight contamination was still evident on all GLC traces.

Faunal disturbance was minor and significant effects were mainly recorded from the high-dose plots where the initial effect of oiled cuttings addition was to depress faunal density, species richness and diversity relative to the controls. Population reductions were observed in a number of species but no expansion of opportunists occurred. In the high dose plots, total faunal density and the abundance of certain species remained depressed for the duration of the experiment. Faunal disturbance occurred more rapidly following diesel treatment than with low-tox treatment; reflecting the greater acute toxicity of the former. After 1 or 2 months, however, the longer term effects of low-tox OBM's became indistinguishable from those of the diesel treatment.

1. INTRODUCTION

In recent years increasing use has been made of oil based muds (OBM's) by the offshore oil industry in the North Sea. In 1981 approximately 36% of wells drilled on the UK continental shelf were drilled with OBM's leading to an estimated 7,000 t of diesel being discharged with drill cuttings. By 1983 approximately 65% of wells were drilled using OBM's, with a loss of 18,100 t of oil to the environment. This corresponded to 90% of the total oil lost by discharge or spillage offshore (Davies *et al*., 1984 and Department of Energy, 1985).

Kuiper, J., Van den Brink, W.J. (eds), Fate and Effects of Oil in Marine Ecosystems. ISBN 90-247-3489-4

OBM's were originally formulated on a diesel base, but concern over the probable environmental impact of large quantities of diesel entering the sea led to the development and increasing use of base oils with reduced toxicities (low tox oils). These are more highly refined products with a much lower aromatic hydrocarbon content than diesel. In standard 96 hour LC_{50} toxicity tests, low-tox base oils appear to be up to 1,000 times less toxic than diesel (Blackman _et al_., 1982) while in a comparison of whole muds, they vary between 10 and 100 times less toxic than diesel based muds (Blackman _et al_., 1983). However comparative 96 hour LC_{50} tests on cuttings coated with used diesel and low-tox muds carried out by Dow (1984) indicated that the low-tox discharge was at least 10 times more toxic to _Crangon crangon_ than the diesel cuttings. Previous tests on that particular unused whole mud system had indicated a toxicity of around 10 times _less_ than diesel OBM's. The enhanced toxicity of this mud following use, in spite of its lower aromatic content, may have arisen from the usual practice of adding a varied selection of chemical agents (some of them toxic) to the mud system during drilling operations. The lack of a clear cut correlation between base oil aromatic content and acute toxicity, or sub-lethal effects, has also been pointed out by Blackman _et al_. (1982) and Payne _et al_. (1985) respectively.

Although low-tox oils have been shown to be less acutely toxic than diesel, there is currently insufficient evidence with which to distinguish their chronic or long-term ecological effects when discharged with cuttings (Davies _et al_., 1984; Matheson _et al_., 1986). In fact the field data available indicate that changes to the macrobenthos resulting from the use of both types of OBM share several characteristics. These have been summarised, in relation to distance from the discharge point, as concentrically arranged zones of impact which are usually elongated in the direction of the locally prevailing tidal currents (Davies _et al_., 1984).

As Addy _et al_. (1984) point out, there are a number of likely mechanisms through which the environmental consequences of oily cuttings discharge are mediated. These include direct physical smothering of the seabed and benthic fauna under piles of cuttings; organic enrichment of the sediment by petrogenic hydrocarbons; and the toxic effects of base oils and other mud additives. The way in which the relative importance of each of these factors changes with distance from the platform has never been tested, and has proved impossible to determine from the limited field data available. It is likely therefore that reliable information on the environmental effects of different OBM's will only be possible following long-term comparative studies under experimental conditions. At the Department of Agriculture and Fisheries for Scotland (DAFS) a series of long-term investigations into the chemical, microbiological and meiofaunal aspects of OBM impact have been conducted in experimental tank systems (e.g. Dow, 1984; Leaver _et al_., in press). The opportunity was taken to complement these tank studies with a field experiment in which the same stocks of OBM contaminated cuttings would be used and similar, or overlapping, dose rates employed. The aim of the field investigation described here was to compare the effects on the macrofauna of single applications of used low-tox and diesel OBM's, and to monitor subsequent recovery over a period of twelve months.

2. METHODS

2.1. Experimental Design

Given that both types of OBM can cause equally severe damage to the benthos through the effects of heavy smothering close to the discharge point, it was decided that any differences in the environmental impact of diesel and low-tox oils due to toxicity would be most apparent at lower hydrocarbon levels. Therefore experimental treatments were chosen that would give sediment hydrocarbon conditions approximating to those found in the 'transition zone in benthic diversity and community structure' (Davies et al., 1984). For logistical reasons only two treatment levels were studied; these being 5000 ppm and 400 ppm, which corresponded to approximately 100 x and 10 x local hydrocarbon concentrations respectively. The oiled cuttings tested were supplied from two North Sea drilling platforms, one using a diesel OBM and the other using a low-tox mud formulation with the trade name 'Enviromul'. The oil content of the diesel cuttings supplied was half that of the Enviromul - coated cuttings (7.15% and 15% by weight respectively) which, in terms of treatment application, would have meant applying twice as much of one type to the seabed to obtain a hydrocarbon dose equivalent to the other treatment. In order to avoid differential smothering effects, each of the treatments was mixed with a quantity of clean afaunal sediment sufficient to bring them all up to the same weight. This enabled the inclusion of a single smothering control in the treatments, consisting of an application of clean afaunal sediment only. For this purpose, sediment was collected by divers from nearby the experimental site and dried in the sun for six weeks. Just prior to treatment-application, the dried sediment was thoroughly re-mixed with seawater in a cement mixer. Treatments were then made up in bulk in the mixer and stored in capped plastic buckets. A total of six treatments, including controls, were investigated and each was replicated three times in a randomised block experimental design. Treatments were as follows:

A. Control	
B. Clean afaunal sediment (C.A.S.)	
C. Enviromul-cuttings + C.A.S.	LOW DOSE
D. Diesel-cuttings + C.A.S.	LOW DOSE
E. Enviromul-cuttings + C.A.S.	HIGH DOSE
F. Diesel-cuttings + C.A.S.	HIGH DOSE

Calculation of the quantity of cuttings required to achieve each of the treatment hydrocarbon concentrations was based on two assumptions; (i) that the specific gravity of the sediment in the experimental area was approximately 2.0, and (ii) that sediment samples for hydrocarbon analysis taken during the course of the experiment would be taken from the upper 5 cm of seabed sediment. Field tests conducted prior to the experiment demonstrated that application rates based on these assumptions did, in fact, achieve the required hydrocarbon concentrations.

2.2. Fieldwork

The experiment was conducted sublittorally (7 m below chart datum) in Milford Haven at a site sheltered from prevailing winds and also remote from industry and shipping (Figure 1). At this site the seabed was flat, consisting of poorly sorted fine muddy sand with a diverse infaunal community numerically dominated by maldanid and cirratulid polychaetes, and bivalves (Abra alba and Mysella bidentata). Salinity is greater than 33‰ at all tidal states and no stratification occurs.

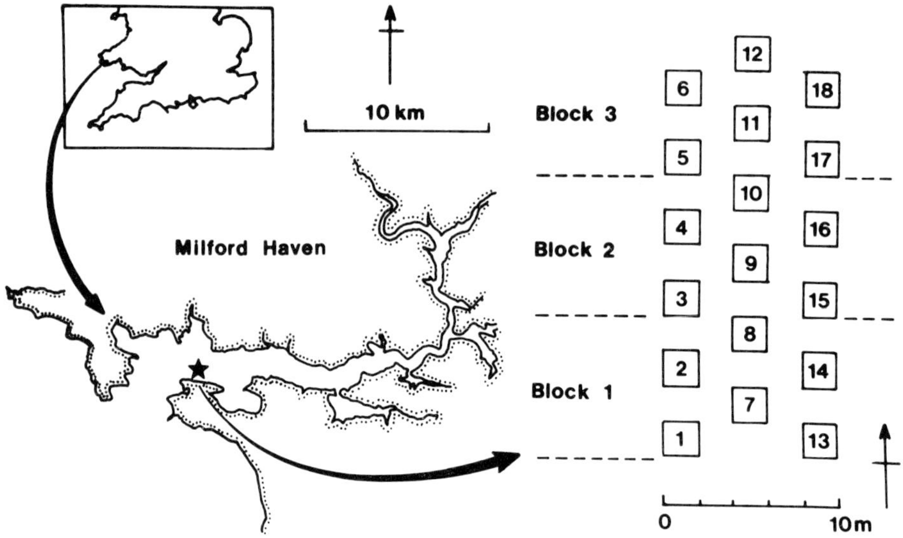

FIGURE 1. Location of experiment in Milford Haven, and plan of experimental site showing plot numbering system.

Eighteen 2 m x 2 m plots (three blocks of six treatments) were marked out on the seabed using corkscrewed steel rods twisted into the sediment as corner posts. The plots were confined to as small an area of seabed as possible (to minimise the effects of sediment/faunal heterogeneity and also to aid in plot-relocation and sampling in low visibility). They were spaced with a minimum separation of 2 m, and positioned in a manner designed to minimise cross-contamination via the prevailing east-west tidal currents (maximum spring velocity of ca. 1.0 knot).

Within each block, treatments were assigned randomly to each of the plots. After pre-treatment sampling, divers applied the treatments to the plots by hand in September 1984, sprinkling even layers of material over the sediment surface.

2.3. Sampling

The experiment followed the following schedule:

June 1984	Pretreatment samples
September 1984	Treatment application
October 1984	2-week samples
October 1984	1-month samples
November 1984	2-month samples
January 1985	4-month samples
March 1985	6-month samples
May 1985	8-month samples
July 1985	10-month samples
September 1985	No samples due to disturbance of plots by fishing gear

On each sampling occasion, four biological and four hydrocarbon samples were taken (sediment for particle size and organic matter analysis was sub-sampled from the hydrocarbon cores). Biological cores were taken with a 5.2 cm diameter corer to a depth of 10 cm, and hydrocarbons were sampled with a cleaned 5 cm diameter alumninium can to a depth of 5 cm. Core co-ordinates in each plot were selected prior to sampling by reference to random number tables, and the same spot was never sampled more than once. In addition, the peripheral 25 cm of each plot was not sampled in order to minimise edge-effects.

2.4. Laboratory analysis

2.4.1. Sediment particle size and organic matter content. Particle size distributions were determined using a combination of wet and dry sieving (Buchanan and Kain, 1971) for particles larger than 4.5 Ø and by electronic particle counting for material smaller than 4.5 Ø (Little *et al*, 1984). Results from both techniques were combined and analysed using the method of moments analysis (Krumbein, 1936).

The organic matter content of sediment was estimated by determining loss of weight on ignition of samples at 480°C for 24 hours. Calcium carbonate standards had previously been shown to undergo no loss of weight under this treatment so that any weight loss due to combustion of sediment carbonates was assumed to be minimal.

2.4.2. Sediment Hydrocarbons. The four cores taken from each plot were pooled and then stored at $-20^{o}C$ until required.

The analysis followed the methods of Little *et al*. (in press), with hydrocarbon material being solvent-extracted from the sediments and separated into aliphatic and aromatic fractions by column chromatography. After weighing each fraction, the aliphatic extract was analysed by capillary gas-liquid chromatography (GLC).

2.4.3. Fauna. Cores were preserved in a *ca*. 5% buffered solution of formaldehyde in seawater to which a vital stain, eosin, had been added. In the laboratory, samples were sieved on a 0.5 mm mesh and the animals sorted from the remaining sediment under a dissecting stereo-microscope. All fauna were identified to the lowest practicable taxon and counted. The level of significance of observed differences between treatments on each sampling occasion was determined by two way analysis of variance on transformed data. A $\log_{10}(x+1)$ transformation was used to normalise the data and to correct slight heterogeneity of variance as indicated by Cochran's test and Hartley's maximum F test. On one occasion, variance heterogeneity was not corrected in spite of transformation and the non-parametric Friedman two-way analysis of variance was employed instead. Significant differences ($p<0.05$) between the means were further examined by the Least Significant Range test (LSR).

Because the analysis of variance indicated no between-block differences in any of the faunal parameters tested, the results from the thrice-replicated treatments were pooled. Examination of treatment effects on both community structure and on individual taxa therefore proceeded using the averaged results from the replicates for each treatment.

3. RESULTS

3.1. Sediment particle size and organic matter content

In June 1984, prior to treatment application, the sediment over the experimental site consisted predominantly of poorly-sorted fine to very fine sand with a mud fraction of 10-20%. A slight gradient was discernible in all parameters such that sediments in block 1 contained most gravel (particles > 2.0 mm diameter), least mud, and were more poorly-sorted than sediments in the block 3 plots.

By October 1984, two weeks after treatment application, this pattern had altered to one where no such gradients existed. This was the result of sediments in block 1 (and to a lesser extent in block 2) becoming muddier with less gravel, and better sorted than previously, with respect to those in block 3 which had shown the least net change.

In July 1985, 10 months after treatment addition, the sedimentary gradients were re-established although not to such a marked extent as in June 1984. Again this was the result of net changes in the sediments of block 1 rather than in block 3.

Mean estimates of sediment organic matter content for the site as a whole varied from 3.3% (SD 0.6) in June 1984, to 2.0% (SD 0.5) two weeks after treatment in September, and then to 4.53% (SD 0.8) at the termination of the study in July 1985. No gradients were evident in the data at any stage and no other sediment or treatment-related patterns of variation were found.

The sedimentary changes noted occurred in treatment and control plots alike and were probably seasonal in nature. It is concluded that treatment additions to the seabed did not detectably alter sediment particle size distributions.

Prior to treatment, and throughout the study, sediments appeared to be well oxygenated with no visible redox potential discontinuity developing in any of the plots.

3.2. Sediment hydrocarbons

The variation over time of total hydrocarbon (THC) values is shown for each plot in Figure 2. Before treatment application, THC levels varied between 17 and 107 ppm and no spatial concentration patterns over the site were discernible, in spite of the sedimentary gradients noted above.

3.2.1. Untreated controls, and clean sediment-treated plots. Throughout the experiment, THC concentrations stayed mainly within the range (17-107 ppm) shown over the whole site by the pre-treatment samples. Nevertheless slight cross-contamination from adjacent oiled plots was evident on GLC traces from all the untreated controls, and also in one of the clean sediment-treated plots (plot 13), at various times up to six months after treatment. In general, instances of cross-contamination were minor and transitory both in relation to the background hydrocarbon trace and to the oil doses applied in September.

3.2.2. Oiled cuttings - treated plots. THC levels in oiled plots usually followed a pattern of initially high loss-rates over the first two weeks post-treatment, followed by an extended period of more gradual decline (Figure 2).

In all Enviromul low-dose plots THC concentrations were reduced from around 400 ppm to less than 150 ppm during the first two weeks after treatment and, after four to six months, fell to within the range of background hydrocarbon values. At this stage evidence of Enviromul treatment was no longer visible on the aliphatic GLC traces.

THC concentrations in diesel low-dose plots showed a greater variation between replicates than in the Enviromul plots (Figure 2). In contrast to the other plots, THC levels in plot 6 were maintained at over 400 ppm for two months following oil-addition. In all plots however, diesel contamination was not visible on the aliphatic GLC traces after four to six months.

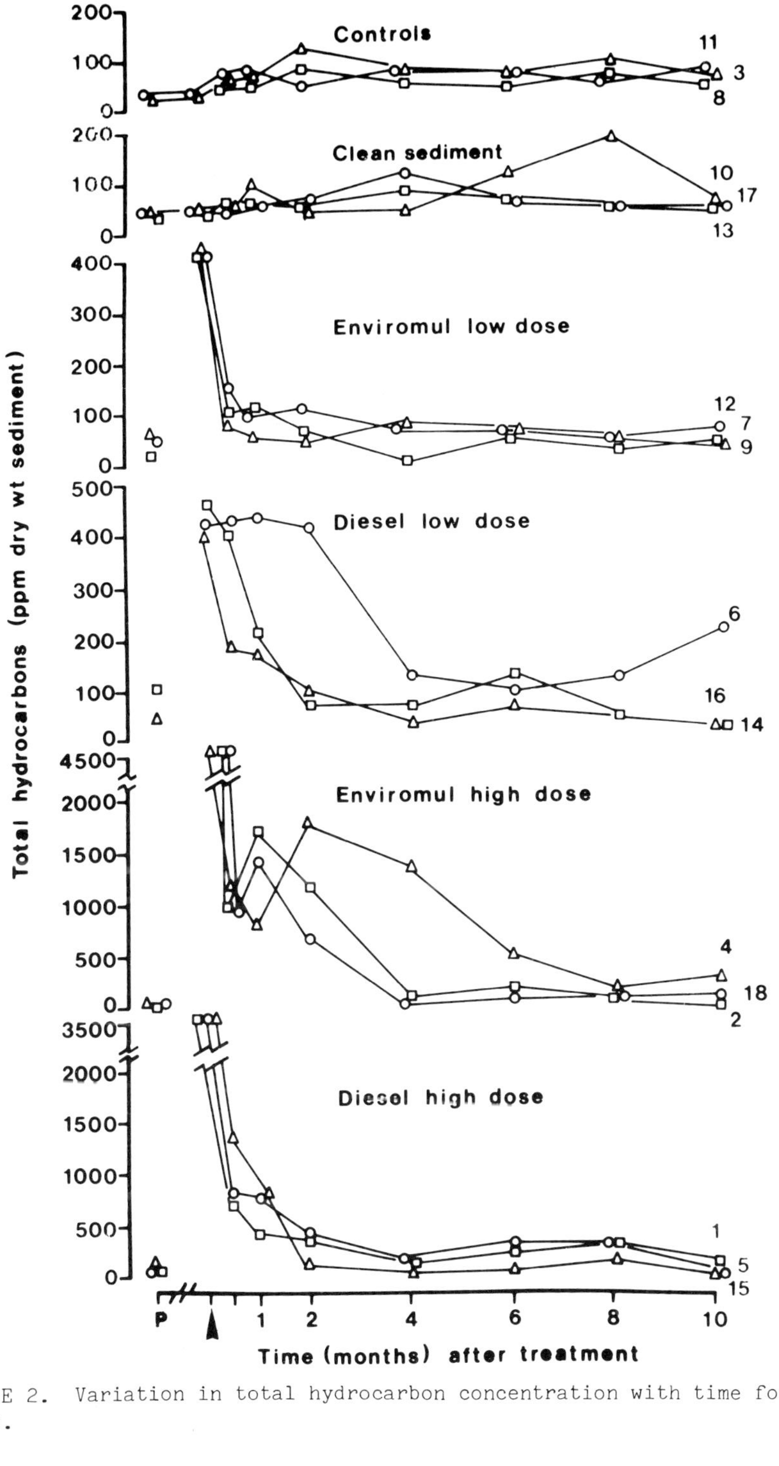

FIGURE 2. Variation in total hydrocarbon concentration with time for all plots.

TABLE 1. Summary of faunal indices at each sampling occasion. Figures given are means of 3 replicates ± standard deviation. N = Density 85 cm^{-2}; S = Number of species 85 cm^{-2}; H(S) = Shannon-Wiener information function; J = Equitability.

A. CONTROL	N		S		H(S)		J	
Pre-treatment	125.0 ±	4.4	23.7 ±	1.0	3.34 ±	0.03	0.73 ±	0.05
2 weeks	177.7	27.0	39.7	4.7	4.54	0.20	0.86	0.00
1 month	166.0	31.0	36.3	6.0	4.29	0.25	0.83	0.10
2 months	121.3	16.3	30.0	4.3	4.13	0.20	0.84	0.02
4 months	109.3	10.5	26.7	1.5	4.06	0.15	0.86	0.03
6 months	94.0	12.0	26.7	3.2	3.85	0.27	0.81	0.03
8 months	116.3	29.4	29.7	2.5	4.16	0.17	0.85	0.05
10 months	216.0	39.0	38.3	2.5	3.81	0.20	0.73	0.03

B. CLEAN SEDIMENT

	N		S		H(S)		J	
Pre-treatment	147.0 ±	36.0	25.7 ±	7.6	3.31 ±	0.60	0.71 ±	0.06
2 weeks	151.0	79.0	32.7	10.0	3.99	0.40	0.80	0.10
1 month	166.7	57.0	38.0	8.0	4.38	0.20	0.84	0.00
2 months	128.3	55.0	28.3	5.9	3.84	0.20	0.80	0.01
4 months	109.0	36.4	25.0	4.6	3.60	0.10	0.78	0.04
6 months	125.3	32.7	31.3	1.5	3.94	0.30	0.79	0.07
8 months	142.7	1.5	31.3	1.5	4.11	0.15	0.83	0.03
10 months	199.0	58.0	35.0	7.2	3.84	0.30	0.75	0.04

C. ENVIROMUL LOW DOSE

	N		S		H(S)		J	
Pre-treatment	131.3 ±	26.0	27.0 ±	3.0	3.47 ±	0.30	0.73 ±	0.04
2 weeks	143.0	60.0	30.0	11.0	3.87	0.40	0.80	0.05
1 month	162.0	27.0	36.3	3.0	4.09	0.20	0.79	0.00
2 months	88.0	10.5	28.7	3.1	4.17	0.20	0.86	0.03
4 months	81.3	11.6	27.3	2.3	4.30	0.08	0.90	0.01
6 months	79.0	27.0	26.7	4.2	3.98	0.07	0.85	0.06
8 months	102.7	20.5	29.3	5.8	4.10	0.28	0.84	0.04
10 months	135.0	30.0	27.3	2.9	3.47	0.30	0.73	0.10

D. DIESEL LOW DOSE

	N		S		H(S)		J	
Pre-treatment	141.3 ±	8.5	25.3 ±	2.3	3.37 ±	0.20	0.72 ±	0.04
2 weeks	116.7	9.5	31.7	3.0	4.10	0.10	0.82	0.00
1 month	87.7	24.0	27.0	3.6	4.11	0.10	0.87	0.00
2 months	87.0	17.0	27.0	4.4	3.98	0.40	0.84	0.05
4 months	80.3	57.7	21.0	3.5	3.64	0.36	0.84	0.12
6 months	86.0	19.7	23.3	5.5	3.67	0.64	0.81	0.09
8 months	82.3	16.6	24.7	8.0	3.80	0.50	0.83	0.04
10 months	144.7	39.0	32.0	5.2	3.76	0.50	0.75	0.10

E. ENVIROMUL HIGH DOSE

	N		S		H(S)		J	
Pre-treatment	99.7 ±	26.0	25.0 ±	1.7	3.33 ±	0.10	0.71 ±	0.04
2 weeks	129.3	10.0	31.7	4.0	4.17	0.20	0.84	0.00
1 month	92.0	23.0	28.0	3.0	4.27	0.20	0.89	0.00
2 months	60.0	4.4	24.7	4.6	4.09	0.30	0.89	0.02
4 months	60.0	14.7	21.0	1.0	3.88	0.07	0.88	0.01
6 months	48.0	4.6	18.3	2.9	3.50	0.40	0.83	0.07
8 months	89.0	19.8	24.5	0.7	3.80	0.13	0.83	0.02
10 months	98.3	36.0	28.7	5.7	4.07	0.20	0.85	0.03

F. DIESEL HIGH DOSE

	N		S		H(S)		J	
Pre-treatment	110.0 ±	10.0	23.3 ±	4.2	3.22 ±	0.40	0.71 ±	0.08
2 weeks	56.0	16.0	19.3	5.0	3.72	0.20	0.88	0.00
1 month	80.3	7.0	24.3	2.0	3.80	0.10	0.83	0.00
2 months	52.3	15.4	21.3	7.5	3.91	0.60	0.89	0.05
4 months	43.0	12.0	21.0	2.6	4.10	0.36	0.94	0.04
6 months	57.3	6.6	23.3	2.5	4.09	0.24	0.90	0.03
8 months	65.3	15.0	24.7	3.8	3.99	0.30	0.86	0.06
10 months	96.3	18.0	33.3	9.1	4.34	0.50	0.87	0.04

In the case of the Enviromul high-dose treatments, THC concentrations fell by approximately 75% to around 900-1200 ppm within the first two weeks. The apparent rise in THC values shown in all three plots over the next six weeks, and the subsequent variability between plots can at best be ascribed to a patchy cuttings distribution. Apparent increases in sediment hydrocarbon load over time were also noted by Bakke *et al.* (1986) in another field experiment with OBM's, and put down to the same cause.

Hydrocarbon loss rates from the diesel high-dose plots following treatment were similar to those from the Enviromul plots, although greater consistency between the three replicates was shown.

By four months into the investigation, both high-dose treatments (with the exception of plot 2) had declined in concentration to a more or less constant level at between 1 and 3 times the maximum background THC values. This condition persisted until the end of sampling at ten months. All aliphatic GLC traces indicated that treatment inputs still contributed a major proportion of the sediment hydrocabon burden at 8 months but by 10 months, contamination was relatively minor.

3.3. Fauna

Mean values from each treatment for the main descriptive community indices are shown in Table 1. The significant differences resulting from two way analysis of variance, and the sources of these differences as determined by the LSR test are shown in Table 2.

TABLE 2. Faunal indices: Results of *a posteriori* comparison of means (Least Significant Range) between treatments exhibiting significant variance (2 way ANOVAR). A = Control; B = Clean sediment; C = Enviromul low dose; D = Diesel low dose; E = Enviromul high dose; F = Diesel high dose.

Parameter	Time	Pairs of treatment means differing at P<0.05					
		A	B	C	D	E	F
N	2 wk	F	F	F	-	F	-
	1 mth	F	F	F	-	-	-
	2 mths	F	F	-	-	-	-
		E	E	-	-	-	-
	6 mths	-	F	-	-	-	-
		E	E	-	-	-	-
	8 mths	F	F	-	-	-	-
		-	D	-	-	-	-
	10 mths	F	F	-	-	-	-
		E	E	-	-	-	-
S	1 mth	F	F	F	-	-	-
	4 mths	-	-	D	-	-	-
		-	-	E	-	-	-
		-	-	F	-	-	-
	6 mths	E	E	E	-	-	-
H(S)	2 wks	F	-	-	-	-	-
	1 mth	F	F	-	-	-	-

Prior to treatment there were no discernible gradients in faunal density (N), species richness (S), diversity (H(S)) or equitability (J) across the site, and both 'between block' and 'between treatment' differences were not significant. In general the effect of adding oiled cuttings to the seabed was to depress N, S and H(S) relative to the controls (Figure 3). In spite of temporal fluctuation, community indices in the clean sediment-treated plots were never significantly different from those in the untreated control plots. Whereas significant treatment differences in H(S) and S were no longer detectable after one month and six months respectively, N was still significantly reduced in both high-dose treatments (relative to the control plots) at 10 months (Table 2).

No evidence was obtained to suggest any difference in the degree of impact between Enviromul and diesel in the long term but, within the first two months following treatment, significant reduction in N, S, and H(S) (relative to controls) occurred quickest in the diesel high-dose plots. This phenomenon is most clearly illustrated by N where, by two weeks after treatment, mean density in the diesel high-dose plots was already significantly lower than in both the control and the Enviromul high-dose plots. Faunal numbers in the Enviromul high-dose plots began to fall only after two weeks but, by two months, were comparable to those in the diesel high-dose plots. A similar lag in effect between diesel and Enviromul treatments appeared to occur in the low-dose plots where temporal variations virtually paralleled those of the corresponding high-dose treatments.

Treatment effects on macrofaunal diversity and equitability were not significant, on the whole, with the exception of the diesel high-dose treatment which caused a reduction in H(S) values (relative to controls) at two weeks and at one month (Table 2, Figure 3). During the course of the experiment, seasonal population fluctuations were evident in most species. Against this variable background, significant differences between oil treatments and controls were noted for some of the more abundant species; particularly in the high-dose treatment plots. The fluctuations in density over time for four species are shown in Table 3 and in Figures 4 to 7.

In the case of *Abra alba*, populations in the diesel high-dose plots were significantly lower than in controls by between two weeks and one month after treatment (Figure 4, Table 4). However, it is also evident that over the same period, all oil treatments caused a similar accentuation of the background seasonal population decline. Thus no differential treatment effects were observed and, after two months, the treatment plot means were not distinguishable from those in the control plots.

Differences between Enviromul and diesel-treated plots were however noted in the case of *Amphiura filiformis* (Figure 5). As noted previously for the community indices, the diesel high-dose caused the most immediate population changes, but after one or two months the population means in the Enviromul high-dose plots tended towards similarity with these. A significant reduction in numbers was still apparent in both high-dose plots for *A. filiformis* at 8 months into the experiment (Table 4). At 10 months numbers were not significantly different from the controls, although there was little indication of population recovery having started (Figure 5). No significant low-dose treatment effects were recorded for this species, although the trends shown in these plots were similar to those in the high-dose plots with the diesel treatment inducing the most immediate changes. Signs of recovery were also evident after 2 months.

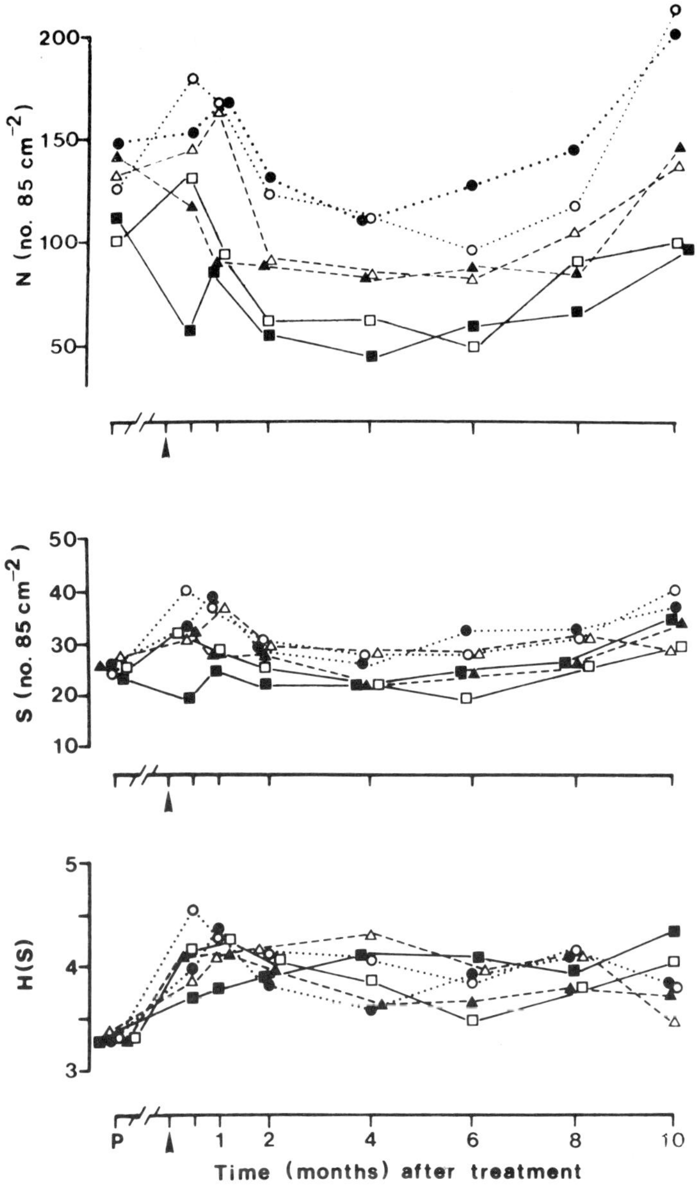

FIGURE 3. Variation in total faunal density, species richness and diversity with time for each treatment (means of 3 replicates). o = Control; ● = Clean sediment; △ = Enviromul low dose; ▲ = Diesel low dose; □ = Enviromul high dose; ■ = Diesel high dose.

TABLE 3. Abundance data for selected species at each sampling occasion. Figures given are means of 3 replicates ± standard deviation.

	Abra alba	Amphiura filiformis	Harpinia antennaria	Pseudocuma longicornis
A. CONTROL				
Pre-treatment	45.3±10.3	3.7±0.5	7.0±4.5	0.3±4.5
2 weeks	20.3 5.7	4.3 1.0	3.7 0.5	4.3 2.5
1 month	16.7 5.5	3.3 0.5	4.0 3.0	1.3 0.5
2 months	7.3 5.5	3.7 0.5	7.0 3.5	1.7 1.5
4 months	3.0 0.0	4.0 0.0	5.3 1.0	0.0 0.0
6 months	5.3 3.8	2.3 0.5	2.3 0.5	1.3 0.5
8 months	18.0 10.4	2.3 1.5	2.3 0.5	0.3 0.5
10 months	19.3 9.9	6.0 2.5	4.3 1.5	11.7 4.5
B. CLEAN SEDIMENT				
Pre-treatment	54.7± 9.1	2.3±2.5	5.7±1.5	0.7±0.5
2 weeks	11.3 8.3	2.7 4.0	3.3 5.0	3.0 3.0
1 month	17.0 10.1	3.7 3.0	4.0 1.0	2.0 0.0
2 months	13.0 6.2	3.0 1.0	1.3 1.5	0.3 0.5
4 months	3.0 1.8	2.0 1.0	4.0 3.5	1.3 2.0
6 months	8.3 8.5	1.7 0.5	2.7 2.0	1.0 1.5
8 months	12.0 2.6	4.3 1.5	2.3 2.5	1.3 1.5
10 months	16.7 7.4	4.7 1.5	5.0 3.0	7.7 5.5
C. ENVIROMUL LOW DOSE				
Pre-treatment	50.0± 6.0	2.7±1.0	6.0±2.0	0.3±0.5
2 weeks	1.7 2.1	5.0 1.5	0.0 0.0	3.7 3.5
1 month	4.7 4.0	3.3 2.0	0.7 0.5	4.7 2.0
2 months	2.7 0.6	3.0 2.5	1.7 1.0	1.0 1.0
4 months	5.7 2.1	2.3 1.5	3.7 4.0	1.0 1.0
6 months	4.3 3.5	3.3 1.5	0.7 1.0	0.7 0.5
8 months	11.7 7.2	2.0 1.0	2.0 0.0	0.3 0.5
10 months	11.0 4.4	3.7 3.5	3.7 2.5	5.7 5.5
D. DIESEL LOW DOSE				
Pre-treatment	54.3± 6.5	4.0±1.0	5.7±1.0	0.0±0.0
2 weeks	7.7 11.5	1.7 1.5	1.7 2.0	6.0 3.5
1 month	2.7 1.5	0.7 0.5	0.3 0.5	4.7 1.5
2 months	3.0 1.7	1.7 1.0	1.0 1.5	0.3 0.5
4 months	3.7 3.8	1.0 1.0	0.7 0.5	0.7 1.0
6 months	4.7 3.2	3.0 1.0	0.3 0.5	2.3 0.5
8 months	18.3 5.8	1.3 1.5	0.7 0.5	1.3 1.0
10 months	17.0 3.6	2.3 0.5	1.3 1.0	8.3 4.5
E. ENVIROMUL HIGH DOSE				
Pre-treatment	41.0±10.6	2.0±1.0	3.7±5.5	0.7±0.5
2 weeks	3.0 3.0	4.3 1.0	0.0 0.0	8.3 3.0
1 month	4.0 4.6	2.3 2.0	0.0 0.0	6.0 1.5
2 months	3.7 3.2	1.0 1.0	0.0 0.0	2.0 0.0
4 months	4.3 4.9	1.7 2.0	0.3 0.5	1.0 1.0
6 months	7.7 7.6	0.7 1.0	0.0 0.0	1.3 1.0
8 months	12.5 0.7	1.0 1.5	1.5 2.0	0.7 0.5
10 months	16.3 4.5	1.0 1.0	1.0 0.0	9.3 2.0
F. DIESEL HIGH DOSE				
Pre-treatment	47.0±11.8	2.7±0.5	1.6±0.5	0.7±1.0
2 weeks	0.3 0.6	0.0 0.0	0.0 0.0	8.7 3.0
1 month	0.7 1.2	0.7 0.5	0.0 0.0	15.3 6.0
2 months	1.7 1.5	0.0 0.0	0.3 0.5	2.3 1.5
4 months	3.3 2.3	0.0 0.0	0.3 0.5	2.0 1.0
6 months	4.3 2.1	0.3 0.5	0.3 0.5	3.7 1.5
8 months	14.3 7.5	0.3 0.5	1.7 2.0	2.0 1.0
10 months	16.3 8.0	0.7 0.5	1.0 1.0	7.3 6.5

TABLE 4. Selected species: Results of a posteriori comparison of means (Least Significant Range) between treatments exhibiting significant variance (2 way ANOVAR). A = Control; B = Clean sediment; C = Enviromul low dose; D = Diesel low dose; E = Enviromul high dose; F = Diesel high dose.

Parameter	Time	Pairs of treatment means differing at P<0.05					
		A	B	C	D	E	F
Abra alba	2 wks	F	F	-	-	-	-
		C	-	-	-	-	-
	1 mth	F	F	-	-	-	-
Amphiura filiformis	2 wks	F	-	F	-	F	-
	2 mths	F	F	F	-	-	-
	4 mths	F	-	F	-	-	-
	6 mths	-	-	F	F	-	-
		-	-	E	-	-	-
	8 mths	-	F	-	-	-	-
Harpinia antennaria	1 mth	F	F	-	-	-	-
		E	E	-	-	-	-
	2 mths	F	-	-	-	-	-
		E	-	-	-	-	-
	4 mths	F	-	-	-	-	-
		E	-	-	-	-	-
		D	-	-	-	-	-
	6 mths	E	E	-	-	-	-
	10 mths	F	F	-	-	-	-
		-	E	-	-	-	-
Pseudocuma longicornis	1 mth	F	F	-	-	-	-

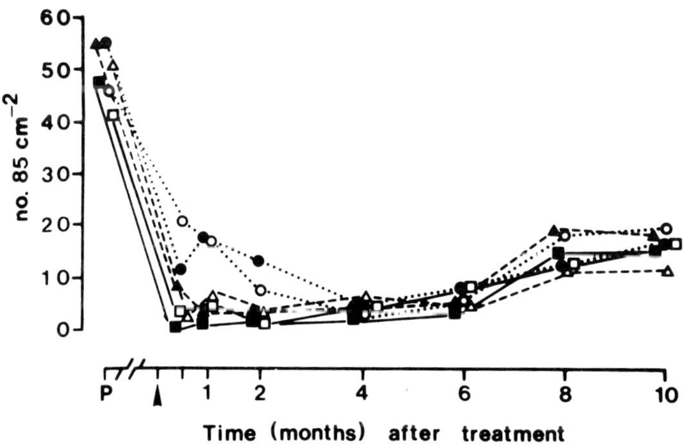

FIGURE 4. *Abra alba*: Variation in density with time for each treatment (means of 3 replicates). o = Control; ● = Clean sediment; △ = Enviromul low dose; ▲ = Diesel low dose; □ = Enviromul high dose; ■ = Diesel high dose

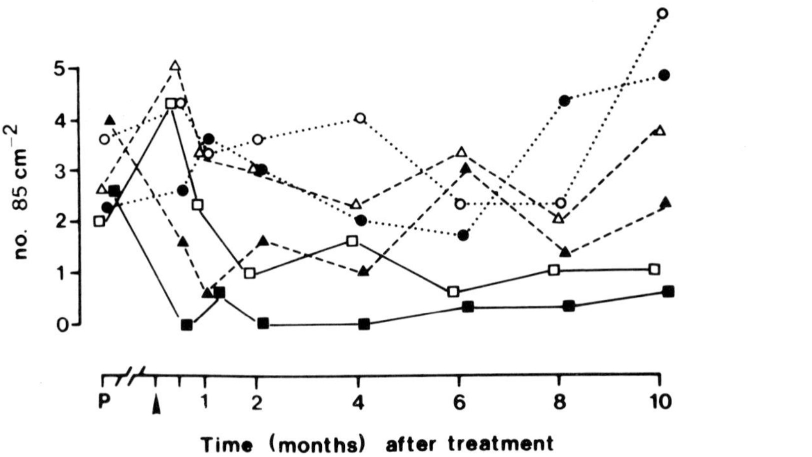

FIGURE 5. Amphiura filiformis: Variation in density with time for each treatment (means of 3 replicates). o = Control; ● = Clean sediment; △ = Enviromul low dose; ▲ = Diesel low dose; □ = Enviromul high dose; ■ = Diesel high dose.

In spite of the low numbers involved, populations of the phoxacephalid amphipod Harpinia antennaria were followed (Figure 6) because of the reported sensitivity of species in this family to pollution and disturbance (Swartz *et al*., 1982, Long and Chapman, 1985). Significant effects were detected in both high dose treatments on the majority of sampling occasions until the end of the experiment (Figure 6, Table 4).

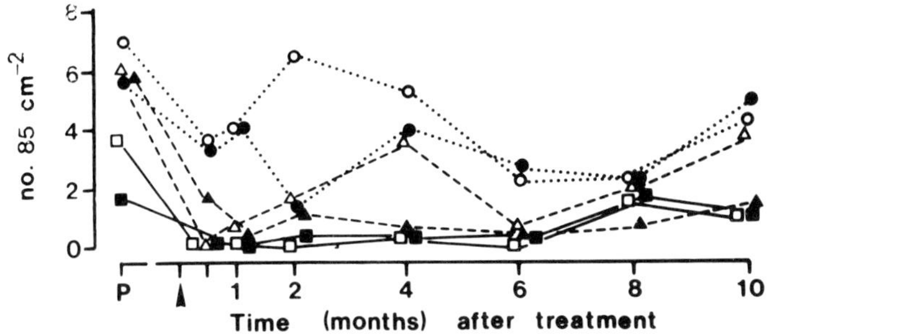

FIGURE 6. Harpinia antennaria: Variation in density with time for each treatment (means of 3 replicates). o = Control; ● = Clean sediment; △ = Enviromul low dose; ▲ = Diesel low dose; □ = Enviromul high dose; ■ = Diesel high dose.

In this instance no time-lag between Enviromul and diesel effect occurred and both levels of each OBM treatment appeared to cause sharp population reductions within two weeks of application. Subsequently, however, the effects of the Enviromul low-dose treatment were not separable from the controls and by 10 months signs of recovery were evident in these plots. In contrast diesel low-dose means showed a continuing similarity to those in the high-dose plots up until 10 months.

The only species for which population enhancement was recorded in the oiled plots, was the cumacean Pseudocuma longicornis (Figure 7). As with H. antennaria, numbers were generally low but a significant peak in the diesel high dose plots occurred at 1 month. Such effects were short-lived, however, and after two months treatment population means merged with the controls again. Similar abundance trends (though not significant) were evident in all plots and therefore were not primarily related to the experimental treatments.

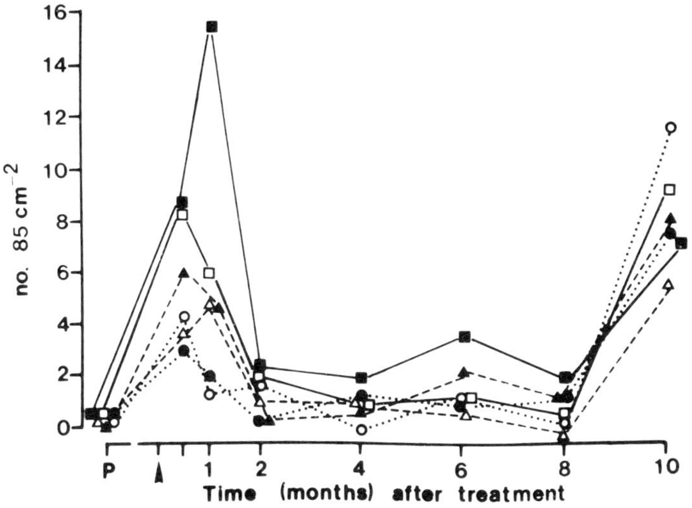

FIGURE 7. Pseudocuma longicornis: Variation in density with time for each treatment (means of 3 replicates). o = Control; ● = Clean sediment; △ = Enviromul low dose; ▲ = Diesel low dose; □ = Enviromul high dose; ■ = Diesel high dose.

4. DISCUSSION

The observed effects of cuttings treatments on the macrobenthos, although statistically significant, were relatively small. In the context of 'zones of impact' as derived from oilfield monitoring surveys (see Davies et al., 1984) data from the high-dose plots could be thought of as equivalent to survey data from somewhere near the outer edge of the 'transition zone' in macrobenthic diversity and community structure. The low-dose plots could then correspond to the outer areas where little or no petrogenic contamination is evident, with no detectable change to the infauna. However it should be emphasised that this study was not intended as a realistic representation of environmental conditions at varying distances from an OBM discharge point. First, only a single application of cuttings was used, and no attempt was made to mimic the continuous or chronic inputs from drilling operations. Second, experimental plots were both small and isolated in comparison with the area of contaminated sediments around a discharge in the North Sea. Thus hydrocarbon removal from the plots was not balanced by fresh inputs, or by inputs resulting from wide scale redistributive processes at the seabed. Therefore the fauna of the experimental plots was not subjected to the constant exposure to pollutant hydrocarbons that takes place around a drilling platform. Furthermore, it might be speculated that in this experiment, faunal recovery was possibly aided by immigration from surrounding unaffected sediments, to a greater extent than would be expected within the much larger contaminated areas in the North Sea.

As already discussed, the environmental effects of OBM-cuttings discharge probably occur through (1) direct physical smothering of the sediments, (2) organic enrichment, and (3) inherent toxicity.

In this study, the plots to which clean sediment was added showed no significant faunal differences from the controls. Since all treated plots received the same bulk of sedimentary material, it may be assumed that the faunal changes recorded in the oiled plots were not the result (primarily) of physical smothering. In addition, the qualitative and quantitative community changes often associated with organic enrichment of sediments (e.g. Pearson and Rosenberg, 1978; Addy *et* al., 1984; Davies *et al*., 1984) were not observed. Concomitantly there was no evidence from diver observations to suggest that significant reduction of sediment Eh was taking place in any of the plots, or that water exchange within the upper layers of sediment was impeded by the oily treatments. Furthermore, no treatment-related increases in sediment organic matter content (percent combustibles) were recorded following cuttings application. Therefore, with the elements of direct smothering and organic enrichment reduced, if not absent, it is likely that treatment effects on the macrofauna were more a reflection of cuttings toxicity.

In general, both Enviromul and diesel treatments caused reductions in faunal density and species richness. At equivalent doses, the only difference noted between the two treatments was the marked tendency for the diesel cuttings to cause faunal change more rapidly than the low-tox cuttings. This could be simply interpreted as reflecting the lower acute toxicity of low-tox OBM's to benthic fauna. This field-test observation is similar in principle to the findings of Dow (1984) who, in toxicity tests using *Crangon crangon*, also recorded a 'lag' in effect between a low-tox and a diesel OBM discharge.

In the cases of total faunal density, and the numbers of the amphipod *Harpinia antennaria*, significant effects (relative to the control plots) were maintained under Enviromul and diesel high-dose treatments for 10 months. In the same plots there was very little evidence of the recovery of *Amphiura filiformis* populations within this period either.

Faunal impact and recovery under the low-dose treatments is more difficult to assess because, overall, few significant treatment effects were recorded. However the trends shown in the first month or two (before treatment hydrocarbons were removed) closely paralleled the changes occurring in the high-dose plots.

5. CONCLUSIONS

Following treatment application there was no difference between the two OBM types in the rate of contaminant hydrocarbon removal from the plots.

At the treatment levels investigated, faunal disturbance was minor, and significant effects were recorded predominantly from the high-dose plots.

The effect of adding oiled cuttings to sediments initially was to depress species richness, faunal density and diversity in relation to the controls.

In the high-dose plots, total faunal density and the abundance of certain species remained depressed for the experiment's duration, showing little sign of recovery for either of the cuttings types.

Faunal disturbance occurred more rapidly following diesel treatment than with low-tox treatment. This probably reflected the greater acute toxicity of diesel OBM's. After one or two months however, the impact of the Enviromul treatment was similar to that caused by diesel cuttings.

Other longer term studies are required before conclusions can be drawn regarding the relative recovery rates of macrobenthos affected by diesel or by various low-tox OBM's. In the meantime, oilfield monitoring studies continue to provide the most valuable information on the effects of OBM discharges and it is hoped that such studies will continue beyond the cessation of drilling activities so that any subsequent environmental recovery may be followed.

6. ACKNOWLEDGEMENTS

I am grateful to the Institute of Petroleum for funding the investigation.

I am also indebted to the many people in OPRU, who gave up much time to assist in fieldwork, mostly in trying wintery conditions; who helped to process the seemingly endless stream of samples; and who assisted in all other aspects of the study right from the planning stages, through data collation and interpretation, to preparation of the final report.

REFERENCES

1. Addy JM, Hartley JP and Tibbets PJC: Ecological effect of low toxicity oil-based mud drilling in the Beatrice oilfield. Marine Pollution Bulletin 15(12), 429-436, 1984.
2. Bakke T, Green NW, Naes K and Pedersen A: Drill cuttings on the sea bed, Phase 3. Field experiment on benthic community response and chemical changes to thin (0.5 mm) layers of cuttings. In: Proc. Symp. Oil Based Drilling Fluids, Cleaning and Environmental Effects of Oil Contaminated Drill Cuttings, Royal Garden Hotel, Trondheim, Norway, 24-26 February 1986. 33-42, 1986.
3. Blackman RAA, Fileman TW and Law RJ: Oil-Based Drill Muds in the North Sea - The use of Alternative Base Oils. ICES CM 1982/E:13 Marine Environmental Quality Committee. 8 pp + tables and figures, 1982.
4. Blackman RAA, Fileman TW and Law RJ: The toxicity of alternative base-oils and drill-muds for use in the North Sea. Internat. Council for the Explor. of the Sea (ICES) CM 1983/E:11. Marine Environmental Quality Committee. 7 pp + tables and figures, 1983.
5. Buchanan JB and Kain JM: Measurement of physical and chemical environment. In: Holme NA and McIntyre AD (eds). Methods for the Study of Marine Benthos. IBP Handbook No. 16. Blackwell: Oxford. 30-58, 1971.
6. Davies JM, Addy JM, Blackman RA, Blanchard JR, Ferbrache JE, Moore DC, Somerville HJ, Whitehead A and Wilkinson T: Environmental effects of the use of oil-based drilling muds in the North Sea. Marine Pollution Bulletin 15(10), 363-370, 1984.

7. Department of Energy: Development of the oil and gas resources of the United Kingdom 1985. HMSO, London. 84 pp + chart, 1985.
8. Dow FK: Studies on the Environmental Effects of Production Water and Drill Cuttings from North Sea Offshore Oil Installations. Ph.D Thesis, University of Aberdeen. 243 pp + figures, tables and appendices, 1984.
9. Krumbein WC: Application of logarithmic moments to size frequency distributions of sediments. Journal of Sedimentary Petrology 6, 35-47, 1936.
10. Leaver MJ, Murison D, Davies JM and Raffaelli D: Experimental studies of the effects of drilling discharges. Proceedings of Royal Society meeting Environmental Effects of North Sea Oil and Gas Developments, Royal Society, London, 19 February 1986, in press.
11. Little DI, Staggs MF and Woodman SSC: Sample pretreatment and electronic particle counting in routine grain size analysis. In: Transfer Processes in Cohesive Sediment Systems. Plenum Press: London. 47-74, 1984.
12. Little DI, Howells SE, Abbiss TP and Rostron D: Some factors affecting the fate of estuarine sediment hydrocarbons and trace metals in Milford Haven 1978-82. In: Pollutant Transport and Fate in Ecosystems, (ed) Coughtrey PJ, British Ecological Society, in press.
13. Long ER and Chapman PM: A sediment quality triad: Measures of sediment contamination, toxicity and infaunal community composition in Puget Sound. Marine Pollution Bulletin 16(10), 405-415, 1985.
14. Matheson I, Kingston PF, Johnston CS and Gibson MJ: Statfjord Field environmental study. In Proceedings from Symposium Oil Based Drilling Fluids, Cleaning and Environmental Effects of Oil Contaminated Drill Cuttings, Royal Garden Hotel, Trondheim, Norway, 24-26 February 1986. 1-16, 1986.
15. Payne JF, Fancey L, Kicenink J, Williams U, Osborne J and Rahimtula A: Mixed function oxygenases as biological monitors around petroleum hydrocarbon development sites: Potential for induction by diesel and other drilling mud base oils containing reduced levels of polycyclic aromatic hydrocarbons. Marine Environmental Research 17, 328-332, 1985.
16. Pearson TH and Rosenberg, R: Macrobenthic succession in relation to organic enrichment and pollution of the marine environment. Oceanography and Marine Biology Annual Review 16, 229-311, 1978.
17. Swartz RC, Deben WA, Sercu KA and Lamberson JO: Sediment toxicity and the distribution of amphipods in Commencement Bay, Washington, U.S.A. Marine Pollution Bulletin 13, 359-364, 1982.

PETROLEUM EFFECTS ON OSMOREGULATION IN THE MARINE HALOPHYTES SPARTINA ALTERNIFLORA AND RHIZOPHORA MANGLE

E.S. GILFILLAN, D.S. PAGE, A.E. BASS, J.C. FOSTER, P.M. FICKETT, W.G. ELLIS, S. RUSK, AND C. BROWN

MARINE RESEARCH LABORATORY AND HYDROCARBON RESEARCH CENTER, BOWDOIN COLLEGE BRUNSWICK, MAINE 04011, USA

1. INTRODUCTION

Salt marshes and mangrove areas are very important in shoreline development and protection (1,2,3). These same areas are highly productive (4,5,6) and are important exporters of detritus to nearby communities (7,8). Both salt marshes and mangrove forests are important to fisheries production beyond their role as exporters of detritus (8,9).

Marshes and mangrove forests are vulnerable to spilled oil because of their location in depositional areas (10) and are also slower to recover from oil spills than are other environments (11).

The effects of oil pollution on survival, growth and development of the saltmarsh grass, *Spartina alterniflora* (12,13) and the red mangrove, *Rhizophora mangle* (14), have been described from experimental systems. Effects of spilled oil on these and related plants have been described from field situations (11). There is, however, little known about the mechanisms whereby these effects are produced.

In this study we examined the effects of oil in sediments on osmoregulatory ability in *Rhizophora mangle* and *Spartina alterniflora* exposed to oil in the field. The working model is that oil in sediments interferes with desalinization of water in the roots of these species. These species exclude salt by a reverse osmosis process involving root membranes (15). Because lipid solvents such as ether and chloroform interfere with salt exclusion it is clear that membrane integrity is vital to this process. The relative amounts of sodium and potassium ions in the plant's tissues have been shown to be an indicator of stress imparted by oil in sediments for mangroves (16). Sodium, the principal cation in seawater, will be elevated in the tissues of plants unable to exclude salt efficiently in their roots. Potassium, a major physiological cation, serves as a reference. The ratio of sodium to potassium will be larger in plants that are unable to efficiently exclude sodium ions from seawater.

2. PROCEDURE

2.1. Leaf Tissue Analysis

Red mangrove leaves were collected from oil spill sites in the Caribbean and West Africa (see Table 1 for sampling locations and dates). At the same time samples of sediments were taken for hydrocarbon analysis. Leaves were wrapped in aluminum foil and packed in ice for transport to the laboratory where they were held at -18°C pending analysis.

Spartina alterniflora leaves were collected from plants that had been experimentally exposed to #2 fuel oil in the field. Fifteen plots of *S. alterniflora*, all at the same height in the intertidal zone, were encased with 30 x 100 cm plastic caissons.

Kuiper, J., Van den Brink, W.J. (eds), Fate and Effects of Oil in Marine Ecosystems. ISBN 90-247-3489-4

TABLE 1. Sediment hydrocarbon content and Na/K ratios for *Rhizophora mangle* leaves from control and oil impacted sites. Hydrocarbon concentrations are given as ppm (dry wt). Na/K ratios given are means of triplicate samples from each plot. Sampling locations are given as PR for Puerto Rico or WA for West Africa. The PR data is from (16).

Sampling Location	O&G PPM	Aliphatic PPM	Aromatic PPM	Weathering Ratio	Na/K Ratio
PR - Oil	64700	12000	13300	2.6	8.5
PR - Oil	13700	2140	2270	3.1	7.4
PR - Oil	187200	72400	36400	1.7	6.0
PR - Oil	122300	28700	34500	1.9	3.3
PR - Oil	144860	28200	14600	3.4	4.7
PR - Oil	368000	91800	98000	3.9	1.9
PR - Oil	343000	94700	79200	2.0	4.9
PR - Oil	526000	135000	145000	1.9	3.2
PR - Oil	33200	3840	4300	4.1	3.0
PR - Oil	136000	25500	23400	2.8	2.9
PR - Ref	26400	201	1240	18.0	4.0
PR - Ref	5430	254	172	12.8	1.8
PR - Ref	24500	1030	354	17.7	1.3
WA - Oil	25000	4130	412	5.5	3.6
WA - Oil	13900	6700	1040	1.9	12.1
WA - Ref	19200	503	410	21.0	0.5

The caissons were filled with sea water and allowed to drain. Three caissons received no treatment; three caissons had sufficient oil added (14 ml) to create a 0.2 mm slick. This thickness corresponds to a moderately heavy oil slick. Three caissons had sufficient oil (28 ml) added to create a 0.4 mm slick. This thickness corresponds to a very heavy oil slick. Three caissons received 14 ml of #2 fuel oil premixed with 10% v/v Corexit 9527; three caissons received 28 ml of #2 fuel oil premixed with Corexit 9527. The slicks were allowed to strand, covering both plants and sediment within the caissons.

The experiment was begun on 5 July 1984. Samples of *S. alterniflora* leaves and samples of sediment were collected prior to the exposure and at intervals thereafter (See Table 2 for sampling dates). Three leaf samples and one sediment sample were taken at each sampling. *Spartina alterniflora* leaves were cut from the plants and rinsed with distilled water, wrapped in foil and frozen (-18 °C) pending analysis.

Mangrove leaf samples were rinsed with distilled water and dried in an oven (180°) to constant weight. They were then digested in HNO_3 and analyzed for Na and K using flame photometric techniques. The data reported in Table 1 are means of three subsamples taken from each field sample.

Frozen *S. alterniflora* leaves were dried to constant weight (80 °C) and digested in HNO_3 and analyzed for Na and K by flame photometry. The data reported in Table 2 are means of the three replicate samples.

2.2. Sediment Analysis

Sediment samples were taken using a coring device. The core at the root zone was transferred to a solvent washed glass container and sealed with a foil lined closure. The sediment samples were returned to the laboratory on ice and frozen until analyzed. Petroleum and biogenic hydrocarbons were determined in the sediment samples by methods described elsewhere (17).

TABLE 2. Sediment hydrocarbon content (ppm. dry wt) and Na/K ratios for the 1984 *Spartina alterniflora* oil exposure experiment. Na/K ratios are means of triplicate samples from each plot. The date is the number of days since the start of the experiment.

Day	O&G PPM	Aliphatic PPM	Aromatic PPM	Total PPM	Weathering Ratio	Na/K Ratio
CONTROL PLOT - NO OIL						
1	4747	68	16	84	57	-
3	5754	138	14	152	38	3.4
7	5922	87	27	114	52	2.5
14	10748	96	45	141	76	3.4
21	7872	86	28	114	69	1.0
28	11073	78	47	125	89	4.4
WHOLE OIL - .2 mm SLICK						
1	12664	172	47	219	58	-
3	9315	43	55	98	95	4.9
7	9035	124	40	164	55	3.3
14	9528	98	18	116	82	4.7
21	9088	86	47	133	68	7.9
28	8436	91	65	156	54	6.6
WHOLE OIL - .4 mm SLICK						
1	8837	165	64	229	39	-
3	8379	113	29	142	59	4.6
7	7820	176	68	244	32	5.6
14	7303	263	77	340	21	5.1
21	8653	174	21	195	44	10.2
28	6891	127	67	194	36	13.7
DISPERSED OIL .2 mm SLICK EQUIVALENT						
1	11618	242	120	362	32	-
3	13280	1068	205	1273	10	3.4
7	11860	496	189	685	17	3.8
14	11244	624	49	673	17	4.3
21	11067	431	85	516	21	3.8
28	11992	590	115	705	17	3.2
DISPERSED OIL .4 mm SLICK EQUIVALENT						
1	13104	163	68	231	57	-
3	9863	207	101	308	32	4.1
7	13679	676	190	866	16	2.9
14	11685	746	188	934	13	4.1
21	9965	341	85	426	23	3.8
28	12648	765	217	982	13	4.2

2.3 Data Analysis

Statistical analysis of the resulting data was carried out using the Regress II software package (Human Systems Dynamics Inc.) on an Apple IIe computer.

3. RESULTS

Results of Na, K and hydrocarbon analyses for the mangrove leaves and sediments are shown in Table 1. Results of Na, K and hydrocarbon analyses for *Spartina alterniflora* leaves and associated sediments are shown in Table 2. Also shown in Tables 1 and 2 is a weathering ratio (WR). The WR is calculated by dividing the total lipid extractable material by the total hydrocarbon content. High values of WR mean that most of the extractable material is not petroleum. Low values of WR mean that most of the lipid

TABLE 3. Results of analysis of variance of Na/K and sediment hydrocarbon data from the 1984 *Spartina alterniflora* oil exposure experiment.

ANOVA Table: Na/K data

Source	Sum of Squares	Degrees of Freedom	Mean Square	F
Time	38.39	4	21.65	8.38**
Oil	207.18	4	51.80	20.06**
Time x Oil	151.68	16	9.48	3.67**
Error	129.1	50	2.58	
Total	574.55	74		

ANOVA Table: Sediment Hydrocarbon Concentration data

Source	Sum of Squares	Degrees of Freedom	Mean Square	F
Oil	1844731.468	4	461182.867	10.60**
Error	1088007.494	25	43520.3	
Total	2932738.962	29		

Treatment Means	Na/K	Total Hydrocarbons
Control	3.42	121.67
.2 mm Dispersed Oil	3.7	702.33
.4 mm Dispersed Oil	3.82	624.50
.2 mm Whole Oil	5.48	147.67
.4 mm Whole Oil	7.84	224.00
Least Significant Difference Between Means (P = .05)	1.87	247.2

** $P < .01$

soluble material present is petroleum. Contaminated sites will have a low WR; clean sites will have a high WR. Table 2 also shows the height of the lower high tide on each date the Spartina samples were taken.

Results of analysis of variance carried out on Na/K data and sediment hydrocarbon data from the 1984 Spartina alterniflora oil exposure experiment are shown in Table 3. Also shown in Table 3 is the least significant difference between treatment means. The results in Table 3 show clearly that the Na/K values change significantly over time ($P < .01$). It is also clear that Na/K values are significantly affected by the oil treatments ($P < .01$). The effects of the oil treatments are different at different times ($P < .01$). It is also clear that significant effects on Na/K are only found as a result of the two whole oil treatments. Sediment hydrocarbon concentrations are also significantly affected by the oil treatments ($P < .01$). It is interesting that sediment hydrocarbon levels are not significantly elevated by treatment with whole oil. Sediment hydrocarbon levels are significantly elevated by both treatments with chemically dispersed oil. No association was observed between any measure of sediment hydrocarbon contamination and Na/K.

Table 4. shows the results of regression of weathering ratio of sediment hydrocarbons on Na/K in Rhizophora mangle leaves. The results of two different regression models are shown. Both models show significant association of elevated Na/K with reduced weathering ratio. The natural log model gives the best fit to the data. It can explain 32.3% of the variation in Na/K in terms of variation in weathering ratio.

TABLE 4. Results of regression of weathering ratio of sediment hydrocarbons on Na/K in Rhizophora mangle leaves.

LINEAR MODEL:

Source	Degrees of Freedom	Sum of Squares	Mean Square	F
Regression	1	36.777	36.777	5.330*
Residual	14	96.607	7.901	
Total	15	133.384		

$R^2 = .2757$

Equation: Na/K = -0.2392 * Weathering Ratio + 5.837

NATURAL LOG MODEL:

Source	Degrees of Freedom	Sum of Squares	Mean Square	F
Regression	1	43.068	43.068	6.676*
Residual	14	90.317	6.451	
Total	15	133.384		

$R^2 = .3229$

Equation: Na/K = -1.9042 * Weathering Ratio + 7.0947

* $P < .05$

4. DISCUSSION

In both Rhizophora mangle and Spartina alterniflora, increased exposure to oil leads to increased Na/K ratios in the leaves. It seems clear that elevated Na/K is a stress response in these plants.

In the case of Rhizophora mangle a significant portion of the variation in leaf Na/K is associated with variation in oil content of sediments. A higher degree of interdependence is obtained with a logarithmic regression model than with a linear regression model. This is to be expected since relationships between stressors and physiological responses are often exponential (18). However, the degree of interdependence is not high. It is clear that there are factors affecting leaf Na/K is this species which were not measured. The most likely extrinsic factor is soil salinity. In any investigation of leaf Na/K in mangroves it will be important to determine soil salinity at the same time that the leaves are collected.

In the case of Spartina alterniflora elevated Na/K in the leaves is associated with treatment with whole oil. No elevation in Na/K in the leaves is associated with dispersed oil treatments.

There are also effects of the oil treatments on sediment hydrocarbon levels. The results show that significantly elevated sediment hydrocarbon levels are associated with the dispersed oil treatments and not with the whole oil treatments. Even though the dispersed oil was incorporated more extensively into the sediment, it was not in a form that exerted any effect on the plants.

Sediment hydrocarbon levels are not significantly associated with changes in leaf Na/K in Spartina alterniflora. Therefore the elevated Na/K observed in plants from the plots treated with whole oil are most likely due to oiling of the above ground portions of these plants. The mechanism of action is most likely membrane damage in the leaves.

However, sediment hydrocarbon levels observed in the dispersed oil test plots, which produced no elevation in Na/K, were in the same range as levels observed to cause sharp elevations in leaf Na/K in West African mangroves. It is likely that the lack of effects observed in the plots with large amounts of dispersed oil in sediments are a result of the fact that in the micellar state, the dispersed oil is no longer lipophilic. Therefore the chemically dispersed oil is not in a form that can disrupt lipid biological membranes. Thus, neither exposure of leaves to dispersed oil nor exposure of roots to chemically dispersed oil caused any measurable damage to the plants.

Support for this conclusion is found in the results of a field experiment in which bivalve molluscs were exposed to a cloud of dispersed oil. Neither Mya arenaria nor Mytilus edulis accumulated any significant quantity of chemically dispersed oil (Page et al., 1984). The conclusion of this 1984 study was that the chemically dispersed oil was not bioavailable because the oil/dispersant micelles were no longer sufficiently lipophilic to be bioaccumulated in lipid structures of the animals.

5. CONCLUSIONS

The overall conclusions of this study may be summarized as follows:

- Elevated Na/K ratios in leaf tissues of halophytes exposed to oil reflect increased stress imparted to the plants by the oil.
- The mechanism involved is similar in both species studied and is related to interference in membrane function.
- Elevated Na/K ratios can also be produced by other factors such as increased soil salinity. Any field investigations should be very careful

to control for other sources of salinity stress.

ACKNOWLEDGEMENT
This work was made possible through support from the Mobil Foundation.

REFERENCES

1. Odum, W.E. and Johannes, R.E. The Response of Mangroves to Man Induced Environmental Stress. In: E.J.F. Wood and R.E. Johannes (eds). Tropical Marine Pollution, pp. 52-62, Elsevier Oceanography Series 12, 1975.
2. Teas, H.J. Ecology and Resotration of Mangrove Shorelines in Florida. Environmental Conservation 4 51-58, 1977.
3. Redfield, A.C. Development of a New England Salt Marsh. Ecol. Monogr. 42 201-237, 1972.
4. Teal, J.M. Energy Flow in a Salt Marsh Ecosystem of Georgia. Ecology 43 614-624, 1962.
5. Krebs, C.T. and Valiela, I. Effects of Experimentally Applied Chlorinated Hydrocarbons on the Biomass of the Fiddler Crab Uca pugnax (Smith). Estuarine and Coastal Marine Science. 6 375-386, 1978.
6. Odum, W.E. and Heald, E.J. Trophic Analysis of an Estuarine Mangrove Community. Bull. Mar. Sci. 22, 671-738, 1975.
7. Odum, E.P. and de la Cruze, A.A. Particulate Organic Detritus in a Georgia Salt Marsh Estuarine Ecosystem. In: G. Lauff (ed). Estuaries pp. 383-388. American Association for the Advancement of Science. Publ. 83, Washington, D.C., 1967.
8. Canestri, V. and Ruiz, O. The Destruction of Mangroves. Mar. Poll. Bull. 4 181-185, 1973.
9. Shenker, J.M. and Dean, J.H. The Utilization of an Intertidal Salt-marsh Creek by Larval and Juvenile Fishes: Abundance, Diversity and Temporal Variation. Estuaries 2 154-163, 1979.
10. Gundlach, E.R. and Hayes, M.O. Vulnerability of Coastal Environments to Oil Spill Impact. Mar. Technol. Soc. J. 12 18-27, 1978.
11. National Academy of Sciences. Oil in the Sea; Inputs Fates and Effects. Chapter 5. National Academy Press, Washington, D.C. (USA), 1985.
12. Alexander, S.K. and Webb, J.W. Effects of Oil on Growth and Decomposition of Spartina alterniflora. In: Proceedings of the 1983 Oil Spill Conference pp. 529-532, American Petroleum Institute, Washington, D.C., 1983.
13. Nestler, J. Interstitial Salinity as a Cause of Ecophenic Variation in Spartina alterniflora. Estuarine and Coastal Marine Science 5 707-714, 1977.
14. Getter, C.D., Ballou, T.G. and Dahn, J.A. Preliminary Results of Laboratory Testing of Oil and Dispersants on Mangroves. In: Proceedings of the 1983 Oil Spill Conference pp. 533-538, American Petroleum Institute, Washington, D.C., 1983.
15. Scholander, P.F. How Mangroves Desalinate Seawater. Physiologa Plantarum. 21 258-268, 1968.
16. Page, D.S., Gilfillan, E.S., Foster, J.C., Hotham, J.R., and Gonzalez, L. Mangrove Leaf Tissue Sodium and Potassium Ion Concentrations as Sublethal Indicators of Oil Stress in Mangrove Trees. In: 1985 Oil Spill Conference Proceedings pp. 391-393. American Petroleum Institute, Washington, D.C. (USA), 1985.
17. Page, D.S., Foster, J.C., Hotham, J.R., Gilfillan, E.S., Vallas, D., Hanson, S.A., and Gerber, R.P., Tidal Area Dispersant Project: Fate of Dispersed and Undispersed Oil in Two Nearshore Test Spills. In:

T.E. Allen (ed). Oil Spill Chemical Dispersants, Research, Experience and Recommendations, pp. 280-298. American Society for Testing and Materials, Philadelphia, PA (USA), 1984.

18. O'Flaherty. E.J. Toxicants & Drugs: Kinetics & Dynamics. John Wiley and Sons, New York. xiii & 398 pp., 1981.

OIL POLLUTION AND PLANKTON DYNAMICS. IV. SUMMARY OF ENCLOSURE EXPERIMENTS IN LINDÅSPOLLENE, NORWAY, WITH SPECIAL EMPHASIS ON THE BALANCE BETWEEN AUTOTROPHIC AND HETEROTROPHIC PROCESSES

H.R. SKJOLDAL[1] AND T.F. THINGSTAD[2]

[1] Institute of Marine Research, Directorate of Fisheries, P.O.Box 1870, N-5011 Bergen-Nordnes, Norway
[2] Department of Microbiology and Plant Physiology, University of Bergen, Allégt. 70, N-5000 Bergen, Norway

1. INTRODUCTION

Oil pollution may have both inhibitory and stimulatory effects on plankton communities. Lethal or sublethal toxic effects belong to the first category, whereas stimulation of heterotrophic processes by oil as an organic carbon source examplifies the latter type of effects. In both cases one would expect primary as well as secondary effects. Secondary effects are being transfered to other components of the dynamic food web system. Examples of secondary effects are growth of species otherwise restrained by competition or predation from species removed due to oil toxicity (17, 55), and depression of primary production caused by a competitive success of bacteria consuming mineral nutrients during degradation of organic substances from the oil (1, 2).

Secondary effects depend on dynamic interactions between populations, and one would therefore expect them to vary extensively with the state of the ecosystem at the time of pollution. Primary effects would also depend on ecosystem status. Species composition is obviously important in this respect since sensitivity to toxic oil components varies among species (21, 31). The degree of nutrient starvation experienced by organisms at the time of pollution may also be important. Environmental factors such as light and temperature influence the state of the ecosystem, and they also affect the properties of oil pollutants. Thus illumination alters the physical (47), chemical (51-54) and toxicity properties of oil (25, 35, 46).

The complexity of possible effects and their dependency on biological and environmental conditions make it difficult to predict effects of oil pollution on plankton communities from simple laboratory toxicity tests. The mesocosm approach represents one possible means of getting information on ecosystem responses towards pollution (32, 37). Such responses must be expected to vary, however, depending on the specific experimental conditions.

We have made three sets of experiments studying the effects of Ekofisk crude oil in seawater enclosures. These three experiments corresponded to different dynamic states of the planktonic ecosystems enclosed. The start of the first experiment (March-April 1980) coincided with the onset of the spring phytoplankton bloom, with mineral nutrient depletion occurring in the course of the experiment. The second and third experiments (June 1980 and 1981, respectively) were conducted with summer plankton in nutrient depleted water. Experimental manipulations other than addition of oil were made on some enclosures in the summer experiments. These included addition of mineral nutrients and addition of dispersant.

Kuiper, J., Van den Brink, W.J. (eds), Fate and Effects of Oil in Marine Ecosystems. ISBN 90-247-3489-4

In this paper we summarize some of the results from these experiments and address more specifically the question of whether addition of oil leads to a shift towards a more heterotrophically dominated pelagic community.

2. METHODS

The experiments were performed in Lindåspollene, a land-locked fjord in western Norway (10). Plastic bag enclosures of the type described by Brockmann et al. (7) were used. The bags, which had a diameter of 1 m and a length of 13.5 m (spring experiment) or 20 m (summer experiments), were filled with naturally stratified water columns. Further details on the experimental systems as well as description of sampling procedures are given elsewhere (40, 42, 52, 54).

Three sets of experiments were performed in March-April 1980, June 1980 and June 1981. The various bags and treatments used in these experiments are summarized in Table 1. Crude oil, dispersant and inorganic nutrients were added to the seawater surface and left to disperse and mix by natural processes.

TABLE 1. Overview of plastic bag enclosures and treatments in oil pollution experiments in Lindåspollene.

Experiment	Time	Bag	Treatment
I	March-April 1980	CB	None (control)
		OB	0.5 l Ekofisk crude oil
II	June 1980	CB	None (control)
		OB	0.5 l Ekofisk crude oil
		ODB	0.5 l crude oil, 100 ml Corexit 9527
		ONB	0.5 l crude oil, inorganic nutrients
III	June 1981	CB	None (control)
		OB	0.5 l Ekofisk crude oil
		NB	Inorganic nutrients
		ONB	0.5 l crude oil, inorganic nutrients

Detailed accounts of the various chemical and biological results have been or will be published in separate reports (oil chemistry: 3, 4, 42, 51, 52, 54; microbiology and phytoplankton: 22, 42, 43, 48-50; zooplankton: 11-13, 41. These should be consulted for descriptions of analytical methods.

3. ENVIRONMENTAL CONDITIONS

Lindåspollene is a semi-enclosed fjord system with a hydrographical regime characterized by a stratified and stable water column (10, 45). Naturally stratified water columns were enclosed in the experimental bags with little disturbance (40, 44). The salinity of the surface layer was generally in the range 28-30 ^{o}/oo whereas the salinity in the bottom part of the bags was 31-31.5 ^{o}/oo. The salinity and density structures were typically characterized by a gradual increase with depth without any marked discontinuity layer.

Due to the stability of the trapped water columns, vertical mixing was low within the enclosures. Estimated values of the coefficient of turbulent diffusivity were in the range 0.01-0.2 $cm^2\ s^{-1}$ (40, 42, 44).

The surface temperature was 2-4 ^{o}C during the first part of the spring 1980 experiment, increasing to 10 ^{o}C by the end of the experiment. In the summer experiments the surface temperature ranged from 12 to 17°C. In the deeper part of the bags the temperature remained fairly constant at about 5°C.

The light conditions within the bags were somewhat reduced in the upper few meters due to the floatation frame and also partly due to the presence of oil (40). Below 5 m this shading effect was small. The bags contained approximately the whole euphotic zone as the 1% light level and nutricline were located close to the bottom of the bags.

4. OIL CHEMISTRY

Photochemistry played a major role in the transformation and water accomodation of the crude oil (3, 51-54). Water soluble polar products from photooxidation accumulated in the surface layer of the oil bags to reach maximum concentrations of 4.8 and 1.7-2.1 mg l^{-1} about 2 weeks after addition of oil in the spring and June experiments, respectively (3, 4, 42). The concentrations of ordinary petroleum hydrocarbons were low.

Addition of the detergent Corexit 9527 resulted in more rapid water accomodation of the oil and a maximum concentration of 21 mg l^{-1} of material derived from petroleum and dispersant was reached at 0.3 m after 1 day (54). Photochemical processes apparently played a major role also in this case.

5. MAJOR BIOLOGICAL EFFECTS

Before treating each experiment separately in terms of the balance between autotrophic and heterotrophic processes, a brief summary of major biological effects will be given. A major effect common to all 3 experiments was a general and pronounced lowering of the phytoplankton biomass and production in oil polluted bags (42, 43). Addition of inorganic nutrients partly relieved the inhibitory effect of oil on primary production (22). This suggests that indirect effects on primary production through reduced nutrients recycling played a role in addition to direct toxic effects on the algae. Addition of oil resulted in a gradual change in phytoplankton species composition towards greater dominance of small and presumably more resistant forms (22, 43, 50). This was most pronounced in the oil and dispersant bag where a few species of tolerant green algae persisted (50).

Planktonic ciliates appeared to be sensitive to oil pollution, and their numbers decreased dramatically in oil polluted bags (11-13, 42). Larger zooplankton was studied in one experiment only (June 1980). The abundance of copepod nauplii was markedly reduced in the upper layer of oil polluted bags. This resulted in reduced recruitment to and decline of the populations of dominant copepods (41).

A typical pattern in most, but not all oil polluted bags was an increase in the number of bacteria. There was also evidence for increased consumption of inorganic phosphate by the bacterial size fraction (<1 µm) relative to that by phytoplankton (>1 µm) (42, 48, 49). The response pattern of various general and hydrocarbon related microbial activity measures showed, however, great variation among the experiments.

By treating response parameters in oil bags relative to those in control bags, mesocosm experiments can provide information on dose - response relationships at an ecological level. The rate of primary production was reduced to about half at a concentration of petroleum derived material of about 100-200 µg l^{-1} (100-200 ppb) (57). Ciliates and nauplii disappeared from the upper layer at concentrations of about 300 µg l^{-1}. No clear ecological effects were observed at concentrations less than 100 µg l^{-1}.

6. HETEROTROPHIC VERSUS AUTOTROPHIC PROCESSES

6.1. Spring 1980 experiment

The experiment was started on 13 March at the onset of the spring phytoplankton bloom. Oil was added to the oil bag (OB) the next day (42). The phytoplankton bloom progressed rapidly after 20 March in the control bag (CB) and culminated by mid April due to inorganic nutrients depletion (Fig. 1A). The phytoplankton development in OB progressed more slowly, especially during the second week, and nitrate was not depleted before the end of April.

pH of seawater is closely connected with the CO_2/carbonate system and reflects the balance between photosynthesis and respiratory processes. pH was lower in OB than in CB, but showed a general increase to high values towards the end of the experiment in both bags (Fig. 1B).

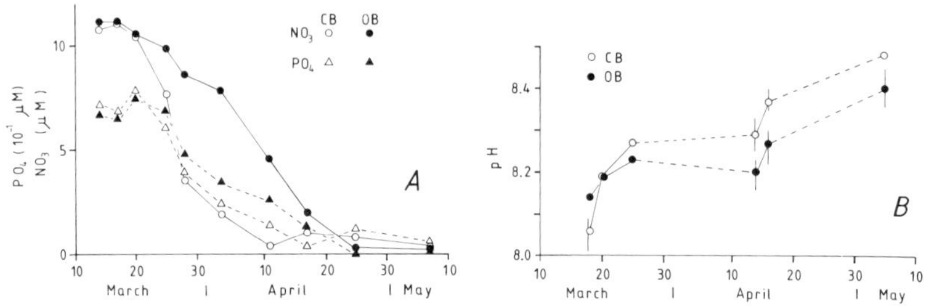

FIGURE 1. Spring 1980 experiment. A. Mean concentrations of nitrate and phosphate at 0.5, 2 and 5 m depths in control bag (CB) and oil bag (OB). B. Mean pH values at 0.5 and 2 m in CB and OB.

Phytoplankton production has been measured by the ^{14}C method and estimated from the decrease in nitrate content and increase in oxygen content (Table 2). The two latter are underestimates since nutrient regeneration, respiration and loss to the atmosphere have not been taken into account. The agreement between measured primary production and production estimated from nitrate was fairly good for the first two week period. Primary production was measured infrequently afterwards and integration for the following period was not attempted.

The estimates based on nitrate consumption and oxygen production were also in fair agreement, especially if the two periods are considered as a whole (Table 2). The estimates based on oxygen tended to be lower than those based on nitrate for the upper layer, but this was at least partly due to loss of oxygen to the atmosphere. The increase in oxygen concentration and the close relation between oxygen and phytoplankton activity (Table 2) are strong evidence that any shift in the balance between autotrophy and heterotrophy caused by the oil was slight, and that autotrophy predominated in OB as it did in CB.

6.2 June 1980 experiment

The bags were filled on 29 May and oil was added to OB and ODB (Table 1) on 31 May. Corexit 9527 was added to CDB one day later. ONB received addition of oil on 3 June and nutrients (agricultural preparation) on 4 June.

A moderate diatom bloom occurred in the surface layer when the bags were filled. This bloom soon collapsed and was in CB followed by a dominance of coccolithophorids (*Emiliania huxleyi*) in the mid-part and small

TABLE 2. Phytoplankton production during the first 4 weeks of the spring 1980 experiment, estimated from decrease in nitrate content, increase in oxygen content, and by the ^{14}C method. Nitrate-N was converted to carbon by assuming an atomic C/N ratio of 6. Oxygen was converted to carbon by assuming a photosynthetic quotient (O_2/CO_2) equal to 1.6 (56).

Depth (m)	Bag	Nitrate consumption ($g\ C\ m^{-2}$) Day 1-14	Nitrate consumption ($g\ C\ m^{-2}$) Day 15-28	Oxygen evolution ($g\ C\ m^{-2}$) Day 1-14	Oxygen evolution ($g\ C\ m^{-2}$) Day 15-28	Primary production ($g\ C\ m^{-2}$) Day 3-14
0-3.5	CB	1.9	0.7	0.6	0.7	1.5
	OB	0.6	1.0	0.3	0.6	0.5
3.5-13	CB	3.7	3.8	2.4	5.2	3.3
	OB	2.0	1.5	0.8	1.9	2.9
0-13	CB	5.6	4.4	3.0	5.8	4.8
	OB	2.6	2.5	1.1	2.5	3.4

flagellates towards the end of the experiment. This development was reflected in the temporal distribution of chlorophyll *a* (chl *a*) at 0.5 m (Fig. 2A). The chl *a* levels were moderately and much lower in OB and ODB, respectively. pH of the upper layer showed a transient decrease in CB, a small decrease during the experiment in OB, and a more pronounced decrease in ODB (Fig. 2B).

Vertical profiles of chl *a* and oxygen 24 days after addition of oil (Fig. 3) reveal a drastic reduction in phytoplankton biomass and a lowered oxygen content in the upper layer of ODB. Apart from chl *a* at 0.5 m and oxygen at 5 m, the profiles in CB and OB were quite similar. The upper layer was slightly supersaturated with oxygen in CB and OB (101-105%) and undersaturated in ODB (86-93%).

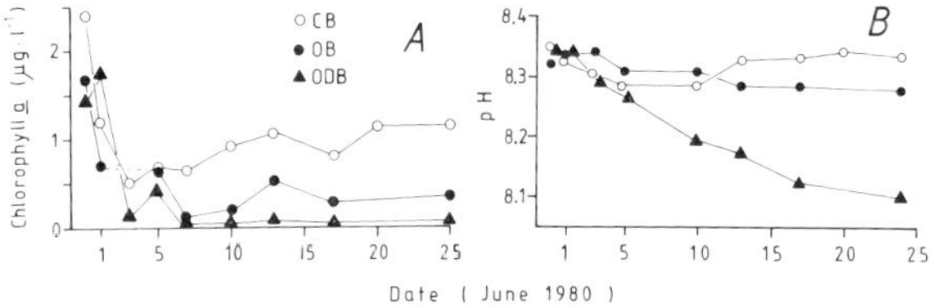

FIGURE 2. June 1980 experiment. A. Mean concentration of chlorophyll *a* at 0.5 and 2 m, and B. pH value at 0.5 m in control (CB), oil (OB) and oil plus dispersant (ODB) bags.

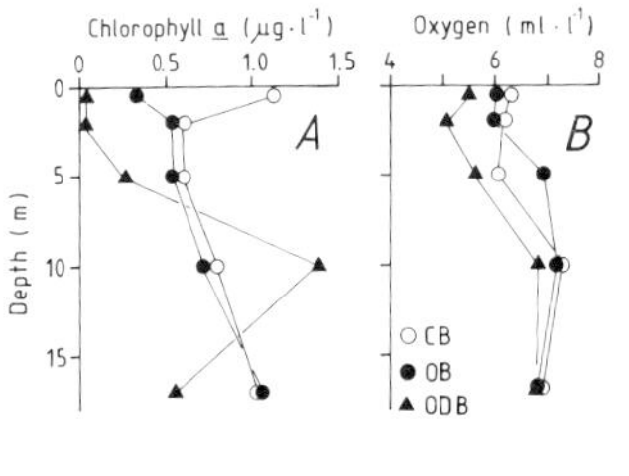

FIGURE 3. June 1980 experiment. A. Vertical profiles of chlorophyll *a* on 24 June, and B. Vertical profiles of oxygen on 25 June in control (CB), oil (OB) and oil plus dispersant (ODB) bags.

A summary of average standing stocks and activities in the upper 3.5 m and the deeper layer is presented in Fig. 4. The amounts of oil-derived material are maximum values obtained on 13 June in OB and 9 June in ODB (3, 4, 54), converted from dry matter by assuming a carbon content of 50%. These amounts exceeded the amounts of particulate organic carbon (POC) in the < 30 μm size fraction in the surface layer.

Phytoplankton C biomass was obtained from POC by subtracting amounts of detritus estimated by regression analysis of POC versus chl a, ATP and POP (43). Oil-derived material contributed to the POC pool in the upper layer of ODB, and phytoplankton biomass was here estimated as chl a multiplied by a C/chl a factor obtained in OB. Bacterial C was estimated from numbers (AODC) (48) assuming a volume of 0.15 μm^3 per bacterium (27) and 5.6 x 10^{-13} g C μm^{-3} (5). Zooplankton C in the size fraction > 63 μm was converted from protein (41) assuming C = 0.7 x protein.

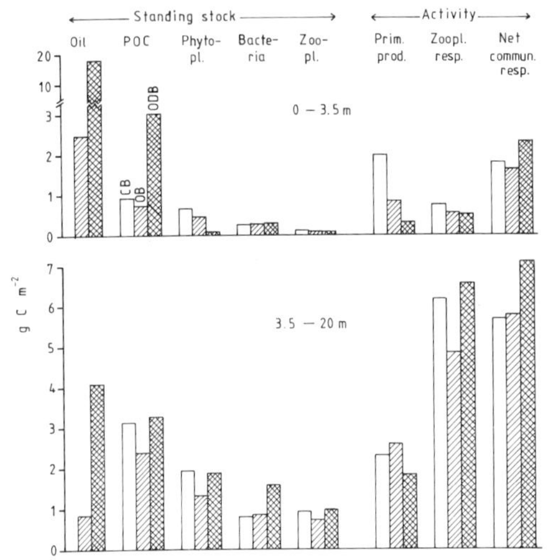

FIGURE 4. June 1980 experiment. Maximum amounts of oil-derived material, average standing stocks of particulate organic carbon (POC), phytoplankton, bacteria and zooplankton, and integrated primary production, zooplankton respiration and net community respiration during the first 17 d following addition of oil. Values for the 0-3.5 m (upper panel) and 3.5-20 m (lower panel) depth intervals in control bag (open columns), oil bag (hatched columns) and oil plus dispersant bag (cross-hatched columns). Further details are given in the text.

Integrated primary production during the first 17 d after oil addition was reduced from 2 g C m^{-2} in the upper layer of CB to 0.8 and 0.3 g C m^{-2} in OB and ODB, respectively. Below 3.5 m there was only a small reduction in primary production in ODB compared to CB and OB. Zooplankton respiration was calculated from protein biomass multiplied by weight specific respiration rate determined for zooplankton from outside the bags (23). Respiration in OB and ODB has possibly been overestimated if oil caused a lowering of the weight specific respiration rate. Net community respiration was taken as the decrease in oxygen content during the experiment and converted to C units assuming a 1:1 molar ratio for CO_2:O_2.

Zooplankton respiration accounted for an appreciable part of the net community respiration in the deeper layer. Its contribution was less in the surface layer, suggesting a possible greater influence of microheterotrophs here. Loss of oxygen to the atmosphere would, however, lead to overestimates of net community respiration, especially for CB and OB which remained supersaturated towards the end of the experiment.

The results in Fig. 4 suggest that a major effect of oil added alone was a reduced primary production, with only a minor change in the autotrophic/heterotrophic balance. Oil dispersed with Corexit 9527 had a stronger effect. The increase in net community respiration was not greater, however, than could be accounted for by the decrease in primary production.

The amount of nutrients added to ONB (Table 1) was high and presumably as a synergistic effect due to ammonium poisoning, primary production was initially strongly suppressed. A phytoplankton bloom developed subsequently and chl a reached 16 $\mu g\ l^{-1}$ in the surface layer on 20 June. pH increased at the same time to 8.38. These results demonstrate that nutrients caused a stronger stimulation of autotrophic than of heterotrophic activity.

6.3 June 1981 experiment

This experiment encompassed 4 bags, two of which were enriched by addition of inorganic nutrients (Table 1). The bags were filled on 1 June, and oil was added to OB and ONB on 2 June. Nutrients (7.5 mmol of nitrate and silicate and 0.6 mmol of phosphate) were added to NB and ONB on 3 occasions: 2, 12, and 15 June.

The level of chl a in the surface layer was generally lower in OB than in CB (Fig. 5A). Chl a increased considerably in response to the additions of nutrients, being consistently higher in ONB than in NB. pH also increased in both of the nutrient enriched bags and was only slightly lower in the oiled bags compared to their respective controls (Fig. 5B). Vertical profiles 3 weeks after addition of oil revealed higher chl a concentrations also at intermediate depths in the nutrient enriched bags and very similar oxygen concentrations in the upper layer of all 4 bags (Fig. 6).

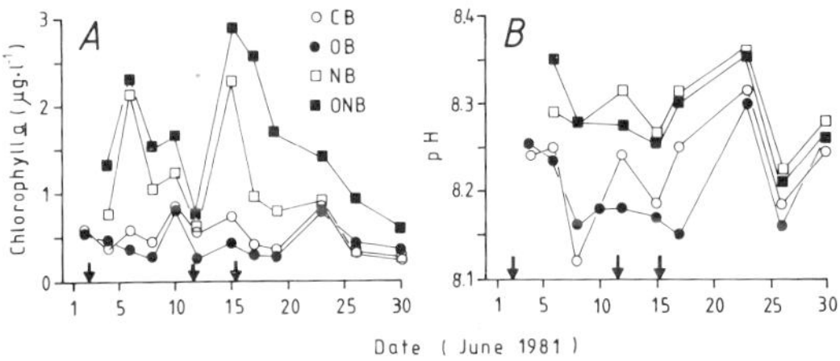

FIGURE 5. June 1981 experiment. A. Mean chlorophyll a concentration at 0.5 and 2 m, and B. mean pH values at 0.5 and 2 m in control (CB), oil (OB), nutrients (NB) and oil plus nutrients (ONB) bags. Arrows show times of nutrients addition.

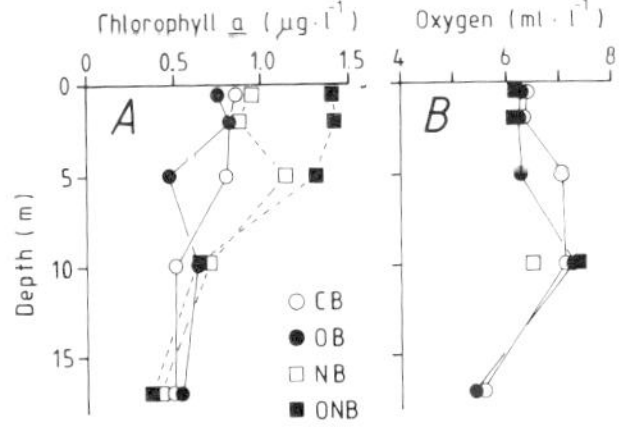

FIGURE 6. June 1981 experiment. A. Vertical profiles of chlorophyll a, and B. vertical profiles of oxygen on 23 June in control (CB), oil (OB), nutrients (NB) and oil plus nutrients (ONB) bags.

A summary of standing stocks and activities is presented in Fig. 7. Oil derived material was determined in OB only, and the maximum amount on 15 June was comparable to that found in the June 1980 experiment. Phytoplankton C was in this case estimated from average levels of chl a, multiplied by average detritus-corrected C/chl a ratios for CB and OB in

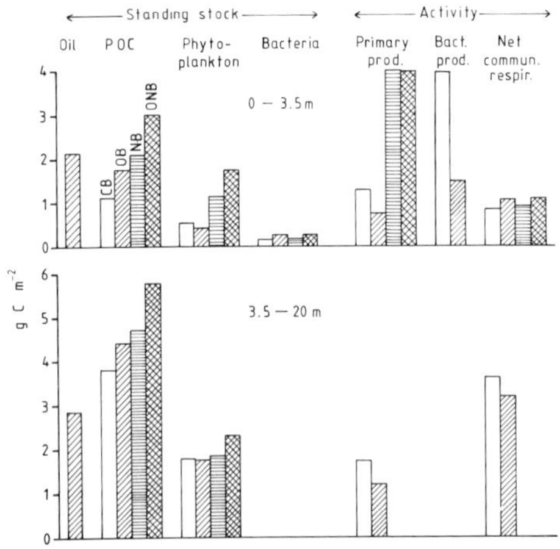

FIGURE 7. June 1981 experiment. Maximum amount of oil-derived material, average standing stocks of POC, phytoplankton and bacteria, and integrated primary production, bacterial production and net community respiration during the first 21 d following addition of oil. Values for the 0-3.5 m (upper panel) and 3.5-20 m (lower panel) depth intervals in control bag (open columns), oil bag (hatched columns), nutrients bag (horizontally hatched columns) and oil plus nutrients bag (cross-hatched columns). Further details are given in the text.

June 1980. The phytoplankton biomass was considerably higher in the upper layer of the nutrient enriched bags, but this trend may have been exaggerated if the nutrients caused increased cellular chl $\underline{a}$ levels. Bacteria was determined at 0.5 m only, and their biomass was almost twice as high in the oiled bags as in their respective controls.

Primary production was reduced to almost half in the upper layer of OB compared to CB. The added nutrients caused a marked increase in primary production which was similar in NB and ONB. Bacterial production was determined from uptake of ^{3}H-thymidine, using a conversion factor of 1.7×10^{18} bacteria produced per mole thymidine (19). The calculated bacterial production, which exceeded the primary production, was considerably lower in OB than in CB. The oxygen content of the bags decreased during the experiment, but there were only minor differences in net community respiration among the bags. Thus, as in the previous experiments, oil did not cause any marked shift in the autotrophic/heterotrophic balance.

7. DISCUSSION

When attempting to generalise from our results, the specific experimental design and environmental conditions must be taken into account. Crude oil was poured onto the sea surface of stable water columns with low vertical mixing and limited wave activity. These conditions favour photochemical influences on the oil (51-53), while mechanical dispersion is limited. Experiments have shown that toxicity (25, 26, 31, 35, 46), dissolution (51, 52) and emulsifying properties (47) of oil are heavily influenced by light. From this one would expect that the behaviour and effects of oil in marine environments are strongly dependent upon the sequence of influences from environmental factors like sun and wind/wave-action following a spill.

A consistent result in our experiments was that addition of oil alone resulted in depression of phytoplankton biomass and primary production. A substantial amount of oil-derived organic carbon was brought into the water column (Figs 4 and 7). Potentially this could have resulted in a strong stimulation of bacterial activity and a shift towards a heterotrophically dominated community. Although there were indications of increased activity of oil degrading bacteria (42, 48, 49), such a stimulatory effect was apparently not of great magnitude in terms of the oxygen and carbon budgets.

There are several possible mechanisms for the low stimulatory effect of oil on heterotrophic processes in our experiments. 1. The oil-derived material was toxic to bacteria. 2. The material was relatively refractile. 3. Growth of oil-degrading bacteria was hindered by predation. 4. Bacterial degradation of oil was limited by inorganic nutrient supply.

The two summer experiments were conducted in a situation with low mineral nutrients content in the upper water layer. Nutrient availability and recycling interact with both autotrophic and heterotrophic processes in such cases. The negative effect of oil on primary production in OB (Figs 4, 7) probably reflected partly a direct toxic effect on the algae and partly a secondary effect due to reduced mineral nutrient supply or availability. The marked decreases in populations of ciliates (12, 13) and juvenile zooplankton (41) suggest that nutrient recycling was reduced. Increased competition between phytoplankton and bacteria for mineral nutrients (6, 8) could also have reduced the nutrients available for phytoplankton.

Relieving the nutrient limitation by addition of nutrients stimulated primary production in the oil bag (ONB) as much as it did in the control bag (NB) (Fig. 7). This reflected the growth of phytoplankton in ONB to biomass levels higher than in NB (Figs 5-7), as well as low consumption of nutrients by stimulated heterotrophic activity. This suggests that bacterial activity in the presence of oil was not controlled by mineral nutrient limitation. The results from the spring experiment, where depletion of nutrients did not occur before at the end of the experiment, support this conclusion. However, ecological interactions should not be ruled out altogether. A combination of inorganic nutrient competition and predation pressure on bacteria has been shown to severely limit heterotrophic activity, even in the presence of a suitable substrate like glucose (36).

It is probable that oil had a more drastic effect on the populations of grazers on phytoplankton than on bacterial grazers. Ciliates were found to be particularly sensitive to oil in our experiments (11-13, 42) as has also been found in another mesocosm experiment (27, 34). The selection of food by ciliates is at present under reevaluation (38, 39), and their diet may be expected to consist of a mixture of small organisms irrespective of the trophic status of the prey. Copepods and other zooplankters are also apparently sensitive to oil pollution (14, 41). Chronic, low-level oil pollution in the MERL tank experiments resulted in an increased phytoplankton biomass due to inhibitory effects on grazer populations (16, 17, 55). Some species of bacterivorous flagellates are, on the other hand, fairly resistant to oil pollution (2, 27, 34, 50). Due to their short generation times (18), they have the potential to rapidly increase the predation pressure in response to an increase in bacterial biomass.

Besides any influence of more complex ecological interactions, toxicity and suitability of the oil-derived material as substrates for bacterial growth remain as likely factors for the restrained bacterial activity. Moderately high concentrations of hydrocarbons (> 300 $\mu g\ l^{-1}$) have been found to inhibit marine bacterial activity (20). It has also been shown that illumination of crude oil results in water soluble products that suppress bacterial activity in natural seawater (35). The chemical nature of the oil-derived material which accumulated in the water column in our experiments has not been characterized in any detail. The major part of this material apparently stems from photooxidation of oil, particularly of the high molecular weight resin/asphaltene fraction (51-53). Such material may be of limited suitability as general carbon substrates for bacterial growth.

In comparing our results with those from other mesocosm oil pollution studies, the experimental designs must be kept in mind. Crude oil floating on the surface has been used in only a few other experiments. It is noteworthy, however, that the strongest negative effects on phytoplankton primary production have been reported from such experiments (15, 25).

In the CEPEX experiments in Canada, water extracts of crude oil, fuel oil or aromatic hydrocarbons were mixed into the water column in large enclosures. Oil resulted generally in a qualitative change in the species composition of phytoplankton, but with less effect on the overall primary production (28-30). Among other effects were increases in numbers of bacterioplankton and tintinnid ciliates and decreases in other zooplankton groups. Increases in bacterial activity were noted which in one experiment resulted in a doubling of the rate of nitrate uptake (30). Even more dramatic effects were observed in a later experiment (SEAFLUXES) where a mixture of crude oil and Corexit 9527 was mixed into the water at 2-4 m depth. This resulted in an initial strong suppression of primary production and increases in bacterial biomass and activity (27, 34). Our results from the bag with dispersed oil (ODB) showed similar trends, and this was the only case where we found a substantial shift towards heterotrophy due to oil. It should be noted that in both the CEPEX and SEAFLUXES experiments were mineral nutrients added to maintain high biological activity.

An experiment with glucose addition to mesocosms (33) forms an interesting comparison with the CEPEX and SEAFLUXES oil pollution experiments. Glucose at concentrations of 1-5 mg l^{-1} caused a more dramatic shift towards heterotrophy and a marked depression of primary production due to bacterial competition for mineral nutrients. This difference reflected probably the more easily degradable nature of glucose as well as the toxic properties of oil.

In oil pollution experiments with the MERL tanks in Rhode Island, USA, water accomodated fractions of fuel oil were used for more long-term dosing with hydrocarbons. A typical pattern in these experiments was an increase in phytoplankton biomass and decreases in zooplankton and zoobenthic populations (16, 17, 55). An increase in bacterioplankton was another common effect, but even so it was estimated that less than 1% of the added oil underwent microbial degradation in the water column (16).

An oil pollution experiment with a design similar to ours was conducted during the spring phytoplankton bloom in Rosfjorden, southern Norway (9). In this experiment there was a rapid bacterial response to addition of oil with a concomitant and marked consumption of mineral nutrients. The spring phytoplankton bloom was depressed from lack of nutrients and possibly also by oil toxicity. These results are markedly different from the ones we found in our spring experiment. Differences in environmental conditions were probably responsible for this. Light intensities were relatively low during the initial part of the Rosfjorden experiment, and photooxidation of oil was assumed to play a minor role (24). Rosfjorden is a more open fjord than Lindåspollene, and the bags were to a greater extent influenced by the forces of waves and currents. This resulted in more mechanical dispersion of oil in the Rosfjorden experiment, especially since much more oil was added than in our case (4 versus 0.5 l). In contrast, photooxidation was a dominant process affecting oil dissolution and toxicity in the spring experiment in Lindåspollene.

8. CONCLUSIONS

Based on our own results and the results from the other mesocosm experiments referred to in the foregoing, we can draw some general conclusions concerning the effects of oil on plankton communities.

The most marked increases in bacterial activity in response to oil hydrocarbons have been found in cases where oil have been mixed or dispersed into the water column. Meteorological conditions with clouds and strong wind, which favours rapid physical dispersion of oil into the water column, would be expected to produce the strongest stimulation of bacterial activity in a real spill situation. In the other extreme of calm and sunny weather, oil-derived material is also brought into the water phase, but the mechanism is based on increased water solubility of photooxidised components. These seem to be more toxic and refractile and less subject to rapid bacterial degradation.

Heterotrophic components of the plankton seem to be equally or more sensitive to oil pollution than is phytoplankton. A common feature of many oil pollution mesocosm experiments has been a stronger decrease in zooplankton populations than in phytoplankton. The latter may even show an increase in presence of oil. Oil will therefore in many situations not lead to a marked shift towards greater heterotrophic dominance.

ACKNOWLEDGEMENTS

The experiments reported here were part of The Norwegian Marine Pollution Research and Monitoring Programme. We thank all our colleagues who participated in these experiments and whose joint effort made this study possible.

REFERENCES

1. Atlas RM: Microbial degradation of petroleum hydrocarbons: an environmental perspective. Microbiol. Rev. 45: 180-209, 1981.
2. Azam F, Fenchel T, Field JG, Gray JS, Meyer-Reil LA, Thingstad F: The ecological role of water-column microbes in the sea. Mar. Ecol. Prog. Ser. 10: 257-263, 1983.
3. Barth T: Weathering of crude oil in natural marine environments: the concentration of polar degradation products in water under oil as measured in several field studies. Chemosphere 13: 67-86, 1984.
4. Barth T: Oil pollution and plankton dynamics. VI. Chemical analysis of dissolved oil components and oxidative degradation products in controlled ecosystem experiments in Lindåspollene, June 1980 and 1981. In prep.
5. Bratbak G: Bacterial biovolume and biomass estimation. Appl. environ. Microbiol. 49: 1488-1493, 1985.
6. Bratbak G, Thingstad TF: Phytoplankton-bacteria interactions: an apparent paradox? Analysis of a model system with both competition and commensalism. Mar. Ecol. Prog. Ser. 25: 23-30, 1985.
7. Brockmann UH, Dahl E, Kuiper J, Kattner G: The concept of POSER (Plankton Observation with Simultaneous Enclosures in Rosfjorden). Mar. Ecol. Prog. Ser. 14: 1-8, 1983.
8. Currie DJ, Kalff J: Can bacteria outcompete phytoplankton for phosphorus? A chemostat test. Microb. Ecol. 10: 205-216, 1984.
9. Dahl E, Laake M, Tjessem K, Eberlein K, Bøhle B: Effects of Ekofisk crude oil on an enclosed planktonic ecosystem. Mar. Ecol. Prog. Ser. 14: 81-91, 1983.
10. Dahl O, Østvedt OJ, Lie U: An introduction to a study of the marine ecosystem and the local herring stock in Lindåspollene. FiskDir. Skr. Ser. HavUnders. 16: 148-158, 1973.

11. Dale T: Oil pollution and plankton dynamics. II. Abundance pattern of ciliates inside and outside enclosures and the responses of ciliates to oil during the 1980 spring bloom in Lindåspollene, Norway. In prep.
12. Dale T: Oil pollution and plankton dynamics. IX. Controlled ecosystem experiment in Lindåspollene, Norway, June 1980: microplankton. In prep.
13. Dale T: Oil pollution and plankton dynamics. XIII. Controlled ecosystem experiment in Lindåspollene, Norway, June 1981: microplankton. In prep.
14. Davies JM, Baird IE, Massie LC, Hay SJ, Ward AP: Some effects of oil-derived hydrocarbons on a pelagic food web from observations in an enclosed ecosystem and a consideration of their implications for monitoring. Rapp. P.-v. Réun. Cons. int. Explor. Mer 179: 201-211, 1980.
15. de la Cruz AA: Effects of oil on phytoplankton metabolism in natural and experimental estuarine ponds. Mar. environ. Res. 7: 257-263, 1982.
16. Elmgren R, Frithsen JB: The use of experimental ecosystems for evaluating the environmental impact of pollutants: a comparison of an oil spill in the Baltic Sea and two long-term, low-level oil addition experiments in mesocosms. In Grice GD, Reeve MR (ed), Marine mesocosms. New York: Springer-Verlag, pp 153-165, 1982.
17. Elmgren R, Vargo GA, Grassle JF, Grassle JP, Heinle DR, Langlois G, Vargo SL: Trophic interactions in experimental marine ecosystems perturbed by oil. In Giesy JP Jr. (ed), Microcosms in ecological research. DOE Symposium Series 52, CONF-781101. Springfield: National Technical Information Service, pp 779-800, 1980.
18. Fenchel T: Ecology of heterotrophic microflagellates. II. Bioenergetics and growth. Mar. Ecol. Prog. Ser. 8: 225-231, 1982.
19. Fuhrman JA, Azam F: Thymidine incorporation as a measure of heterotrophic bacterioplankton production in marine surface waters: evaluation and field results. Mar. Biol. 66: 109-120, 1982.
20. Hodson RE, Azam F, Lee RF: Effects of four oils on marine bacterial populations: controlled ecosystem pollution experiment. Bull. mar. Sci. 27: 119-126, 1977.
21. Hsiao SIC: Effects of crude oils on the growth of arctic marine phytoplankton. Environ. Pollut. 17: 93-107, 1978.
22. Kristiansen S, Pengerud B, Skjoldal HR, Thingstad TF: Oil pollution and plankton dynamics. XI. Controlled ecosystem experiment in Lindåspollene, Norway, June 1981: phytoplankton biomass and activities in relation to nutrients addition. In prep.
23. Kvalsund RS: Respirasjon, ammoniumekskresjon og elektron transport system (ETS) aktivitet på størrelsefraksjonert zooplankton i Lindåspollene. Cand.real. thesis. Univ. of Bergen, 1985.
24. Laake M, Tjessem K, Rein K: Fate of a tritiated Ekofisk crude oil in a controlled ecosystem experiment with North Sea plankton. Environ. Sci. Technol. 18: 641-647, 1984.
25. Lacaze JC, Villedon de Naïde O: Influence of illumination on phytotoxicity of crude oil. Mar. Pollut. Bull. 7: 73-76, 1976.
26. Larson RA, Bott TL, Hunt LL, Rogenmuser K: Photooxidation products of a fuel oil and their antimicrobial activity. Environ. Sci. Technol. 13: 965-969, 1979.
27. Lee K, Wong CS, Cretney WJ, Whitney FA, Parsons TR, Lalli CM, Wu J: Microbial response to crude oil and Corexit 9527: SEAFLUXES enclosure study. Microb. Ecol. 11: 337-351, 1985.

28. Lee RF, Takahashi M: The fate and effect of petroleum in controlled ecosystem enclosures. Rapp. P.-v. Réun. Cons. int. Explor. Mer 171: 150-156, 1977.
29. Lee RF, Takahashi M, Beers J: Short term effects of oil on plankton in controlled ecosystems. In Bates CC (ed), Proceedings of the Conference on Assessment of Ecological Impact of Oil Spills. Arlington: American Institute of Biological Sciences, pp 634-650, 1978.
30. Lee RF, Takahashi M, Beers JR, Thomas WH, Seibert DLR, Koeller P, Green DR: Controlled ecosystems: their use in the study of the effects of petroleum hydrocarbons on plankton. In Vernberg FJ, Calabrese A, Thurberg FP, Vernberg WB (ed), Physiological responses of marine biota to pollutants. London: Academic Press, pp 323-342, 1977.
31. Østgaard K, Hegseth EN, Jensen A: Species-dependent sensitivity of marine planktonic algae to Ekofisk crude oil under different light conditions. Bot. Mar. 27: 309-318, 1984.
32. Parsons TR: The use of controlled experimental ecosystems: a review. J. oceanogr. Soc. Japan 37: 294-298, 1981.
33. Parsons TR, Albright LJ, Whitney F, Wong CS, Williams PJleB: The effect of glucose on the productivity of seawater: an experimental approach using controlled aquatic ecosystems. Mar. environ. Res. 4: 229-242, 1981.
34. Parsons TR, Harrison PJ, Acreman JC, Dovey HM, Thompson PA, Lalli CM, Lee K, Guanguo L, Xiaolin C: An experimental marine ecosystem response to crude oil and Corexit 9527: part 2 - biological effects. Mar. environ. Res. 13: 265-275, 1984.
35. Pengerud B, Thingstad F, Tjessem K, Aaberg A: Photoinduced toxicity of North Sea crude oils towards bacterial activity. Mar. Pollut. Bull. 15: 142-146, 1984.
36. Pengerud B, Skjoldal EF, Thingstad TF: The reciprocal interaction between degradation of glucose and ecosystem structure. Studies in mixed chemostat cultures of marine bacteria, algae and bacterivorous nanoflagellates. Mar. Ecol. Prog. Ser., in press.
37. Pilson MEQ, Nixon SW: Marine microcosms in ecological research. In Giesy JP Jr. (ed), Microcosms in ecological research. DOE Symposium Series 52, CONF-781101. Springfield: National Technical Information Service, pp 724-741, 1980.
38. Rivier A, Brownlee DC, Sheldon RW, Rassoulzadegan F: Growth of microzooplankton: a comparative study of bactivorous zooflagellates and ciliates. Mar. microb. Food Webs 1: 51-60, 1985.
39. Sherr EB, Sherr BF, Fallon RD, Newell SY: Small, aloricate ciliates as a major component of the marine heterotrophic nanoplankton. Limnol. Oceanogr. 31: 177-183, 1986.
40. Skjoldal HR: Oil pollution and plankton dynamics. V. Controlled ecosystem experiment in Lindåspollene, Norway, June 1980: the experiment and the physical-chemical environment. In prep.
41. Skjoldal HR: Oil pollution and plankton dynamics. X. Controlled ecosystem experiment in Lindåspollene, Norway, June 1980: mesozooplankton. In prep.
42. Skjoldal HR, Dale T, Haldorsen H, Pengerud B, Thingstad TF, Tjessem K, Aaberg A: Oil pollution and plankton dynamics. I. Controlled ecosystem experiment during the 1980 spring bloom in Lindåspollene, Norway. Neth. J. Sea Res. 16: 511-523, 1982.
43. Skjoldal HR, Haldorsen H, Kristiansen S, Pengerud B, Sakshaug E, Thingstad TF: Oil pollution and plankton dynamics. VII. Controlled

ecosystem experiment in Lindåspollene, Norway, June 1980: phytoplankton biomass, activities, and physiological condition. In prep.
44. Skjoldal HR, Johannessen P, Klinken J, Haldorsen H: Controlled ecosystem experiment in Lindåspollene, western Norway, June 1979: comparisons between the natural and two enclosed water columns. Sarsia 68: 47-64, 1983.
45. Skjoldal HR, Wassmann P: Sedimentation of particulate organic matter and silicium during spring and summer in Lindåspollene, western Norway. Mar. Ecol. Prog. Ser. 30: 49-63, 1986.
46. Sydnes LK, Hemmingsen TH, Skare S, Hansen SH, Falk-Petersen I-B, Lønning S, Østgaard K: Seasonal variations in weathering and toxicity of crude oil on seawater under arctic conditions. Environ. Sci. Technol. 19: 1076-1081, 1985.
47. Thingstad TF, Pengerud B: The formation of "chocolate mousse" from Statfjord crude oil and seawater. Mar. Pollut. Bull. 14: 214-216, 1983.
48. Thingstad TF, Pengerud B, Bratbak G, Jelmert A, Dundas I: Oil pollution and plankton dynamics. VIII. Controlled ecosystem experiment in Lindåspollene, Norway, June 1980: microbial population levels and activities. In prep.
49. Thingstad TF, Pengerud B, Dundas I: Oil pollution and plankton dynamics. XII. Controlled ecosystem experiment in Lindåspollene, Norway, June 1981: microbial population dynamics. In prep.
50. Throndsen J: Oil pollution and plankton dynamics. III. Effects on flagellate communities in controlled ecosystem experiments in Lindåspollene, Norway, June 1980 and 1981. Sarsia 67: 163-169, 1982.
51. Tjessem K, Aaberg A: Photochemical transformation and degradation of petroleum residues in the marine environment. Chemosphere 12: 1373-1394, 1983.
52. Tjessem K, Aaberg A, Kobberstad O: Chemical aspects of the water soluble material of Ekofisk crude oil under field and laboratory conditions. Toxicol. environ. Chem. 7: 131-159, 1984.
53. Tjessem K, Kobberstad O, Aaberg A: Photochemically induced interactions in Ekofisk crude oil. Chemosphere 12: 1395-1406, 1983.
54. Tjessem K, Pedersen D, Aaberg A: On the environmental fate of a dispersed Ekofisk crude oil in sea-immersed plastic columns. Water Res. 18: 1129-1136, 1984.
55. Vargo GA, Hutchins M, Almquist G: The effect of low, chronic levels of no. 2 fuel oil on natural phytoplankton assemblages in microcosms: 1. Species composition and seasonal succession. Mar. environ. Res. 6: 245-264, 1982.
56. Williams PJleB, Raine RCT, Bryan JR: Agreement between the ^{14}C and oxygen methods of measuring phytoplankton production: reassessment of the photosynthetic quotient. Oceanol. Acta 2: 411-416, 1979.
57. Zafiriou OC, Joussot-Dubien J, Zepp RG, Zika RG: Photochemistry of natural waters. Environ. Sci. Technol. 18: 358A-371A, 1984.

THE EFFECTS OF OIL AND CHEMICALLY DISPERSED OIL ON NATURAL PHYTOPLANKTON COMMUNITIES

M. SCHOLTEN, J. KUIPER, H. VAN HET GROENEWOUD, G. HOORNSMAN AND E. VAN DER VLIES.

Division of Technology for Society TNO, Department of Biology, Laboratory for Applied Marine Research, P.O. Box 57, 1780 AB Den Helder, Netherlands.

1. INTRODUCTION

The potential functioning of a marine ecosystem strongly depends on the quantity and quality of the primary production of the phytoplankton (37). Effects of oil on phytoplankton can thus cause changes in the structure and functioning of the entire ecosystem.
A lot of information about oil sensitivity of marine algal species is gained from single species laboratory tests (10, 13, 30, 34, 36, 39). However, it is not generally possible to predict the functioning of a natural phytoplankton community after an oil spill from these experiments. This is primarily due to interactions between primary production and other ecological processes.
Availability of light and nutrients, influenced by competition between algal species and bacterial mineralization, as well as grazing by filter-feeding herbivores strongly effect the structure and functioning of the phytoplanton community.
Mesocosms have proved to be useful tools in analyzing ecosystems, as they possess the essential properties of a natural ecosystem, which cannot be integrally studied in the laboratory (9, 12, 15, 16, 18, 33). It is considered reliable to extrapolate the information gained from mesocosms to natural field situations. The simultaneous measurements of a number of interdependent parameters allows for the deduction of ecologically relevant cause-effect relationships.
In this paper a recapulation is given on the effects of oil or chemically dispersed oil on the functioning of phytoplankton communities as seen in several mesocosms experiments using model plankton systems (MOPs) or model tidal flat systems (MOTIFs). Primary production will be considered in terms of nutrient economy or grazing pressure. Observations on a North Sea plankton community during an oil spill are used as field verification of these mesocosm experiments.

2. MATERIAL AND METHODS

2.1 The mesocosms

At the Applied Marine Research Laboratory of the TNO Division of Technology for Society in Den Helder model shallow-water ecosystems are used in study the fate and effects of pollutants in marine coastal ecosystems.
A vertical water column of the North Sea or other natural plankton systems can be simulated by means of so-called MOPs (model plankton ecosystems) (15, 16). A MOP is a large plastic bag that is filled with natural seawater (1 to 30m^3) in which plankton organisms are present in natural densities. A MOP is suspended in natural seawater.
A typical intertidal ecosystem of the Wadden Sea or other estuaries can be simulated by means of so-called MOTIFs (model tidal flat ecosystems) (7, 25, 26). A MOTIF is composed of two concrete reservoirs coupled to

Kuiper, J., Van den Brink, W.J. (eds), Fate and Effects of Oil in Marine Ecosystems. ISBN 90-247-3489-4

each other. Each reservoir measures 6 metre lenght x 3.5 metre width x 1.2 metre height.
The bottom of the reservoir is covered with a 0.5 metre-thick layer of natural marine sediment. Seawater containing natural plankton is continually supplied from the Wadden Sea, so that every week 20% of the water is renewed.
To simulate tidal movements, water is pumped from one reservoir into the other every six hours. Thus, the sediment in the former reservoir is caused to be left dry for some hours.
At the start of experiments macro- and meiofauna, such as molluscs, bivalves, lugworms and nematodes, are added to the MOTIFs in natural densities.

2.2. The experiments
In tabel 1 an overview is given of eight experiments using MOPs and MOTIFs in order to study the fate and ecotoxicological effects of oil on marine ecosystems. Various types of oil spills were tested. In some experiments the effects of chemical dispersion of the oil was also tested. The sampling programme includes time series measurements of physio-chemical parameters, oil concentrations in biotics and abiotics, and biotic parameters such as bacteria, phytoplankton, zooplankton and if present also phytobenthos and benthic macro- and meiofauna.
The phytoplankton measurements are described in more detail in the next section.

2.3. Phytoplankton measurements
The phytoplankton measurements include:
a) Spectrophotometric determination of chlorophyl concentrations (45)
b) Total number and size frequency determination of algal cells using Particle Data Ltd. electronic particle counter (42).
c) Primary production measurements based on oxygen production or consumption in light or dark incubated water samples. Oxygen concentrations were determined using a sensitive Winkler titration with photometric end point determination (40, 49).
d) Inspection of species destribution of lugol-iodine preserved phytoplankton samples (48) using a Zeiss inverted microscope.

3. RESULTS
A verbatim synopsis of the main phytoplankton results is presented in this paper. Figures and tables involving more detailled information are inserted in seperate reports (see table 1).
3.1. MOPs exposed to water soluble oil fraction (40).
In this preparatory experiment the most elementary interactions between the oil and the phytoplankton were observed. The primary productivity per chlorophyll unit was reduced by the oil to only 15-65% of the productivity in the control MOPs from one day after the oil application till the end of the experiment four weeks afterwards.
On the first day following the oil application the productivity in the oil treated MOPs was up to three times that of the control MOPs.
In spite of the inhibited productivity the algal biomass in the oil treated MOPs became higher compared to the control ones, especially after a small phytoplankton bloom. This seemed to be caused by a reduction of the grazing pressure as a consequence of the inhibited activity of copepods.
In the control MOPs a flagellate dominated community, with only small

mesocosm	date	oil type	oil concentration	dispersion	report
1. MOPs with North Sea water	May-June 1982	watersolube fraction of fuel oil	100 $\mu g.l^{-1}$	-	Scholten, 1983 b
2. MOPs with North Sea water	April-May 1983	Nigerian light crude oil: toplayer or watersoluble fraction	0.5 l m^{-2} or 60-500 $\mu ug.l^{-1}$	-	Kuiper et al, 1985
3. MOPs with Grevelingen water	June-July 1984	Topped Forties: North Sea crude oil	0.5 l m^{-2}	-	Kuiper et al, 1985
4. MOPs with North Sea water	April-May 1984	Topped Forties	0.5 l m^{-2}	Finasol OSR5	Kuiper et al, 1983
5. MOPs with	March-April 1985	F3: oil	o.4-2 l m^{-2}	-	Kuiper et al, 1986
6. MOTIFs	October 1981-December 1982	Oil mousse: fuel oil gas oil and seawater	1 l m^{-2}	-	Kuiper et al, 1983
7. MOTIFs	March 1984-'85 March 1985-'86	Topped Forties	0.1 l m^{-2} 0.5 l m^{-2}	Finasol OSR5	Kuiper et al, 1986 b

Table 1: Overvieuw of experiments

diatoms present, succeed to the initial diatom dominated one. In the oil dominated MOPs a more rapid decline of the diatoms led to a dominance of very small microflagellates.
As a consequence micro-flagellate consuming protozoans reached high densities in the oil treated MOPs.

3.2. MOPS exposed to a Nigerian light crude oil (22).
Effects of the addition of Nigerian crude, either as a water solube fraction or as a floating layer, on the primary productivity could not be ascertained due to large variation between replicates.
Due to fact that microflagellates (mainly *Phaeocystis poucheti*) dominated the control MOPs as well, no shift to microflagellate dominance due to oil pollution was recorded.
This seemed to be the reason that chlorophyll concentrations of the oil treated MOPs resemble the control ones, despite the inhibited development of copepods (e.g. *Temora longicornis*) by oil, as the copepods are mainly feeding on diatoms.

3.3. MOPs exposed to a topped Forties crude oil in nitrogen poor seawater (22).
No effects of oil on the biomass production of the phytoplankton could be detected in this experiment. In the control MOPs large diatoms (e.g. *Chaetoceros* spp. and *Rhizosolenia* spp.) were abundant, where as in the oil treated MOPs microflagellates dominated the phytoplankton community.

3.4. MOPs exposed to topped Forties and chemically dispersed topped Forties (19).
Addition of Forties alone did not result immediately in perceptible differences in the fuctioning of the phytoplankton compared to control MOPs. Only after one week elevated chlorophyll levels were recorded in the oil treated MOPs, although no affection of the productivity per unit chlorophyll by oil could be detected. However, the enhanced chlorophyll concentrations correlated well with the inhibition of the copepod production, primarely due to a lower copepod egg production in oil contaminated MOPs.

In the contrary to the former MOP experiments oil pollution did not result in a more rapid succession of the phytoplankton community to a microflagellate dominated one. Large flagellates (e.g. *Prorocentrum sp.*) and diatoms (e.g. *Chaetoceros debilis* and *Nitzschia spp.*) remained dominant, whereas microflagellates became more important in the control MOPs. Addition of the dispersant alone resulted in an immediately and strong decrease of the phytoplankton biomass. However, due to the high mortality of copepods long-term effects were comparable to that in oil-treated MOPs: i.e. high chlorophyll concentrations in spite of a non-affected productivity per chlorophyll unit.

Treatment of the oil with dispersant immediately resulted in a very strong effect, exceeding the sum of the separate effects of the oil and the dispersant. Phytoplankton activity was severely inhibited, resulting in very low chlorophyll densities throughout the experiment. The zooplankton suffered very high mortality.
Complete dispersion prevented recovery of the phytoplankton, despite the low grazing rates. In the MOPs in which the effectiveness of the dispersion was least, as a consequence of insufficient agitation, a phyto-

plankton recovery was seen after some weeks in the absence of copepod grazing.
The obvious explanation of the severe effects is the fact that very high oil concentrations were found in the water following dispersion of an oil-spill.
3.5 MOPs exposed to an F3-oil (20).
Phytoplankton biomass was reduced by the F3-oil in a dose-related way, resulting in a delayed and lower maximum biomass level during the phytoplankton bloom.
Although in the control MOPs copepod activity was relatively lowe, an oil-induced mortality and inhibited reproduction of copepods (e.g. *Temora longicornis*) was seen.
It was obvious that even the primary productivity per unit chlorophyll was substantually inhibited by the oil. As a consequence there was no benefit observed from the reduced grazing rates. *Skeletonema costatum* was the dominant species in all MOPs, but *Chaetoceros spp.* and large flagellates seemed to be also relatively tolerant of the oil-gas pollution. Most of the diatomspecies that occurred abundant in the control MOPs were absent in the oil treated MOPs (i.e. *Asterionella glacialis* and *Rhizosolenia delicatula*), whereas microflagellates (e.g. *Phaeocystis poucheti*) became dominant in oil treated MOPs.

3.6. MOTIFs exposed to an oil mousse (23, 25, 26, 40).
The results of this experiments resemble those of the MOP-experiments. Long-term inhibition of the primary production per unit chlorophyll is recorded from the oil treated MOTIFs up to the end of the experiment 10 months after the artificial oil spill. These long term effects of oil, despite the continual flow through of fresh seawater, may be caused by leakage of oil from the sediment. Part of the oil that was left behind at low tide was covered with sediment coming to the surface due to bioturbation activities of worms (e.g. *Arenicola marina*). This sediment-bound fraction caused a prolonged exposure of the ecosystem to oil components. The inhibited primary production did not result in lower chlorophyll concentrations. Phytoplankton biomass even became higher in the oil treated MOTIFs relative to the uncontaminated ones. Just as in the MOP-experiments this was probably due to lower grazing pressure as a consequence of lower copepod activity in the oil treated mesocosms. Moreover the high mortality of filter-feeding macrofauna species (e.g. *Cerastoderma edule*, *Corophium volutator*) introduced into the MOTIFs caused an even stronger grazing effect in the MOTIFs compared to the MOPs. As a consequence of the reduced growth of diatoms such as *Nitzschiaceae*; *Naviculaceae*; *Plagiogramma spp.*; *Chaetoceros spp.*; *Rhizosolenia* spp. and *Skeletonema spp.*, especially during diatomblooms, microflagellates became dominant in the oil treated MOTIFs on the long-term.
Just as in the first MOP-experiment a stimulation of the primary productivity of the phytoplankton was seen at the first day following the oil spill.

3.7. MOTIFs exposed to Forties crude oil spills and chemical dispersion of these spills (7, 21).
In the first experimental year the MOTIFs were exposed to a relative large oil spill. After oil exposure, the primary productivity per unit chlorophyll was inhibited throughout the summerperiod and again during the winter. This long-term oil effect has its origin in the transportation of oil into the sediment by bioturbation, just as in the first

MOTIF-experiment.
In the second year a small oil spill was simulated, and no inhibition of the primary productivity could be detected.
As the chlorophyll concentrations of incoming water exceeded that of the water that was already been in the control MOTIFs, a high potential grazing rate of algae consuming filterfeeders is indicated.
Reduced grazing rates as a consequence of a large spill resulted in elevated chlorophyll levels compared to these in control MOTIFs. Such an elevation of chlorophyll concentrations could not be observed in MOTIFs exposed to a small spill in combination with the removal of floating oil one day after the oil application. Removing the floating oil two days afterwards, resulted in an elevation of the chlorophyll concentrations compared to the control MOTIFs not before September, four months after the oil spill.

Application of the dispersant alone did not result in differences with the control MOTIFs, except for the first few days after the application, due to a strong dilution of the dispersant as a consequence of the continual water throughflow. Dispersion of the large oil spill caused stronger effect than oil alone. In spite of the high mortality of filterfeeding organisms, no elevated chlorophyll concentrations occured, indicating the strong inhibition of the primary productivity by dispersed oil.
On the contrary, elevated chlorophyll concentrations were observed after dispersion of the small oil spill, as a consequence of reduced algae consumption by filterfeeders.
Again, dispersion led to high oil concentrations in the water phase, explaining the more severe effects compared to those of the oil alone.
The control MOTIFs showed a dominance of large flagellates and a few diatom species up to June, and microflagellates during the following summer peak and afterwards. After a large spill diatoms became more important, e.g. <u>Nitzschia closterium</u> and <u>Skeletonema costatum</u>, whether the spill is dispersed or not. The maximal difference in species composition compared to the control MOTIFs is observed during the summer period.
After a small spill large flagellates remained dominant, preventing a dominance of micro flagellates as seen in the control MOTIFs. Dispersion led to an abundance of diatoms (e.g. <u>Nitzschia closterium</u>) within the phytoplankton community.

4. DISCUSSION

4.1. <u>Mesocosm experiments</u>

At first the experimental data did not show clear overall effects of oil on the functioning of the phytoplankton.
Nevertheless, the outcomes of the various experiments gives a clear indication about the way that oil pollution can alter the primary production. Under natural field conditions a diatom dominated phytoplankton community is present during springs. Diatoms seem to have a competitive advantage over flagellates under these conditions. Durring summer the diatoms become less abundant, making room for a flagellate bloom. This can be caused both by an exhaustion of the dissolved silicate (as in the MOPs) or by an increase of the grazing pressure by filterfeeders (as in the MOTIFs). Diatoms need silicate for their growth, whereas flagellates do not, while on the other hand, diatoms suffer heavier grazing by filterfeeders. Microflagellates especially are not easily be grazed.

In the control MOTIFs a similar phytoplankton dynamics occured as well as in the control MOPs. However in the latter case, the succession is somewhat faster than in natural field situations (15, 16, 18, 24, 25, 26). When concidering the overall impact of oil pollution on phytoplankton, direct effects such as growth inhibition, can not alone explain the observed phenomena.
Indirect effects such as reduced grazing, altered competition or differences in available nutrients must also be conciderd. Inhibition of the growth of algae, measured as primary productivity per unit chlorophyll, due to oil pollution seems to appear in most of the experiments, and is related to type and conentration of oil. Differences in oil sensitivity between the various species can cause change in the species composition of the phytoplankton community. Thus relatively tolerant species often can be replace the more sensitive ones (40, 49).
Many laboratory studies on the relative sensitivety of algal species show that diatoms seem in general be more susceptible to oil indiced growth reductions than do microflagellates (4, 31, 39). However, the mesocosm experiments showed some diatom species to be tolerant to oil pollution (i.e. Chaetoceros debilis, Nitzschia closterium, Skeletonema costatum). Therefore, the relatively high densities of microflagellates observed in some of the oil treated mesocosms, as well as in other mesocosm experiments and field investigations concerning oil pollution (4, 31, 35, 39, 43, 46, 47), can not completely be explained in terms of differences in oil sensitivity.
Moreover, some of the mesocosm experiments showed even higher diatom densities after an oil exposure compared to control conditions. Variations in grazing upon algae by filterfeeding organisms seemed to be a key factor determining the amount of biomass present under conditions when direct inhibition of phytoplankton growth is moderate or absent. Irrespective of grazing it is expected that inhibited productivity per unit chlorophyll will result in reduced standing stock algal biomass and reduced algal production. However, this is never observed in the mesocosm experiments. Futhermore, elevated algal biomass concentrations were found in oil-treated mesocosms in most of the experiments, resulting in higher total primary production despite the inhibited productivity per unit chlorophyll. This phenomenon can be explained in terms of reduced grazing upon algae by filter feeders due to oil pollution (24, 29, 32). From the mesocosm experiments it is clear that copepods are very suspectible to oil pollution (i.e. Temora longicornis). The same is reported by other authors (2, 3, 5, 8, 32, 35). Copepods feed mainly on algae (especially larger forms such as diatoms). Thus, oil pollution indirectly causes reduced grazing rates.
From MOTIF experiments high mortality of filter feeding bivalves is observed (7, 25, 26). This also reduces the grazing pressure upon phytoplankton, when the algal densities of the removal water are compared with those of the standing water, a high grazing pressure is indicated in the control MOTIFs. If the reduction of the grazing pressure exceeds the inhibition of algal productivity by oil, the equilbrium level of the algal standing stock will be elevated compared to control situations. As a consequence the daily algal production can increase, and thus nutrients will be more rapidly exhausted. As diatoms are heavily grazed, grazing reduction can lead to an initially higher diatom production and therefore to a silicate deficiency. In the long term diatom dominated algal communities shift earlier to flagellate or even microflagellate dominated situations.

As silicate is not becoming deficient a prolonged abundance of diatoms is usually observed. Especially in combination with lower grazing pressure upon diatoms this can lead to higher diatoms densities in oil treated mesocosms compared to control ones in some of the experiments.
Prelonged diatom dominance after an oil spill can be caused by the following:

a) High mineralization rates can prevents silicate deficiency.
b) An extremely low production (also in the controls) or a severe inhibition of the productivity per unit chlorophyll can result in a less rapid exhaustion of silicate.
c) Small reductions in grazing pressure, probably due to a relative lower toxicity of oil to filter feeding organisms or to an also low grazing pressure under control situation can cause no elevated chlorophyll levels, and therefore no rapid silicate exhaustion. This can also be the case if opportunistic diatom eating organisms (e.g. nematodes and harpacticids) can replace the inhibited ones.

Inhibited algal productivity combined with high algal densities does not result in a reduced total primary production after an oil spill, but reduced transfer of energy to secondary producers can occur. This can be caused by both a direct oil inhibition of the activity of secondary producers, as well as a dominance of less consumable algae (i.e. microflagellates) resulting from the oil pollution.
Sometimes a sudden increase of small, short living opportunistic organisms feeding on the remaining algae is observed e.g. microflagellate consuming protozoans (1, 8, 31) the consequence of the oil induced alteration of the secondary producers on the functioning of orgnisms from higher trophic levels within the ecosystems (11) is still obscure. The initial stimulation of the primary productivity immediatly after an oil spill, as observed in some of the experiments, seemed to be caused by the low concentrations of oil dissolved in the water shortly after the oil spill (10, 36, 38).An explanation for this phenomenon is not yet available, but may be explained by hormesis (44).

4.2. Recommendations for combating oil spills at sea

While revieuwing the effects of oil on phytoplankton it is clear that type and concentration of oil compounds disolved in the water determines the severeness of the effects (50). After a relatively large oil spill (>o,1 l. m^{-2}) a rapid mechanical removal of the oil from the water surface is recommended. Treatment of the oil with dispersant generally will result in a aggravation of the effects, caused by high dissolved oil concentrations in the water as a direct result of dispersion. Alternative combat strategies ared needed, i.e. mechanically removal of the oil. In order to prevent long term effect transportation of oil to the sediment must be prevented.

4.3 Fieldverification

The results of mesocosm experiment can be compared to field observations on a North Sea plankton system after an oil spill, when a compatable sampling programme is carried out. On june 7th, 1982 the oil-tanker "Katina" leakt oil, due to a collision at sea nearby Hoek van Holland. It was estimated that 1600 m^3 fuel oil escaped into the North Sea, creating a 0.2 mm mousse-layer (14).

Although hardly any dissolved oil was found in the seawater some effect on the phytoplankton could be observed. Comparing the primary productivity of the 0-1 metre toplayer with that of 2-3 metre depht-layer it was obvious that at seawater where the mousse had drifted past one day before sampling, a somewhat inhibited production was found compared to control stations. At a station were the oil mousse had just arrived a stimulated production was seen (17).
These observations resemble those of the mesocosms, thus indicating the reliability of mesocosm experiments in predicting effects in natural fiels situations. Similair effects of an oil on natural phytoplankton communities in field situations and mesocosms were described elsewhere (4, 5, 6, 8, 27, 28, 29, 31, 32, 35, 43, 46).

5. ACKNOWLEDGEMENTS

Thanks are due to our colleageus N. Admiraal, T. Bowmer, M.C. Hendriks, W. Chr. de Kock, M. van der Meer, P. Roele and B. Schrieken for their assistance and comments, as well as R. Dekker and G. van Moorsel (NIOZ) and Chr. Brouwer (RIN) for the collaboration in the MOTIf-experiments.
Parts of the experiments were funded by:
*Commision of the European Community
- Directorate - General XI
- Directorate - General XII

*Netherland Ministery of Transport and Public Work (Rijkswaterstaat).
*United duth oil industries (VNA): Vereniging van de Nederlandse Aardolieindustrie.
*Nederlandse Aardolie Maatschappij B.V. (NAM).

6. REFERENCES

1. W. Admiraal and L.A.H. Venekamp (1986)
Significance of tintinnid grazing during blooms of Phaeocystis pouchetii (haptophyceae) in dutch coastal waters.
Neth. J. Sea Res. 20: 61-66.
2. Bernem, K.H. van (1982)
Effect of experimental crude oil contamination on abundance, mortality and a resettlement of representative mud flat organisms in the mesohaline area of the Elbe estuary.
Neth. J. Sea Res. 16: 538-546.
3. Connell, D.W. and G.J. Miller (1980)
Petroleum hydrocarbons in aquatic ecosystems - Behaviour and effects of sublethal concentrations. Part II.
CRC Critical Reviews in environmental Control 11: 105-163.
4. Dahl, E., M. Laake, K. Tjessem, K. Eberlein en B. BØle (1983)
Effects of Ekofisk crude oil on an enclosed planktonic ecosystem.
Mar. Ecol. Prog. Ser. 14: 81-91.
5. Davies, J.M., I.E. Baird, L.C. Massie, S.J. Hay en A.P. Ward (1980)
Some effects of oil-derived hydrocarbons on a pelagic food web from observations in an enclosed ecosystem and a consideration of their implications for monitoring.
Rapp. P.v. Réun. Cons. int. Explor. Mer. 179: 201-211.
6. Davies, J.M., R. Hardy and A.D. McIntyre (1981)
Environmental effects of North Sea oil operations.
Mar. Poll. Bull. 12: 412-416.

7. R. Dekker and G.W.N.M. van Moorsel (1987)
Effects of different oil doses, dispersant and dispersed oil on macrofauna in model tidal flat ecosystems.
This proceedings.

8. Elmgren, R. and J.B. Frithsen (1982)
The use of experimental ecosystems for evaluting the environmental impact of pollutants: a comparison of an oil spill in the Balthic Sea and two long term low level oil addition experiments in mesocosms.
In: Marine Mesocosms, G.D. Grice and M.R. Reeve (eds.).
Springer, New York. pp. 153-166.

9. Gamble, J.C. en J.M. Davies (1982)
Application of enclosures to the study of marine pelagic systems.
In: G.D. Grice and M.R. Reeve (eds.), Marine Mesocosms, Springer, New York, pp. 25-84.

10. Gordon, D.C. en N.J. Prouse (1973)
The effects of three oils on marine phytoplankton photosyntheses.
Mar. Biol. 22: 329-333.

11. Greve, W. and Parsons, F.R. (1977)
Photosynthesis and fish populations: Hypothetical effects of climatic change and pollution.
Helgolander Wiss. Meeresunters. 30: 666-672

12. Grice, G.D. en M.R. Reeve (1982)
Introduction and description of experimental ecosystems.
In: G.D. Grice en M.R. Reeve (eds.). Marine Mesocosms.
Springer Verlag, New York.

13. Hegseth, E.N. en K. Ostgaard (1985)
Application of in situ dialysis cultures in studies of phytotoxicity of North Sea crude oils.
Water Res. 19: 383-391.

14. W. Koops, F.J. Sanders and J.M. Gubbens (1985)
The Katina oil spill 1982, combating operation at sea.
Proceedings at the 1985 Oil Spill Conference.
American Petroleum Institute, Washington.

15. Kuiper, J. (1977)
Development of North Sea coastal plankton communities is separate plastic bags under identical conditions.
Mar. Biol. 44: 97-107

16. Kuiper, J. (1982)
The use of enclosed plankton communities in aquatic ecotoxicology.
Thesis Wageningen. 256 pp.

17. Kuiper, J. (1983)
Planktonwaarnmeningen en olie-analyses in het kustwater na de olielozing door de tanker Katina - juni 1982.
TNO report nr. R 83/35, Delft NL.

18. Kuiper, J. (1984b)
Marine ecotoxicological tests: multispecies and model ecosystem experiments.
In: G. Persoone and C. Claus (eds.).
Ecotoxicological testing for the marine environment.
Proc. MARTOX Symposium, Ghent, Vol 1: 527-588.

19. Kuiper, J. (1984b)
The fate and effects of dispersants and dispersant-treated crude oil in marine model ecosystems with different mixing regimes.
Report MT-TNO R 84/167 Delft 79 pp.

20. Kuiper, J., H. v.h. Groenewoud, N. Admiraal, G. Hoornsman, M. v.d. Meer, F. Schulting and E. Verkoelen.
Fate and effects of F3-oil in an enclosed wadden sea plankton community.
21. Kuiper, J. H.v.h. Groenewoud, N. Admiraal, M. Scholten, P.A.W.J. de Wilde, G. van Moorsel, R. Dekker, W.J. Wolf and C. Brouwer (1986).
The influence of dispersants on tha fate and effects of oil in model tidal flat ecosystems.
TNO report nr. R86/182; Delft NL.
22. Kuiper, J., H. van het Groenewoud, G. Hoornsman, M. van der Meer, F. Langeweg and U. Brockmann (1985)
Fate and effects of crude oil in natural marine plankton ecosystems.
Report MT-TNO, Delft. R 85/660, 117 pp.
23. Kuiper, J., H. van het Groenewoud, G. Hoornsman, P.A.W.J. de Wilde, E. Berghuis, A. Kok, W.J. Wolff en H. Schoonheden (1983)
A study of marine pollution in outdoor model ecosystems representing a tidal flat (OPEX).
Eindrapport MT-TNO no. 84/14, 103 pp.
24. Kuiper, J. and A.O. Hanstveit (1984)
Biodegradation rates of xenobiotics in laboratory die-away and in natural plankton communities.
Proc. 4th Disposal Symposium. Plymouth 1983 (In Press).
25. Kuiper, J., P.A.W.J. de Wilde en W.J. Wolff (1984a)
Effects of an oil spill in outdoor model tidal flat ecosystem.
Mar. Pollut.Bull. 15: 102-106.
26. Kuiper, J., P.A.W.J. de Wilde en W.J. Wolff (1984b)
Oil Pollution Experiment (OPEX).
I. Fate of an oil mousse and effects on macrofauna in a model ecosystem representing a Wadden Sea tidal mud flat.
In: Ecotoxicological testing for the marine environment. G. Persoone, E. Jaspers and C. Claus (eds.).
State University of Ghent and Inst. mar. Scient. Res. Bredene Belgium, Vol. 2: 331-359.
27. Laake, M., K. Tjessem en K. Rein (1984)
Fate of tritiated Ekofisk crude oil in a controlled ecosystem experiment with North Sea plankton.
environm. Sci. Technol. 18: 641-647.
28. Lacaze, J.C. (1974)
Ecotoxicology of crude oils and the use of experimental marine ecosystems.
Mar. Poll. Bull. 5: 153-156.
29. Lännergren, C. (1978)
Net- and Nanoplankton: Effects of an Oil Spill in the North Sea.
Botanica Marina 21: 353-356.
30. Lee, R.F., W.s. Gardner, J.W. Anderson, J. Blaylock en Barwell-Clark (1987)
Fate of polycyclic aromatic hydrocarbons in controlled ecosystem enclosures.
Environ. Sci. & Technol. 12, 832-838.
31. Lee, R.F. en M. Takahashi (1977)
The fate and effects of petroleum in controlles ecosystem enclosures.
Rapport P.V. Réun. const. Int. Explor. Mer. 171: 150-156.

32. Linden, O., R. Elmgren en P. Boehm (1987)
The Tsesis oil spill.
Ambio 8: 244-253.
33. Menzel, D.W. en J. Case (19770
Concept and design: Controlled ecosystem pollution experiment.
Bull. Mar. Sci. 27: 1-7.
34. Østgard, K., I. Eide and A. Jansen (1984)
Exposure of phytoplankton to Ekofisk crude oil.
Mar. Environm. Res. 11: 183-200.
35. Parsons, T.R., P.J. Harrison, J.C. Acreman, H.M. Dovey, P.A. Thompson en C.M. Lalli (1984)
An experimental marine ecosystem response to crude oil and Corexit 9527: part 2 - biological effects.
Mar. environm. Res. 13: 265-275.
36. Parsons, t.R., W.K.W. Li en R. Waters (1976)
Some preliminary observations on the enhancement of phytoplankton growth by low levels of mineral hydrocarbons.
Hydrobiologia 51: 85-89.
37. Parsons, T.R. en M/ Takahashi (1977)
Biological oceanographic processes.
Pergamon Press, Oxford.
38. Prouse, N.J., D.C. Gordon and P.D. Keizer (1976)
Effects of a low concentration of oil accomodated in seawater on the growth of uni algal marine phytoplankton cultures.
J. Fish. Res. Bd. Can. 33: 810-818.
39. Pulich, W.M.; K. Winters en C. van Baalen (1974)
The effects of a No. 2 fuel oil and two crude oils on the growth and photosynthesis of microalgae.
Mar. Biol. 28: 87-94.
40. Scholten, M. (1983)
Een vergelijking van de Winklen zuurstofmethode voor de bepaling van de primaire produktie van marien fytoplankton, met de C14 metode.
TNO rapport R 83/6, 86 pp.
41. Scholten, M. (1983)
De effecten van contaminanten op de primaire productiviteit van marien fytoplankton in modelecosystemen en in het veld.
TNO report nr. R83/7; Delft NL.
42. Sheldon, R.W. and T.R. Parsons (1967)
A practical manual on the use of the coulter counter in marine science.
Coulter Electronics corp., Toronto, Canada, 66 pp.
43. Skjoldal, H.R., T. Dale, H. Haldorsen, B. Pengerud, T.F. Thingstad, J. Tjessen en A. aaberg (1982).
Oil pollution and plankton dynamics. I. Controlled ecosystem experiment during the 1980 spring bloom in Lindaspollene, Norway.
Neth. J. Sea Res. 16: 511-523.
44. Stebbing, A.R.D. (1981)
Hormesis stimulation of colony growth in Campanularia flexuosa (Hydrozoa) by copper, cadmium and other toxicants.
Aquatic Toxicol. 1: 227-311.
45. Strickland, J.D.H. and T.R. Parsons (1986)
A practical handbook of seawater analysis.
Bull. Fish. Res. Bd. Canada 167: 1-311 pp.

46. Throndsen, J. (1982)
Oil pollution and plankton dynamics. III. Effects on flagellate communities in controlled ecosystem experiments in controlled ecosystem experiments in Lindaspollene, Norway, June 1980 and 1981.
Sarsia 67: 163-169.

47. Vargo, G.A., M. Hutchins en G. Almquist (1982)
The effect of low, chronic levels of no. 2 fuel oil on natural phytoplankton assemslages in microcosms:
I. Species composition and seasonal succession.
Mar. Environm. Res. 6: 245-264.

48. Vollenweider, r.A. (ed.) (1969)
A manual on methods for measuring primary production in aquatic environments.
Blackwell Sci. Publ., 213 pp.

49. Williams, P.J.B. le and N.W. Jenkinson (19820
A transportable microprocessor-controlled precise winkler titration, suitable for field situation and shibboard use.
Limnol. Oceanogr. 27: 576-584.

50. Winters, K.; R. O'Donell; J.C. Batterton en C. van Baalen (1976)
Water soluble components of four fuel oils: characterization and effects on growth of microalgae.
Mar. Biol. 36: 269-276.

OIL AND THE MANGROVES OF THE NORTHERN RED SEA

B. DICKS and S. S. C. WESTWOOD

Oil Pollution Research Unit, Field Studies Council,
Orielton Field Centre, Pembroke, Dyfed, U.K.

ABSTRACT

Avicennia marina mangroves on South Geisum Island in the northern Red Sea were heavily oiled by a series of oil spills in 1982 and 1983. The oil was viscous and weathered, and formed a complete coating over pneumatophores. Although death of the mangroves was anticipated, the majority of trees survived. Field observations suggest sediment drainage characteristics influence pneumatophore density for this species and play a major role in determining the ability of the trees to survive heavy oiling. In order to test these hypotheses, further studies were carried out in April 1985 at four sites on South Geisum Island, which included obtaining measurements of Eh, interstitial water oxygen content and salinity, hydraulic conductivity, particle size distribution and hydrocarbon contamination as well as sampling infaunal communities. The data are presented and discussed.

It is postulated that pneumatophore density, combined with field assessment of sediment grain size, drainage characteristics and, where necessary, a rapid measure of hydraulic conductivity, may provide a simple index for identifying areas of high susceptibility to oil spill damage. With this knowledge, the highest priorities for protection can be identified and suitable protective measures taken in the event of a spill. It is recommended that the first step should be a more exhaustive investigation of the relationships between breathing root density and the sediment drainage/grain size/hydraulic conductivity, in order to confirm the hypotheses put forward from the South Geisum Island studies.

1. INTRODUCTION

In the northern Red Sea, mangal stands are rare and mono-specific. Only the black mangrove, *Avicennia marina* has been recorded this far north; at Ras Muhammed on Sinai, the Geisum islands and Abu Minqar in the mouth of the Gulf of Suez, and the outer regions of the Gulf of Aqaba. This species lives around the middle and high water mark in these micro-tidal habitats, and also extends into areas of nearshore lagoon in embayments or inshore of fringing coral reefs, in areas which may or may not dry out at low water. Such mangals are characteristically associated with extreme conditions of high salinity and low winter temperature. Por *et al*. (1977) describe the mangals of the Gulf of Aqaba as 'oases of high biomass wedged between the desert of Sinai and the blue oligotrophic waters of the Red Sea'. Nevertheless, the community of the mangal is not exceptionally rich in species and tends to comprise hardier elements of shore animals and plants, albeit in exceptional numbers. However, in view of the rareness of this type of habitat in the Red Sea, the undoubted local importance of the mangal in terms of productivity and the bird

Kuiper, J., Van den Brink, W.J. (eds), Fate and Effects of Oil in Marine Ecosystems. ISBN 90-247-3489-4

communities which they support (Jennings, 1983 and unpublished data), it deserves both further study and protection from the activities of man.

It is well established that mangroves are sensitive to the effects of spilled oil which causes defoliation and death of trees by coating breathing roots and penetrating into sediments. Only slow recovery follows such damage. Although many mangrove species can survive light oilings, (Baker, 1982; Getter et al., 1984; Lewis, 1983) heavy oiling over the pneumatophores almost invariably results in death.

The exponential growth of coastal and offshore oil exploration and production in the Gulf of Suez, and more recently amongst the Ashrafi, Gubal and Geisum Islands in the northern Red Sea, has given widespread cause for concern regarding damage to habitats of both coral and mangrove. Wennink and Nelson-Smith (1979) reported heavy oiling along considerable lengths of Red Sea coastline from shipping, refining and oil production, with the worst affected areas near coastal oilfields extending from Ras Gharib south to Ras Shukeir. More recent surveillance of this coastline (Edwards, pers. comm; Dicks, in press; Jennings, pers. comm.) reveals that the situation may be deteriotating rather then improving. Of particular concern have been a series of large spills during 1982 and early 1983 from oil operations in the southern third of the Gulf of Suez, which resulted in heavy oiling of a considerable length of more southerly coastline, including the islands noted above (Hanna, 1983; Dicks, in press).

Visits to the Hurghada areas in June 1983 and then to Zeit Bay, the Ashrafi, Gubal, Geisum and Abu-Minqar Islands in August and September/October 1983 and March and August 1984, showed that the mangroves of South Geisum and Abu Minqar had been heavily oiled. Whilst some areas of mangrove had been killed on the south shore of South Geisum, equally heavily oiled mangal nearby, along the west shore and on Abu Minqar Island had not died, in spite of a 100% cover of oil on breathing roots and heavy (up to 5 cm thick) accumulations on sediments. As one of the main reported mechanisms of effect of oil is the restriction of oxygen exchange to the root system from breathing roots, the reasons for survival of some heavily oiled trees was further investigated. Preliminary observations (reported in Dicks, in press) suggested that trees which had survived tended to have low densities of breathing roots and live in somewhat coarser and better drained sediments, whilst those which had died lived in muddier sediments and had higher densities of breathing roots. It was concluded that sediment drainage characteristics influence pneumatophore density for this species and play a major role in determining the ability of the trees to survive heavy oiling. It was also postulated that pneumatophore density may provide a simple index for identifying areas of high susceptibility to oil spill damage. Consequently, the sediment characteristics of four areas of mangrove were investigated in greater detail during a further visit to South Geisum Island in April 1985, with the aim of determining the factors which result in survival or death of the trees.

2. METHODS

2.1. Field measurements

The field visit to South Geisum was made in April, 1985. The four experimental areas were:

- Oiled mangroves which had survived (2 areas, LM1 and LM3)
- Oiled mangroves which had died (DM2).
- Unoiled mangroves (LM4).

The experimental sites (Fig. 1) were selected to be comparable in terms of tree height and density (assessed visually) and vertical height on the shore (all were within a vertical range on the shore of 10 cm, measured at high water). At all sites the mangroves were growing in a layer of sediment 15 - 20 cm deep overlying consolidated coral limestone. The main root mat from which the pneumatophores extended vertically was found at 8 - 12 cm depth. Measurements made at each location were:

- o - a depth profile of soil redox potential (Eh) at 2 cm intervals;
- o - salinity of interstitial water, measured in 3 to 5 small pits 5 cm deep using a portable refractometer;
- o - oxygen content of interstitial water, measured in 3 to 5 small pits 5 cm deep with a portable O_2 meter;
- o - sediment hydraulic conductivity, measured in undisturbed 3.8 cm diameter cores. Time and equipment constraints prevented replication of these measurements. The cores were sealed with rubber bungs and returned undisturbed to the field laboratory for measurement of rates of infiltration and permeation of water through the core, using a constant head permeameter.
- o - breathing root density, measured in 5 random 0.1 m^2 quadrats;
- o - oil layer thickness.

In addition sediment samples were taken for particle size, hydrocarbon and infaunal analysis. In order to assess both sediment structure with depth and penetration by hydrocarbons, two 6.5 cm diameter cores were taken to the depth of bedrock (15 - 20 cm) and subsectioned at 2 cm intervals and stored in separate containers (clean plastic pots for grain

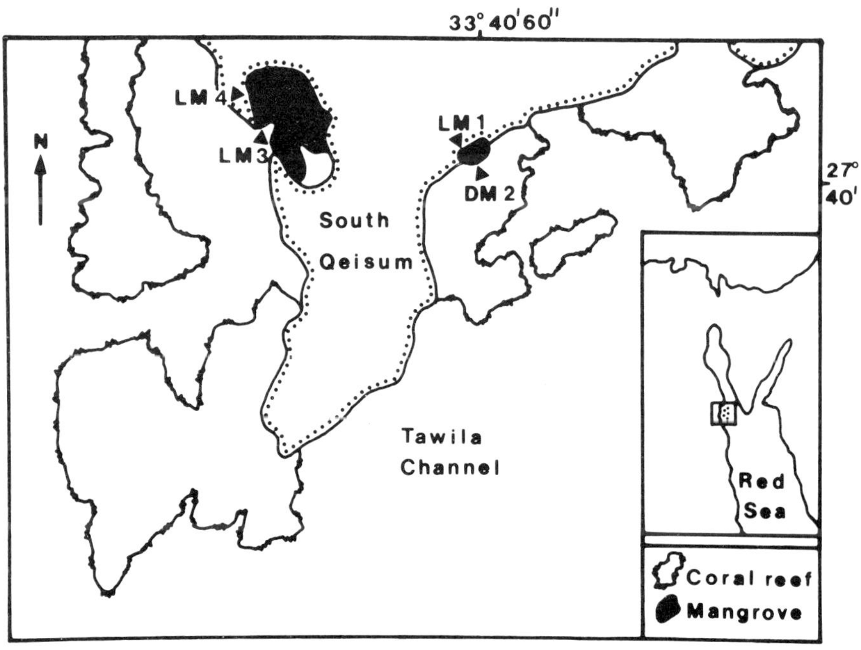

Fig. 1. The location of experimental sites on South Geisum Island.

size analysis and solvent cleaned aluminium cans for hydrocarbon samples). Sections for hydrocarbon analysis were stored at -20°C. Three 10 cm diameter cores were taken to 15 cm depth for subsequent analysis of the infauna. They were preserved unsieved in glass jars with a 5% formalin in seawater mixture. The stain eosin was added to facilitate laboratory sorting. A photographic record was made of each of the four sites.

2.2. Laboratory analysis

2.2.1. *Sediment grain size*. Particle size analysis was conducted by a combination of wet/dry sieving (Buchanan, 1984) and, where a significant silt/clay fraction was present (i.e. more than 2% by weight) Coulter Counter analysis was employed (Little *et al*, 1984). The particle size data were analysed by a computerised method of 'moments' (Krumbein, 1936). The 'moment' measures define the particle size distribution in terms of 0 mean, 0 standard deviation, 0 skewness and 0 kurtosis. This method has the advantage over graphic techniques in that all of the distribution data are used in the computation and not just a few points on the cumulative curve. Mud content is a sediment characteristic of importance in faunal studies and is often referred to in isolation from the rest of the particle size distribution. It is defined as that fraction of a sediment which passes a 4.0 0 (63 um) sieve by wet sieving, and includes both silt and clay. It is expressed on a percentage dry mass basis.

2.2.2. *Organic matter*. Organic matter content was determined by combustion analysis. This required a separate subsample. Organic matter content was calculated as percent weight loss on ignition of oven dried material and is approximately equivalent to organic carbon content.

2.2.3. *Hydrocarbons*. The procedure for analysis of hydrocarbons can be summarised as follows:

i) extraction of the solvent-soluble components of the sample.

ii) partitioning of the hydrocarbons from the other organic components (e.g. alcohols, esters, fatty acids) of the solvent extract, which were discarded.

iii) separation of the hydrocarbons into aliphatic and aromatic fractions by column chromatography, and determination of the concentrations of these fractions in relation to dry mass of sediment by gravimetry.

iv) analysis of the separated fractions by capillary gas liquid chromatography (GLC).

Full details of the methods can be found in Baker *et al*, (1984). Gravimetric analysis provides a relatively simple and standard procedure for determining the concentration of the aliphatic and aromatic fractions. The total hydrocarbon concentration (THC) is obtained from the sum of the two values. Results are expressed in ppm relative to dry sediment mass. Analysis of the hydrocarbon fractions by GLC provides a trace or 'fingerprint' of the analysed fraction, from which information on the nature of the hydrocarbons and degree of degradation can be obtained. In most cases the GLC traces obtained from the aliphatic fractions consist of an envelope of unresolved complex mixture (UCM) upon which are superimposed resolved peaks. The relationship between resolved peaks and the UCM allows assessment of the degree of weathering of the oil.

2.2.4. *Infauna*. On returning to the laboratory, the biological samples were washed over a stack of 2 sieves (1.0 and 0.5 mm) to separate the fractions and remove the preservative. The washed fractions were then placed in a white sorting tray and washed again with running water to decant light-bodied organisms on to the appropriate mesh size. The decanted fraction was examined under a binocular microscope and the animals were picked out. The remaining portion was hand sorted under bright illumination, the eosin dye added at preservation facilitating this task by staining organisms red. Both portions of the 0.5 mm samples were sorted under a binocular microscope. Organisms were then identified to the lowest practical taxon within the constraints of the available taxonomic literature and time. In some cases organisms have been identified only to genus or broader taxa as a consequence of the paucity of taxonomic literature for some groups. A complete collection of reference material and laboratory notes has been retained.

3. RESULTS

The sediment and interstitial water characteristics for each of the experimental sites are shown in schematic form in Figures 2 to 5.

3.1. Heavily oiled sites where the mangroves survived - LM1 and LM3

3.1.1. LM1 - Sediment characteristics. Sediments at all depths were poorly sorted and multimodal, with the 0 mean lying in the sand mode (Table 1). With the exception of the 14-16 cm section (which was solvent extracted to remove oil and consequently lost an undefined proportion of fine sediments) all showed a significant gravel fraction, sand mode or modes, and a significant mud mode. Percentage mud averaged 19.8% with a range through the core of 13.3 - 32.5% (excluding the 14 - 16 mm section).

Site	LM1			DM2			LM3			LM4		
	Type	%mud	%org	Type	%mud	%org	Type	%mud	%org	Type	%mud	%org
Depth:												
0 - 2 cm	FS	28.7	19.8	FS	15.6	32.0	MS	7.8	12.5			
2 - 4 cm	MS	14.0	4.5	VFS	31.4	10.9	MS	15.9	6.0			
4 - 6 cm	MS	13.6		VFS	35.2	8.3	FS	22.9	3.7	MS	6.7	1.9
6 - 8 cm	CS	14.6	4.5	VFS	30.3	4.5	FS	21.1	2.2			
8 - 10 cm	MS	19.2	7.8	VFS	26.8	5.7	MS	15.3	2.1	MS	8.1	2.9
10 - 12 cm	CS	15.3	6.4	VFS	27.1	5.7	MS	15.9	2.9			
12 - 14 cm	MS	20.7	4.0	VFS	28.7	4.7	MS	9.7	2.1			
14 - 16 cm		3.4	4.3	VFS	29.2	4.6	MS	11.9	3.2			
16 - 18 cm	FS	32.5	4.6				MS	12.4	3.2			
18 - 20 cm							MS	15.2	2.2			

Table 1. Sediment characteristics with depth (sediment type, percentage mud and percentage organics) of the four experimental sites. VFS = Very fine sand, FS = Fine sand, MS = Medium sand, CS = Coarse sand on the Wentworth scale. LM4 was analysed at 2 sections only (0 - 5 cm and 5 - 10 cm).

The hydraulic conductivity (water permeability) was 13.4 ml min^{-1} in the 2-7 cm section, and 4.0 ml min^{-1} in the 7-12 cm section which contained the main mangrove root mat. The conductivity coefficients (k), calculated from:

$$k\ (cm\ sec^{-1}) = \frac{\text{flow rate (ml } sec^{-1})}{(\text{head (cm) x cross sectional area})/\text{length of core (cm)}}$$

were 9.86 x 10^{-3} cm sec^{-1} and 2.94 x 10^{-3} cm sec^{-1} respectively. These readings indicate a high potential for water movement in the soil and rapid drainage, and these sediments were observed to be well drained on the shore. The reduced hydraulic conductivity in the lower core section may reflect the presence of a binding root mat at 8 cm. Redox potential readings at this site (Fig. 2) were all highly positive (mean 155 mV, range 40 - 210 mV), indicative of a state of rapid re-oxidation of reducing products from biological and chemical processes. Interstitial water salinity was high (53 ppt), indicating evaporative concentration and/or lateral movement of hypersaline water from terrestrial sources and resulting in considerable salinity stress on organisms. The % oxygen load in the water was also high at 58% (range 49-70%).

3.1.2. LM1 - Hydrocarbons. Surface sediments were grossly contaminated with crude oil, which penetrated to 14 cm in high concentrations (THC 890 ppm-190,000 ppm, see Fig.2). Unexpectedly, a band of low concentration of THC (210 ppm) was found at the depth of the root mat (8-10 cm). Below 14 cm, THC concentrations were lower (35-130 ppm). Even though the samples were taken approximately 2-3 years after the spills, the oil was only partially weathered, as shown by the homologous series of resolved *n*-alkane peaks in sample LM1 at 10-12 cm (Fig. 3).

3.1.3. LM1 - Mangrove breathing root density. Mean breathing root density was 17.2 per 0.1 m^2 with a range of 12-22 per 0.1 m^2.

3.1.4. LM1 - Biological community A highly impoverished community of only 10 taxa and 1 - 12 individuals per core (34 individuals in total) was present. Characteristic species were opportunists, well known from heavily polluted or stressed locations; *Capitella capitata*, oligochaetes and nematodes (Pearson and Rosenburg, 1978; Raffaelli and Mason, 1981; Warwick, 1981). The polychaete *Augeneriella lagunari*, an indicator of high salinity (Gitay, 1970), was also recorded.

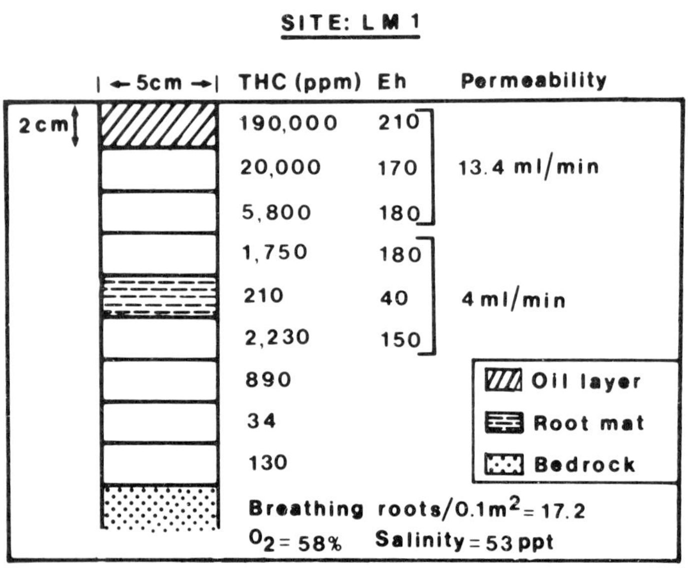

Fig. 2. A schematic profile of the sediment cores at site LM1. Eh is given as mV potential corrected to the H_2 electrode. % oxygen and salinity refer to interstitial water.

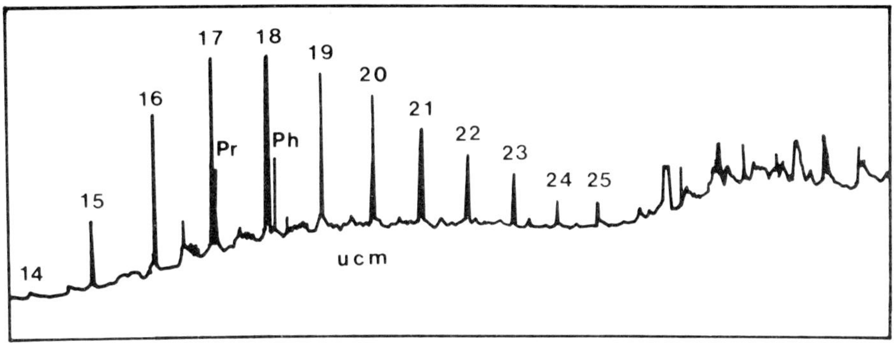

Fig. 3. Gc trace of the aliphatic hydrocarbons in the 10 - 12 cm dewpth section. n-alkanes are numbered. Pr = pristane, Ph = phytane.

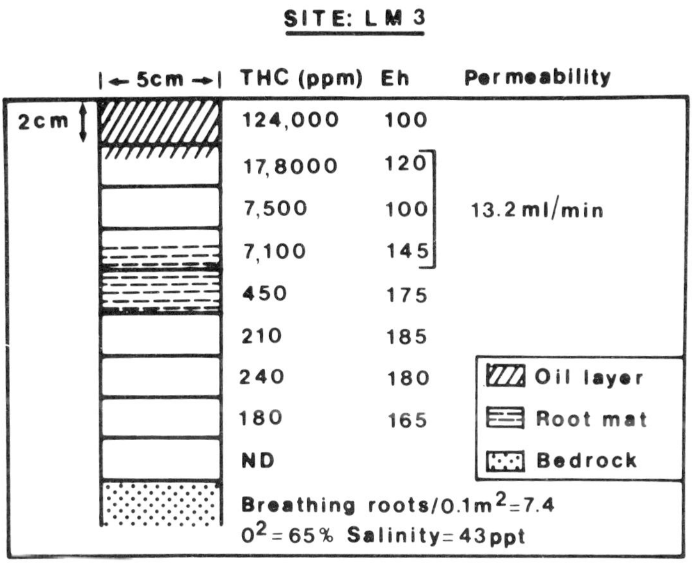

Fig. 4. A schematic profile of the sediment cores at site LM3. Eh is given as mV potential corrected to the H_2 electrodc. % oxygen and salinity refer to interstitial water.

3.1.5. LM3 - Sediment characteristics. Sediments down to 10 cm depth showed two dominant sand modes, a coarse and a fine, plus small gravel and mud modes (Table 1). 0 mean was in the medium/fine sand mode. Below 10 cm a single mode was dominant in the medium sand range, associated with a significant mud mode. Overall, % mud was elevated in all sections (mean 15%, range 7.8 - 22.8%) and similar to that found at LM1. The % gravel was lower than that found at LM1, and similar to that at DM2 (4%).

The hydraulic conductivity in the 2-7 cm core section was 13.2 ml min^{-1}, giving a coefficient (k) of 9.45 x 10^{-3} cm sec^{-1}. This indicates a high potential for water movement and drainage, and is consistent with shore observations which showed the soil to be well drained.

Redox potential readings at this site were all highly positive (mean 146 mV, range 100 - 185 mV, see Fig. 4), indicative of a well oxygenated soil in a state of rapid re-oxidation of reducing products from biological and chemical processes. Interstitial water oxygen content was high at 71% (range 58-90%), and salinity was very close to that of seawater (42 ppt) at 42.6 ppt.

3.1.6. LM3 - Hydrocarbons. Surface sediments (0 - 8 cm depth) were grossly contaminated with crude oil (THC 124,000 - 7,100 ppm, see Fig. 4). The surface oil was heavily weathered, much more so than at LM1 or DM2, and may be older oil, or have weathered more rapidly for reasons which are not clear. Deeper core sections (8 - 16 cm) showed the presence of relatively fresh crude as well as biogenic components from leaf waxes. Of the 3 oiled sites, this was the least heavily contaminated at these depths (THC 180 - 450 ppm). No detectable hydrocarbons were found in the deepest 2 sections (Fig. 4).

3.1.7. LM3 - Mangrove breathing root density. Mean breathing root density was very low at this site at 8 per 0.1 m^2 (range 5-11 per 0.1 m^2).

3.1.8. LM3 - Biological community Although highly impoverished, the community was slightly more diverse than that of LM1. Like LM1, the most common species were opportunists, well known from heavily polluted or stressed locations (*Capitella capitata*, oligochaetes and nematodes). There was considerable variability between cores, with numbers of individuals ranging from 4 - 144 per core. More specimens were found in the 0.5 - 1.0 mm fraction than in the 1.0 mm fraction.

3.2. Site at which mangroves were heavily oiled and died - DM2

3.2.1. DM2 - Sediment characteristics. Sediments at all depths were moderately to poorly sorted with the 0 mean lying in the very fine sand mode (Table 1). Most core sections showed multimodal grain size distributions, with a dominant very fine sand mode, but also significant mud modes and a small gravel mode. Percentage mud was high in all sections, ranging from 15.6 - 35.2%, with a mean of 28% (Table 1).

The hydraulic conductivity in the 2-7 cm core section was very low at less than 1 ml min^{-1}, giving a conductivity coefficient (k) of 0.4×10^{-3} cm sec^{-1}. This indicates a very low potential for drainage which is consistent with shore observations showing the sediment to be waterlogged.

Redox potential readings were all strongly negative (mean -230 mV, range -180 to -285 mV, see Fig. 5), indicative of a state of highly reduced conditions. Interstitial water salinity was slightly elevated (46 ppt) above that of nearby seawater (42 ppt). The % oxygen load in the water was low at 27.5% (range 11 - 40%). Both salinity and % oxygen may have been elevated by small contributions from surface water run-off into the measuring pits at this site.

3.2.2. DM2 - Hydrocarbons. The hydrocarbons at DM2 were very similar in concentration, distribution and type to that reported for LM1, with gross contamination of surface layers (Fig. 5). Weathering was less in the surface sediments where concentrations were very high (THC 145,000 ppm) and a thick layer of oil was present. A second significant band of oil was found just below the mangrove root mat (at 8-10 cm), where concentrations reached 21,500 ppm. This oil was heavily weathered and may be the result of spills earlier than the 1982/83 series. The lowest

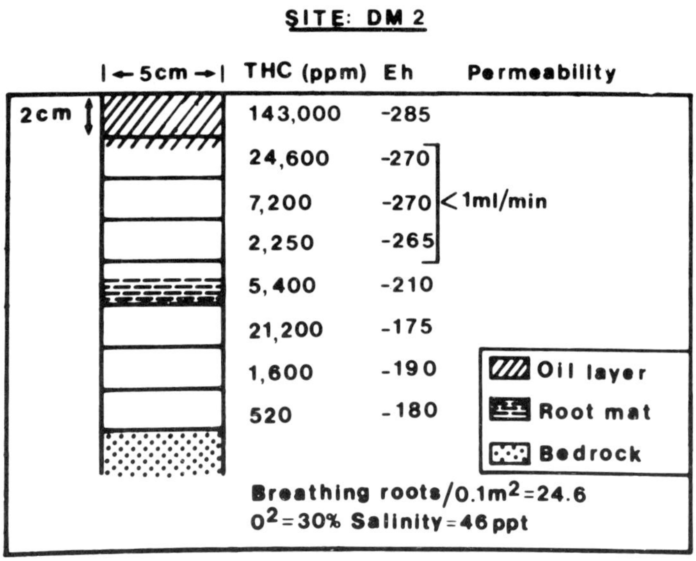

Fig. 5. A schematic profile of the sediment cores at site DM2. Eh is given as mV potential corrected to the H_2 electrode. % oxygen and salinity refer to interstitial water.

concentration of THC was found at 14 - 16 cm (520 ppm, Fig. 5), but all other sections were contaminated by 1,600 ppm or more.

3.2.3. DM2 - Mangrove breathing root density. Breathing root density was 24.6 per 0.1 m^2 with a range from 14 - 37 per 0.1 m^2.

3.2.4. DM2 - Biological community Virtually no macrofauna was found, except for a single nematode and one specimen of *Capitella capitata*. This probably reflects severe pollution stress.

3.3. Unoiled and healthy site - LM4

3.3.1. LM4 - Sediment characteristics. Only 2 core sections (0 - 5 cm and 5 - 10 cm) were analysed at this site (Table 1). The sediments were moderately sorted and unimodal with 0 mean lying in the medium sand mode. A mud tail was present in both sections, although both % mud and % gravel concentrations were small at 6.6 - 8.0% and 0.1 - 0.9% respectively. The sediment at this site was similar to that found in the deeper sections at LM3.

The hydraulic conductivity was less than 1 ml min^{-1} in the surface 5 cm and 1.6 ml min^{-1} in the 5-10 cm section, giving coefficients (k) of 0.7×10^{-3} cm sec^{-1} and 1.4×10^{-3} cm sec^{-1} respectively. Although higher than the coefficient at DM2, these relatively low values indicate a relatively poor drainage potential, which is again consistent with field observations that the sediments were only poorly drained.

Redox potential readings at this site were highly positive down to 20 cm (mean 208 mV, range 190 - 220 mV, see Fig. 6), indicating a state of

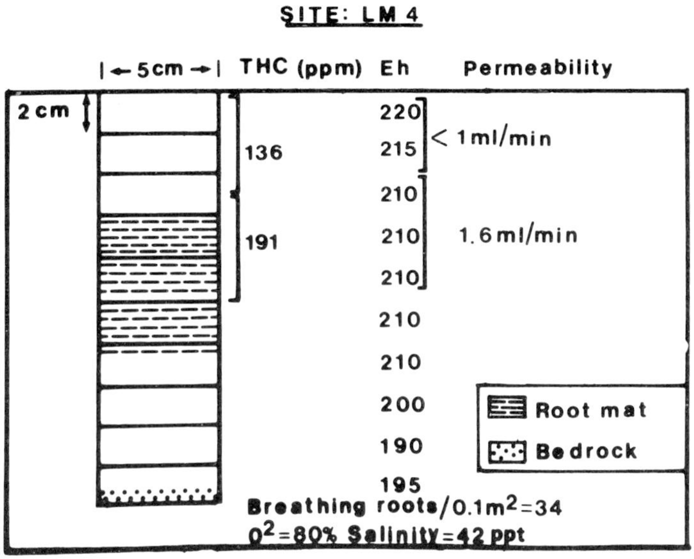

Fig. 6. A schematic profile of the sediment cores at site LM4. Eh is given as mV potential corrected to the H_2 electrode. % oxygen and salinity refer to interstitial water.

rapid re-oxidation of reducing products from biological and chemical processes, and the presence of plenty of oxygen. The oxygen content of interstitial water was high at 81% (range 76-90%) and salinity, at 42 ppt, was identical to local seawater.

3.3.2. LM4 - Hydrocarbons. Sediments were only lightly contaminated by crude oil (135 - 190 ppm, Fig. 6). The gc traces showed weathered material as well as small components of relatively fresh oil and a clear biogenic input of plant waxes.

3.3.3. LM4 - Mangrove breathing root density. This site showed the highest breathing root density at 34 per 0.1 m^2 (range 27-50 per 0.1 m^2).

3.3.4. LM4 - Biological community The community found at LM4 was more diverse than at any of the other experimental sites, and may be regarded as typical of upper shore mangroves in the Geisum Island area. Like the heavily oiled sites, commonest species were indicators of stress (*Capitella capitata* and oligochaetes) or hypersaline conditions (*Augeneriella lagunari*), but their numbers and the presence of an additional 18 taxa, including both molluscs and crustacea which were found only at this location, reflect the absence of the heavy oil pollution at the other sites.

4. DISCUSSION

4.1. Heavily oiled sites where the mangroves survived - LM1 and LM3

At both locations the sediments were seen to be well drained in the field although the laboratory analyses showed them to differ somewhat in grain size characteristics. Both had similar % mud contents (18 and 15%

respectively) and 0 mean values (1.25 and 1.67 respectively) but very different % gravel components (31% and 4% respectively). The hydraulic conductivity coefficient was also high and similar at each site (k = 9.87 and 9.45 respectively) indicative of good potential for drainage. Whilst it is not possible to accurately predict hydraulic conductivity from grain size distributions (or vice versa) because factors such as particle shape, degree of sorting and modality of grain sizes come into play, there is a general relationship in that coarser sediments tend to drain more quickly than finer. Although there are some differences in sediment characteristics with depth, both sites can be classed as medium sands which are likely to drain well. Irrespective of sediment characteristics, whether or not good drainage occurs depends on shore topography, as both drainage and re-flooding depend upon suitable inlet and exit routes for the water.

The presence of oil may affect the ability of sediments to drain, either by blocking subsoil exit or inlet routes or preventing surface exchange. The field observations showed that in both cases the sites drained freely at low water in spite of the presence of thick layers of oil.

Interpretation of the relationships between pore water salinity and oxygen content, the drainage characteristics discussed above, and the ability of mangrove to acquire oxygen from pore waters or via breathing roots is not simple. It might be expected that well drained soils which regularly reflood as a result of tidal rise and fall and which may be permeated easily by air would have well oxygenated pore water, from which the mangrove root system could obtain an adequate supply. Under such conditions breathing roots may play a secondary role, except perhaps during 'crises' of low oxygen content, and therefore be produced in lower numbers. Both LM1 and LM3 have the lowest numbers of breathing roots of all 4 experimental sites, but the density at LM3 was significantly lower than LM1 ($p = 0.05$ by one way ANOVAR test on log transformed data) in spite of similar drainage characteristics and coefficients of hydraulic conductivity. An additional major difference between the sites was the salinity of the pore water, with much higher values at LM1 than LM3 (53 ppt vs. 42.6 ppt). Increasing salinity is reported by Parsons et al. (1984) to reduce solubility of oxygen, although his observations only extend up to salinities of 34 ppt. This may present a partial explanation for increased numbers of breathing roots at LM1. The elevated salinity at LM1 indicates poor mixing between interstitial pore water and tidal water, but the cause of the high salinity has not been established. It may result from evaporative concentration or from brine seepage from nearby terrestrial groundwater sources. If the latter is the case, water is likely to be depleted in oxygen which may place additional oxygen stress on the mangrove. Although the % oxygen saturation was lower at LM1 than LM3, little can be inferred from this due to potential interference from surface water. Eh values at these two sites were all highly positive (96 to 196 mV), strongly indicative of oxidising conditions, but this may result, at least in part, from the mangroves because they pass oxygen from their root system into the pore water and thus aid physical oxygenation processes.

4.2. Heavily oiled site where mangroves had died - DM2

The sediment characteristics of this site were different to those at LM1 and LM3, with the 0 mean in the very fine sand mode and a mud content of 28%. No significant gravel mode was found, and the sediments appeared muddy and waterlogged in the field. The hydraulic conductivity coefficient was very low at 0.42×10^{-3} cm sec^{-1} in the sediments immediately below the oil. Together these factors indicate poor drainage and low potential for natural aeration of interstitial water. Eh values were indicative of highly reducing conditions (-180 to -288 mV), partly a reflection of the naturally low water movements and high organic content of the sediment, as well as the absence of any O_2 input from the dead mangroves. Oxygen saturation of pore water was the lowest recorded at 27%. Interstitial salinity was only slightly elevated above sea salinity at 46 ppt. Breathing root density was higher than at either LM1 or LM3, but was only significantly different from LM3 in a log-transformed one way ANOVAR test. Together, these factors support the hypothesis that the number of breathing roots produced is dependent upon the natural potential for aeration of pore water. At this location, the trees would be more dependent on the pneumatophores than at LM1 and LM3. This may explain the death of the mangroves on oiling of the breathing roots.

4.3. Unoiled mangroves, Site LM4

The sediments were moderately sorted medium sands with a small mud content (6.6 - 8.0%), somewhat similar to those found in the deeper sections at LM3. In spite of this similarity, hydraulic conductivity of the sediments was low ($k = 0.74 - 1.47 \times 10^{-3}$ cm sec^{-1}). Breathing root density was the highest recorded (34 per 0.1 m^2), and Eh and percentage O_2 saturation were both high (187-213 mV and 81% saturation). Salinity was close to that of seawater. These sediments were observed to drain only poorly in the field, which is consistent with the low hydraulic conductivity, and their potential for natural aeration is thus likely to be low. The high density of breathing roots probably reflects the low natural aeration potential of these soils and the high oxygen content and Eh values found were a consequence of the action of breathing roots in aerating the soil. It is likely that in the event of heavy oiling of these breathing roots, death of the trees would result.

5. CONCLUSIONS

Dicks (in press) put forward the following hypotheses: a) in muddy sediments where drainage is limited and reducing conditions are likely to prevail, *A. marina* produces greater densities of pneumatophores, presumably in order to keep root systems provided with sufficient oxygen. b) the converse applies in sandy sediments which drain well: i.e. fewer pneumatophores are produced and the root systems are able to utilise at least some oxygen from interstitial water or air in the pores at low water.

The evidence presented above supports these hypotheses and we conclude that the key to explaining the survival of heavily oiled mangroves lies in the ability of the root mat to obtain sufficient oxygen once the breathing roots are put out of action. The only route available is to obtain oxygen from pore water, and the ability to achieve this appears to be related to the drainage characteristics of the soil. These are not dependent on any single sediment parameter, but relate to a number of features, notably grain size characteristics, hydraulic conductivity and shore topography. The presence of the mangrove root mat may also modify these characteristics as the mangal develops and grows.

These observations suggest that it may be possible to identify in advance those areas of mangrove likely to be susceptible to damage by oil. This could be achieved by a field assessment of drainage characteristics, breathing root density, soil grain size and, where necessary, a rapid measure of hydraulic conductivity.

It is recommended that the first step should be a more exhaustive investigation of the relationships between breathing root density and the sediment characteristics of drainage/grain size/hydraulic conductivity, in order to confirm the hypotheses put forward on the basis of the 4 sites on South Geisum Island.

We believe that the potential to identify susceptibility of mangroves to oil and thus apply suitable protection is more widely applicable than the northern Red Sea, for *Avicennia marina* is well adapted to cope with a wide range of salinity and soil conditions, and occurs widely throughout the Indo-Pacific as a dominant species. It is also possible that such predictive ability can be applicable to species other than *Avicennia marina*, and in view of the worldwide importance of mangrove the diversity, productivity and economy of the tropics would repay further study.

6. ACKNOWLEDGEMENTS

We are deeply grateful to the Geisum Oil Company, Egypt for providing the opportunity to carry out these studies, and to British Petroleum (London) and the Shell Internationale Petroleum Matschappij (Den Haag) who kindly supported the analytical programme. Our thanks are also due to our colleagues in OPRU, particularly Nicola Dodd and Jan Smith who carried out the sediment grain size and hydrocarbon analyses.

7. REFERENCES

1. Baker JM: Mangrove swamps and the oil industry. Oil and Petrochemical Pollution 1(1): 5-22, 1982.
2. Baker JM, Crothers JH, Little DI, Oldham JH and Wilson CM: Comparison of the fate and ecological effects of dispersed and non-dispersed oil in a variety of intertidal habitats. In: Oil Spill Chemical Dispersants, ed Allen TE. 239-279. ASTM, Philadelphia, 1984.
3. Buchanan JB: Sediment analysis. In: Methods for Study of Marine Benthos. Eds. Holme, NA and McIntyre, AD. IBP Handbook No. 16, 41 - 65, 1984. Blackwell, Oxford.
4. Dicks B: Oil pollution in the Red Sea - environmental monitoring of an oilfield in a coral area, Gulf of Suez. Deep-Sea Research 31(6-8A), 833-854, 1984.
5. Dicks B: The Red Sea - Pollution. Key Environment handbook series, Pergamon, in press.
6. Dicks B: Oil and the Black Mangrove, *Avicennia marina* in the northern Red Sea. Mar. Poll. Bull, in press.
7. Getter CD, Cintron G, Dicks B, Lewis RR, and Seneca ED: The recovery and restoration of saltmarshes and mangrove following an oil spill. In: Restoration of Habitats Impacted by Oil Spills, 65-113. Butterworth, Boston, 1984.
8. Gitay A: A review of *Augeneriella* (Polychaete:Sabellidae) and a new species from Northern Sinai. Israeli. Journ. Zool. 19(2), 105-109, 1970.
9. Hanna RM: Oil pollution on the Egyptian Red Sea coast. Mar. Poll. Bull, 14, 268 - 271. 1983
10. Jennings MC: An interim report on the ornithological expedition to the Red Sea province of Egypt in spring 1983. OSME Bull. 11, 8-9, 1983.
11. Krumbein WC: Application of logarithmic moments to size frequency distributions of sedimnents. J. Sed. Pet. 6, 35-47, 1936.

12. Lewis RR: Impact of oil spills on mangrove forests. Tasks for Vegetation Science, Vol 8, 171-183, 1983. Dr. W. Junk Publishers, The Hague.
13. Little DI, Staggs MF and Woodman SSC: Sample pretreatment and size analysis of poorly-sorted cohesive sediments by sieve and electronic particle counter. In: Parker WR and Kinsman DJJ (eds.), Transfer Processes in Cohesive Sediment Systems. Estuarine and Brackish Water Sciences Association. London: Plenum Press. 47-74, 1984.
14. Parsons TR, Maita Y and Lalli CM: A manual of chemical and biological methods for seawater analysis. Pergamon, Oxford. 173 pp. 1984.
15. Pearson THJ and Rosenberg R: Macrobenthic succession in relation to organic enrichment and pollution of the marine environment. Oceanogr. Mar. Biol. Ann. Rev. 16, 229-311, 1978.
16. Por FD, Dor I and Amir A: The mangal of Sinai: Limits of an ecosystem. Helgolander wiss. Meeresunters, 30, 295-314, 1977.
17. Raffaelli DG and Mason CF: Pollution monitoring with meiofauna using the ratios of nematodes to copepods. Mar. Poll. Bull. 12(10), 158-163, 1981.
18. Warwick RM: The nematode/copepod ratio and its use in pollution ecology. Mar. Poll. Bull. 12(10), 329-333, 1981.
19. Wenninck CJ and Nelson-Smith A: Coastal Oil Pollution Study for the Gulf of Suez and the Red Sea Coast of the Republic of Egypt. IMO, London, 1979.

EFFECTS OF SEVEN DISPERSANTS ON GROWTH OF THREE SUBTROPICAL/TROPICAL ATLANTIC SEAGRASSES

A. THORHAUG AND J.H. MARCUS

GREATER CARIBBEAN ENERGY AND ENVIRONMENT FOUNDATION, 1121 CRANDON BLVD. MIAMI, FLORIDA 33149, U.S.A.

1. INTRODUCTION

Seagrasses are immediately impinged upon by tankers, pipeline leaks, and in some cases, by offshore drilling spills, barge and harbor spills. Millions of acres of seagrasses throughout the world have already been decimated with a loss of fisheries nurseries. However, there remain throughout the world's tropics and subtropics several million acres. One of the three most important, and probably the most extensive critical habitat in the Caribbean, Gulf of Mexico and Florida is seagrasses, primarily composed of *Thalassia testudinum* (turtle grass), *Halodule wrightii* (shoal grass) and *Syringodium filiforme* (manatee grass). The most important function of seagrass is critical nursery habitat for fisheries, but in addition, seagrasses provide food, shelter, surface for attached animals and plants and anti-erosion functions.

The effects of dispersed oil have only previously been studied for subtropical and tropical seagrasses on the dispersant Corexit 9527. Thorhaug *et al.* (1986) carried out a series of laboratory tests over a range of concentrations (from 0.75 ml dispersant in 7.5 ml oil in 100,000 ml seawater to 100 times these concentrations) and time periods from 5 to 100 hours. Three species of subtropical and tropical seagrasses were used. The results showed high variability between species' tolerance to Corexit 9527 with *Thalassia testudinum* being far more tolerant than *Halodule wrightii*, which was slightly more tolerant than *Syringodium filiforme*. A single dosage time and species was tested by Baca and Getter (1982) for Corexit 9527.

A second study tested Corexit 9527 in a field site in Panama using the most dispersed-oil tolerant seagrass, *Thalassia* (Getter, 1986). A subsequent real spill in the area of the Caribbean coast of Panama, which had Corexit 9527 dispersant applied, appears to corroborate the field and laboratory studies. The results showed concentrations of 5 to 20 ppm for short (5 to 24 hr) time periods are not toxic. Longer time periods (100 hr) can be toxic to *Syringodium*.

No tests of the wide variety of commercially available and usable dispersants have been made on the seagrasses as have been tested on temperate shrimp (Anderson *et al.*, 1985).

The use of dispersants may impinge on seagrasses (Indonesia and Bahrain as real life examples) especially in developing nations during nearshore spills (harbors, ports, pipelines, oil terminals) as well as for coastal spills from boats and drilling rigs. Knowledge of which dispersants can be safely used on which seagrass is of great importance to set standards or guidelines in developing nations.

2. METHODS

The methods duplicate those described by Thorhaug and Marcus (1985),

Kuiper, J., Van den Brink, W.J. (eds), Fate and Effects of Oil in Marine Ecosystems. ISBN 90-247-3489-4

and Thorhaug et al. (1986). Out-of-door aquaria with sand and 100,000 ml natural seawater were planted with 15 healthy sprigs each of 3 species (*Thalassia testudinum*, *Halodule wrightii*, and *Syringodium filiforme*). Treatments included dispersed oil and control. Variables included: concentration of dispersed oil and type of dispersant. The dispersants tested included Corexit 9527, Corexit 9550, OFC-D-609, Conco K (K), Jansolv-60, Cold Clean 500 and Finasol OSR-7. Oil was weathered in shallow pans exposed to sunlight for 24 hours prior to experimentation. Dispersants were vigorously mixed with oil at time 0. Aeration throughout the experiment kept dispersed oil in a "mousse" phase. After treatment, oil was thoroughly removed and tanks cleaned. The contaminated water was immediately replaced with clean ocean water. Specimens were then measured for 14 days for mortality, sublethal effects and growth of 15 new blades of each species.

The aquaria were protected from rain by a transparent plastic covering. The temperature of the water ranged from 28 to 31°C with a daily fluctuation of 1°C. The seagrasses were illuminated by natural sunlight. Experiments reported here were carried out from March to September, 1986. Seagrasses were acclimatized to the aquaria for a minimum of 48 hours prior to each experiment. Each aquarium was designated for treatments by random sampling techniques. The experiments were designed to test the effect of 75 ml and 125 ml of Louisiana Crude oil combined with one of seven dispersants (10:1 or as instructed*for that dispersant); on the seagrasses; and the condition and growth of seagrasses growing in an aquarium with no exposure to oil or dispersant (control).

Statistical analysis of data is described in detail in Thorhaug and Marcus (1985). Values for b were estimated by linear regression analysis of (usually) n = 5 determinations over a 14 day period for each blade measured. All the growth rates are expressed as specific growth, G, % day^{-1}, obtained from G = b(100). The mean specific growth rate $\bar{G} \pm$ 95% CL, was then determined for 15 blades of each species.

3. RESULTS

When exposed to 125 ml and 75 ml of Louisiana Crude oil combined with each of seven dispersants for 100 hours, the three test species reacted in the following manner (see Table 1):

3.1 *Thalassia testudinum*

Mean growth was highest using the dispersants Finasol OSR-7, Jansolv-60 and Cold Clean 500. Growth ranged $\bar{G}$ = 4.22 - 3.54 % day^{-1} when dispersants were combined with 75 ml of oil and from $\bar{G}$ = 4.11-2.95 % day^{-1} with 125 ml oil. The most toxic dispersants were OFC-D-609 ($\bar{G}$ = 2.73% day^{-1} with 125 ml oil and $\bar{G}$ = 3.10% day^{-1} with 75 ml oil) and Conco K (K) which, when combined with 125 ml oil and 75 ml oil resulted in $\bar{G}$ = 1.74% day^{-1} and 2.10% day^{-1}, respectively. The Corexit dispersants 9550 and 9527 fell in the middle growth range: when combined with 125 ml, mean specific growth ranged from $\bar{G}$ = 2.78 - 2.60 % day^{-1}; when combined with 75 ml, mean specific growth ranged between $\bar{G}$ = 3.35 - 3.28 % day^{-1}.

3.2 *Syringodium filiforme*

Mean growth was highest using the dispersants Finasol OSR-7, Jansolv-60 and Cold Clean 500. Growth ranged from $\bar{G}$ = 3.41 - 3.25 % day^{-1} when these dispersants were combined with 75 ml oil and from $\bar{G}$ = 3.11 - 2.80 % day^{-1} when dispersants were combined with 125 ml oil. Dispersants that affected growth most were Corexit 9550, Corexit 9527, OFC-D-609 and Conco K (K). These, where combined with 75 ml oil resulted in a growth rate

*OFC-D-609 (30:1);Cold Clean (2% dilution, 1:1);Jansolu (6% dilution, 1:1).

TABLE 1. The effect of 75 ml and 125 ml of Louisiana Crude oil combined with one of seven dispersants (10:1 or as instructed for that dispersant) on *Thalassia*, *Halodule* and *Syringodium* for 100 h in 100 l seawater. Mean specific growth rates ($\bar{G}$ % per day), 95% confidence limits, and coefficient of variation in growth rates (CV%) are given on numbers of blades measured (n) for each species of seagrass.

Dosage oil (ml)	Dispersant	*Thalassia testudinum*			*Halodule wrightii*			*Syringodium filiforme*		
		$\bar{G} \pm$ 95% C.L.	CV%	% mortality	$\bar{G} \pm$ 95% C.L.	CV%	% mortality	$\bar{G} \pm$ 95% C.L.	CV%	% mortality
0	----	4.51 ± 0.22	18	0	4.20 ± 0.31	24	7	3.98 ± 0.25	28	7
75	Finasol OSR-7	4.22 ± 0.55	48	0	3.64 ± 0.35	15	7	3.41 ± 0.26	28	7
125	Finasol OSR-7	4.11 ± 0.41	37	7	3.52 ± 0.29	21	10	3.11 ± 0.61	35	10
75	Jansolv-60	3.98 ± 0.31	51	10	2.25 ± 0.61	64	41	3.05 ± 0.41	299	12
125	Jansolv-60	3.76 ± 0.24	18	18	2.11 ± 0.37	41	57	2.86 ± 0.35	22	25
75	Cold Clean 500	3.54 ± 0.11	9	16	3.31 ± 0.28	36	21	3.25 ± 0.25	18	18
125	Cold Clean 500	2.95 ± 0.37	21	26	3.22 ± 0.66	45	27	2.80 ± 0.54	28	30
75	Corexit 9550	3.35 ± 0.31	28	22	2.05 ± 0.45	31	58	1.97 ± 0.35	17	63
125	Corexit 9550	2.78 ± 0.41	35	37	1.69 ± 0.55	65	88	1.49 ± 0.45	41	84
75	Corexit 9527	3.28 ± 0.41	31	26	1.84 ± 0.41	60	76	1.39 ± 0.37	25	84
125	Corexit 9527	2.60 ± 0.33	28	40	0.66 ± 0.21	31	100	0.51 ± 0.45	37	100
75	OFC-D-609	3.10 ± 0.27	24	35	1.66 ± 0.31	24	73	1.42 ± 0.38	46	86
125	OFC-D-609	2.73 ± 0.45	38	51	0.54 ± 0.31	38	100	0.44 ± 0.39	29	100
75	Conco K(K)	2.10 ± 0.41	49	65	0.76 ± 0.75	61	97	0.59 ± 0.35	37	98
125	Conco K(K)	1.74 ± 0.35	37	88	0.45 ± 0.25	28	100	0.41 ± 0.68	52	100

ranging from $\bar{G}$ = 1.97 -1 0.59 % day^{-1}. When combined with 125 ml oil, growth ranged from $\bar{G}$ = 1.49 - 0.41 % day^{-1}.

3.3 Halodule wrightii

Mean specific growth rate was highest using the dispersants Finasol OSR-7 and Cold Clean 500. Growth ranged from $\bar{G}$ = 3.64 - 3.31 % day^{-1} when dispersants were combined with 75 ml oil and $\bar{G}$ = 3.52 - 3.22 % day^{-1} when dispersants were combined with 125 ml oil. The dispersants that most affected growth were Corexit 9550, Corexit 9527, OFC-D-607 and Conco K(K). These, when combined with 75 ml oil resulted in growth ranging from $\bar{G}$ = 2.05 - 0.76 % day^{-1} and growth ranging from $\bar{G}$ = 1.69 - 0.45 % day^{-1} when combined with 125 ml oil. The dispersant Jansolv-60 fell in the middle range: when combined with 75 ml oil, growth reached $\bar{G}$ = 2.25 % day^{-1};when combined with 125 ml oil, growth reached $\bar{G}$ = 2.11 % day^{-1}.

4. DISCUSSION

The dispersants examined exhibited a wide variety of effects on seagrass growth the Greater Caribbean Basin. Guidelines or standards to dispersant use in oil spills as an aggregated group of substances may not be an effective management strategy. Using only the least biologically toxic group of dispersants and building a methodology of spill clean-up using these, appears to us to hold more promise. We recommend publishing dispersant toxicity lists and proceeding with oil spill clean up plans which name the dispersant to be used on each area's seagrass. Our group of preliminary laboratory experiments clearly demonstrate wide differences in toxicity on between dispersants. The results indicate that the most toxic dispersants to *Thalassia* growth are OFC-D-609, and in particular, Conco K(K). These and the Corexits 9550 and 9527 are also toxic to *Halodule* and *Syringodium*, especially at extended exposure periods. The least toxic dispersants to growth of *Thalassia* and *Syringodium* are Cold Clean 500, Jansolv-60 and especially Finasol OSR-7. Jansolv-60 appears to be more toxic to the growth of *Halodule* than to *Syringodium*.

There is a fairly wide difference in toxicity tolerance to the various dispersants and concentrations. Various seagrass species have concentration, time of exposure, and dispersant tolerances which must be taken into account in oil spill contingency plans and application dosages. Oil spill plans must provide information about which seagrasses are present in areas where dispersants are to be applied which are now completely lacking.

We recommend that dispersed oil field testing be done on seagrasses and other critical habitats in tropical and in subtropical locations to ensure both the tolerance range and the critical range for the dispersed oil likely to be used, even if the use of dispersants is only meant to be the third alternative clean-up plan. If the use of dispersants is contemplated anywhere in the plan, knowledge of the biological toxicity on the critical habitat organisms in the locality is necessary.

We should point out that these data are for the coastal, bay and estuarine seagrass communities in the western subtropical Atlantic from Brazil to Florida. Since other species are present in Mediterranean, Red Sea, Arabian Gulf, Indian Ocean, Pacific Ocean, China Sea these results do not extrapolate easily. These results do not extrapolate to coral reefs, which are frequently more sensitive to pollutants than seagrasses are. Testing on species found in other ocean basins with the dispersant to be used in that locality should be carried out for reliability of use of these data in oil spill clean-up plans for other tropical oceans.

5. ACKNOWLEDGMENTS

The authors gratefully acknowledge the support of Mobil Oil Foundation and Mr. Stewart Horn. We are particularly grateful for conversations about dispersant use with the following: Michael Flarety and John Riley, EPA; Captain Charles Corbett and J. A. Broom, U.S. Coast Guard; James Hildrew and John McCullough from Mobil; Duane Levine from Exxon and Drs Page and Gilfilin, Bowdoin College ; Howard Teas, University of Miami, supplied crude oil, and the following persons supplied dispersants: Dr. Gordon Lindblom, Mr. P. D. Turits, Dr. June Siva, Virginia A. Watters, Dr. Larminie, G. D. Fielder, Jerry W. Johnson and Stephen Kaufman.

6. REFERENCES

1. Anderson JW, DL McQuerry, and SL Kiessere: Laboratory evaluation of chemical dispersants for use on oil spills at sea. Environmental Science and Technology, 1985.
2. Baca BJ and CD Getter: The toxicity of chemically dispersed oil on the seagrass Thalassia testudinum using static and flow through bioassays. In: Dispersants: Five years Research. ASTM Press, Philadelphia, 1982.
3. Getter CD: The tropics experiment, designed to give answers for managing tropical oil spills. International Petroleum Environmental Conservation Association (IPIECA) Symposium on Oil spills Research in Tropical Environments and Contingency Planning in the Wider Caribbean, Caracas, Venezuela, 1986.
4. Thorhaug A and J Marcus: Effects of dispersant and oil on subtropical and tropical seagrasses. Proceedings of the 1985 Oil Spill conference, 1985.
5. Thorhaug A, J Marcus and F Booker: Oil and dispersed oil on subtropical and tropical seagrasses in laboratory studies. Marine Pollution Bulletin, 1986.

TOXICITY OF CHEMICALLY DISPERSED OIL IN A FLOW-THROUGH SYSTEM

C. BARDOT and G. CASTAING
Institut Français du Pétrole, BP 311, 92506 RUEIL MALMAISON cédex, France

Even though for a long time it has been included in the arsenal of oil combatting techniques, the treatment of an oil slick by dispersants is still a debatable solution. In addition to the difficulty in assessing the effectiveness of such products on oil, the evaluation of their consequences on marine ecosystems is very delicate and is the subject of numerous research projects.

Offshore tests carried out since 1979 by the Institut Français du Pétrole (PROTECMAR) have revealed the presence of small concentrations of dispersed hydrocarbons in the first few meters of water (less than 100 ppm) as well as the importance of subsurface currents in causing the rapid dilution of oil droplets. This is why we have been working to define a laboratory test which effectively takes such phenomena into consideration. A test organism would be subjected, for a given length of time (6 h), to a decreasing concentration of oil emulsified by an influx of unpolluted water. A description of the equipment as well as the preliminary results have already been published (Bardot, 1984).

Because it has traditionally been used in toxicity tests in Great Britain and the Netherlands, the brown (or common) shrimp Crangon crangon was chosen as the biological species. This organism coming from a natural environment is subjected to its first stress when it is caught and transported to the laboratory. The need to use a test organism that is as well gauged as possible led us to take into consideration all the physiological variations caused by its capture. Indeed, its use requires a length of time for it to become adapted to the test conditions (temperature, salinity, absence of sediment).

To have a test organism available all year round, we were led to use a reference toxic substance to judge the physiological state of the shrimp by thus reducing to a minimum, the effects on the measurement of its great seasonal sensitivity. A cationic surfactant, Noramium DA50 (alkyldimethylbenzylammonium chloride, CECA SA), was chosen for its total solubility in sea water at the test concentrations, its perfect stability, as well as its toxic action at weak doses. On each batch of shrimp caught, a static measurement was made of the lethal dose for 18 hours of the reference toxic (five concentrations, thirty shrimps per concentration). Any batch resulting in death outside of the limits set was then rejected.

Kuiper, J., Van den Brink, W.J. (eds), Fate and Effects of Oil in Marine Ecosystems. ISBN 90-247-3489-4

The measurement of the toxicity of chemically dispersed oil was performed in a tank filled with 16 liters of seawater. Comparative tests were performed, working either in a closed system or in a flow-through system. The closed system is the one used by MAFF in Great Britain (central stirring by a propeller). The flow-through system was tested with two energy variants. The first consisted in using MAFF type stirring. The second, which was more representative of actual sea stirring conditions, used a metal crown beating just underneath the surface of the water. In both cases, the pollutant concentration in the tank was reduced by the influx of unpolluted water. The technique used in the latter case was derived from the method of assessing the effectiveness of dispersants by the dynamic flow-through test (Bocard, 1984).

The advantage of using two tests with different energy lies in being able to compare the toxicities measured when oil dispersion is total (MAFF type stirring) or when it is only partial (IFP type stirring).

When dilution occurs, the variation in the pollutant concentration follows the following mathematical law :

$$x = x_0 \exp(-Dt) \text{ or } \mathrm{Ln}\ x/x_0 = -Dt$$

with x = pollutant concentration at time t

with x_0 = pollutant concentration at time o

with D = rate of dilution

To compare the effect of dilution in relation to the test in a closed medium, the pollutant concentration chosen in this case, called "concentration equivalent to half-exposure" (x_m) is defined as :

$$\mathrm{Ln}\ x_m = \frac{\mathrm{Ln} xo + \mathrm{Ln}\ x}{2}$$

The tests were performed on shrimp. Toxicity is expressed in percentage of dead animals at the end of the experiment (6 h). A comparison of the mortality percentages is made by a test of χ^2.

The principal conclusions of this experiment are :

- A similarity of toxicities is found in the closed and flow-through systems. The lethal toxicity of an average and constant dose is just as great as a much larger initial dose that is reduced by time.

- No significant differences are observed between the mortality occurring in the flow-through system with the two types of stirring. For a dispersant that is not very effective, because the oil concentration at the bottom of the tank where the shrimp are kept (a sieve prevents them from rising to the surface), it can be supposed that toxicity can correspond to the finest oil droplets. The size of the droplets effectively seems to be a determinant factor in the toxicity since mortality can be seen to be all the greater as mean droplet diameter is small (at constant pollutant concentration).

- Dilution can be used to assess the behavior of the products responsible for the toxicity of an oil/dispersant mixture. In this way,the evolution in time of the nature of toxics can be determined (elimination of various constituents by leaching) or, on the contrary, a residual presence of toxic effects can be determined (fast and permanent fixation of some products).

REFERENCES

- Bardot C., Bocard C., Castaing G., Gatellier C. 1984. Proceedings from the 7th Artic Marine Oilspill Program Technical seminar, Environment Canada, Ottawa, pp 179-201.

- Bocard C., Castaing G., Gatellier C. 1984. ASTM Special Technical Publication, 840, pp 125-142.

THE USE OF TESTS IN A WAVE BASIN TO DEFINE DISPERSANT EFFECTIVENESS

H.M.Brown, N.M.To and R.H.Goodman
Esso Resources Canada Limited
Calgary,Canada

INTRODUCTION

The use of dispersant chemicals on accidental oil spills is now an acceptable clean-up option in many areas of Europe and North America. There are a variety of commercial dispersant products available but not all will work equally well under a given set of sea conditions or oil types. Some form of 'effectiveness' grading for each product is needed so that field personnel can quickly choose the dispersant most appropriate for use in a particular spill. Such grading is currently done by means of one of two or three common laboratory tests (1). Unfortunately, none of these tests very realistically reproduces the sea conditions normally encountered in accidental spills and the dispersant results expected from laboratory tests are rarely achieved in spills at sea (2). In order to resolve the discrepancy between laboratory tests and sea spill observations, tests have been carried out in a large wave basin. These have led to a clearer definition of effectiveness and to a proposed method of measuring effectiveness in experimental spills at sea.

DEFINING DISPERSANT EFFECTIVENESS

The measurements which must be made in an experimental spill to determine dispersant effectiveness, are often misunderstood. Those required will be evident from the first figure, which shows the effect of oil removal processes over time on the amount of oil remaining in a spill. Without the addition of dispersant chemicals, oil will be removed from the sea surface by a number of natural processes. Depending on the sea state and other factors, the removal will follow one of a family of curves as shown. When dispersant is added to the oil, the processes of oil incorporation into the water are greatly accelerated, causing the curve to dip toward the origin. Once the dispersant is consumed, natural removal again prevails. To determine dispersant effectiveness, one must measure the oil placed in the water by the dispersant alone, that is, the oil in the water at time t_D when the dispersant is used up, less the oil that would enter the water from natural processes during the same time.

$$\text{Effectiveness (\%)} = 100 - R_D$$

To determine the effectiveness, points along the curve after time t_D must be measured, so that an extrapolation to R_D can be made. At present there is no effective method of measuring the oil remaining on the surface in a spill at sea and R_D can not be determined directly. An alternative would be to measure the oil dispersed beneath the slick. The problems associated with this approach have been studied in wave basin experiments and are discussed below.

Kuiper, J., Van den Brink, W.J. (eds), Fate and Effects of Oil in Marine Ecosystems. ISBN 90-247-3489-4

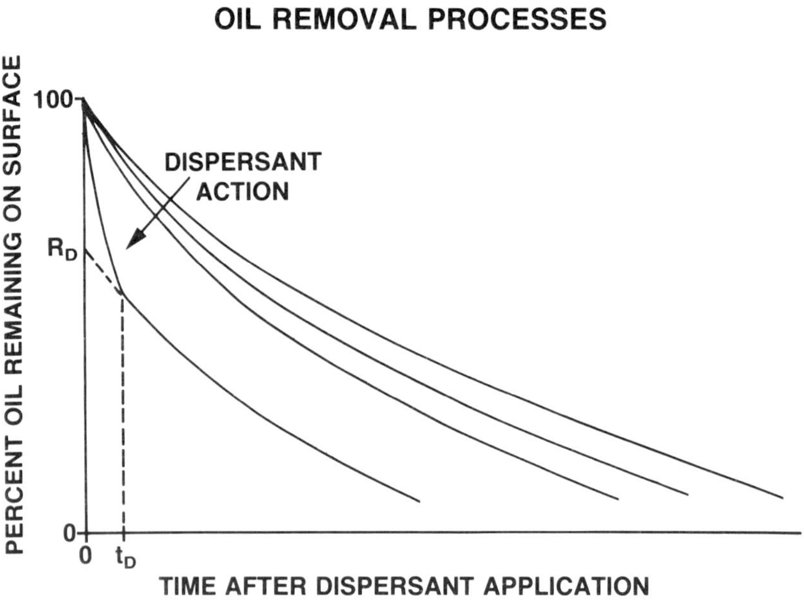

CALCULATION OF EFFECTIVENESS FROM WAVE BASIN DATA

During the past three years a number of dispersant experiments have been conducted in a large wave basin at Esso Resources Laboratory in Calgary, Canada. A description of the facility and the procedures used can be found in Brown et al.(3).

Dispersant effectiveness has been determined by measuring the concentration of dispersed oil under the slick as the experiment progresses. Integrating the concentration over time will yield the amount of oil flowing through the water column as a function of time. The second figure shows a typical concentration curve (time averaged to remove small fluctuations) measured in the wave basin below a slick. For the conditions of this experiment, the concentration normally peaks within a few minutes after dispersant application and then decreases to a constant value (C_D). Two processes appear to be operating. At first, the dispersant is actively introducing oil into the water column and the concentration increases, but as the chemical is consumed, the oil concentration drops to a constant value. This value is due to oil which is uniformly dispersed throughout the wave basin and is maintained in the water column by the wave energy. Oil is not being added to the water nor resurfacing from it. The time t_D at which the dispersant ceases to be active can be determined by extrapolating from the curve as shown. The total amount of oil removed from the surface via the water column is determined by an integration procedure which matches a diffusion model to the observed vertical oil concentration profiles. Details of the procedure can be found in reference 4. After correcting this amount for evaporative losses, the dispersant effectiveness can be immediately determined. Note that t_D, the dispersant activity time, is the same quantity as that shown in the diagram of oil removal processes.

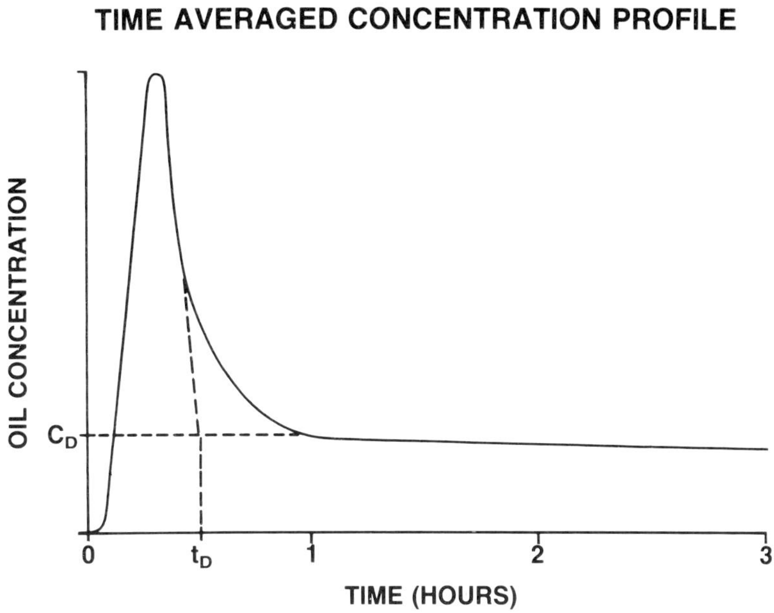

The wave basin experiments show that it is possible to determine dispersant effectiveness by using measurements of oil concentration from beneath the oil slick. A numerical model of the wave basin experiment has been developed to better understand the movement of oil through the water column. This has shown that the total oil in the water can be determined if concentration data as a function of depth anywhere below the slick is known (4). Thus in a sea spill it is not necessary to detect and measure all of the dispersed oil but only to measure a vertical profile of the oil concentration at some point under the slick. This information together with the area of the slick is sufficient to determine the total oil in the water at a given time. Several such measurements should enable one to define an oil removal line as shown on the first graph and then to extrapolate back to find R_D. The dispersant effectiveness follows immediately from the definition.

REFERENCES

1. Meeks DG: A Review of the Laboratory Testing and Assessment of Oil Spill Dispersant Efficiency. Proc. of the 1981 Oil Spill Conf. API, Washington, D.C.,1981, pp. 19-26.
2. Fingas M: The Effectiveness of Oil Spill Dispersants. Spill Technology Newsletter, July/December 1985, pp. 47-64.
3. Brown HM, Canevari GP, and Goodman RH: Dispersant Effectiveness in Cold Water. Proc. of the 8th Annual Arctic Marine Oil Spill Program, Edmonton, Canada 1985, pp. 245-259.
4. Brown HM, To NM, and Goodman RH: Experimental and Theoretical Basin Studies of Dispersant Effectiveness in Cold Water. Proc. of the 8th Annual Arctic Marine Oil Spill Program, Edmonton, Canada, 1986, pp. 635-651.

AN ASSESSMENT OF THE ENVIRONMENTAL HAZARD ASSOCIATED WITH THE DISCHARGE OF PRODUCTION WATER FROM A NORTH SEA OIL PLATFORM BASED ON LABORATORY BIOASSAYS WITH A CALANOID COPEPOD - *ACARTIA TONSA* (DANA)

A.E. GIRLING

SHELL RESEARCH LIMITED, SITTINGBOURNE RESEARCH CENTRE, SITTINGBOURNE, KENT

1. INTRODUCTION

The routine operation of offshore oil production platforms results in the discharge to the sea of water after it has been separated from oil produced from the reservoir. Recent attention has focussed on the need to assess the toxic hazard of this 'production water' to the marine environment. Shell UK Exploration and Production has initiated and implemented a research programme to assess the environmental hazard of discharges from its production platforms in the UK sector of the North Sea.

Acute and chronic laboratory toxicity tests on production water from Shell's Dunlin Alpha platform with the marine calanoid copepod *Acartia tonsa* have formed part of the hazard assessment programme. *A. tonsa* was selected as a test species in view of:-

i) its presence in the fauna of the North Sea.
ii) its ease of culture in the laboratory.
iii) the role of copepods as important links in the marine food web.
iv) the general susceptibility of calanoid copepods to hydrocarbon contamination. (Davies et al. 1980).
v) the availability of documented test methods for determining the acute toxicity of test substances to this species (EPA, 1976).

2. MATERIALS AND METHODS

Samples of production water from the Dunlin Alpha platform have been taken from the water phases in the 3rd stage bulk oil/gas and water separator. Sample pH was measured on collection and then adjusted to pH 4.0 to prevent the development of microbial populations. The samples were transported to the laboratory in sealed 11 litre steel International Air Transport Association cans internally coated with an Epikote resin. On arrival, the samples were transferred to glass aspirators, acidified further to pH 1.5 and then stored in the dark at 4.0°C prior to use.

Subsamples of the bulk samples were analysed to determine salinity, pH, total organic carbon, total hydrocarbons and the concentration of some aromatic hydrocarbons (benzene, toluene, xylene and naphthalene).

The acute toxicity of production water was determined by exposing groups of 10 animals to each of a logarithmic series of concentrations. Mortality was determined in each group at 24, 48, 72 or 96 hours. LC50 values - the concentration lethal to 50% of the exposed organisms - was calculated by probit statistical analysis or by graphical interpolation.

Chronic toxicity was assessed on the basis of a reproduction study in which adult populations in production water dosed test vessels were monitored over 20 days. Test vessels were stocked with adult *A. tonsa* at a density of 20/litre. Each vessel was supplied with synthetic sea water and an algal diet mix via a continuous flow-through system. Production water was added continuously to the pipes delivering the medium at rates required

Kuiper, J., Van den Brink, W.J. (eds), Fate and Effects of Oil in Marine Ecosystems. ISBN 90-247-3489-4

to produce the final desired range of test concentrations. Adult populations - animals older than approximately 14 days - were then monitored at intervals of 5 days to determine the numbers present.

3. RESULTS

Chemical analysis of the subsamples yielded the following typical values:-

pH 6.8 (Measured on collection at the platform)
Salinity 30‰
Total organic carbon 250 µg C ml^{-1}
Total hydrocarbons 30 µg ml^{-1}
Total aromatic hydrocarbons* - 10 µg ml^{-1}.
*(Benzene + Toluene + Xylene + Naphthalene)

The results of the tests show production water to be acutely toxic to A. tonsa at concentrations of approximately 2.0% - 5% (96 and 24 hour LC50 values respectively). Chronic effects on reproduction were only present at concentrations greater than approximately 0.3%.

4. CONCLUSIONS

The results of our laboratory bioassays indicate that production water shows no acute or chronic toxicity to A. tonsa at dilutions greater than 50 and 300 fold respectively.

Production water discharges from the Dunlin Alpha platform averaged 730 m^3 $hour^{-1}$ over a 6 month period in 1986. The behaviour of an effluent of the same density as sea water discharged under North Sea conditions has been modelled; mathematically using standard diffusion equations (Read and Blackman, 1980) and in laboratory flume experiments (Somerville et al, In preparation). The resulting estimates of dilution rates around platform discharge points range from 800 fold within 320 metres (mathematical model) and 1000 fold within 50 metres (flume model).

We can therefore conclude that the toxic effects on calanoid copepods of a production water discharge in the North Sea are likely to be restricted to a small zone close to the point of discharge and that such a discharge will not cause any significant changes at the population level. This conclusion is supported by the results of other experiments we have performed (not reported) in which we have shown biodegradation to considerably reduce the acute and chronic toxicity of production water to A. tonsa.

REFERENCES

Davies, J.M., Baird, I.E., Massie, L.C., Hay, S.J. and Ward, A.P. Some effects of oil-derived hydrocarbons on a pelagic food web from observations in an enclosed ecosystem and a consideration of their implications for monitoring.
Rapp. P-V Reun., Cons. Int. Explor. Mer, 179, 201-211, 1980.

Read, A.D. and Blackman, R.A.A. Oily water discharges from offshore North Sea installations: A perspective. Mar. Pollut. Bull. 11, 44-47, 1980.

US Environmental Protection Agency. Bioassay procedures for the ocean disposal permit program.
EPA-600/9-76-010, 1976.

Somerville, N.J., Wilkinson, T.G., Stephenson, R.R., Bennett, D., Watkinson, R.J., Davenport, J.N., Lynes, A., McCourt, B. and Mahieu A. The environmental effect of produced water from North Sea oil operations (In preparation).

THE EFFECTIVENESS OF A NEW TYPE III DISPERSANT IN THE TREATMENT OF WEATHERED CRUDE AND EMULSIFIED FUEL OILS ON SALTMARSHES AND SANDFLATS

D.I. LITTLE

Oil Pollution Research Unit, Field Studies Council,
Orielton Field Centre, Pembroke, Dyfed, U.K.

and

D. L. SCALES

Leonard Wills Field Centre, Field Studies Council,
Nettlecombe Court, Taunton, Somerset, U.K.

ABSTRACT: The aim of this experiment was to assess the effectiveness of BP Enersperse 1037 chemical dispersant on stranded lightly weathered crude oil and emulsified medium fuel oil (mousse). The oils were applied to field plots on contrasting shorelines (saltmarsh and sandflat) where both oils had previously been persistent (see Figures 1a and b). Type III dispersants were formulated for aerial spraying and are intended to be effective on weathered oils. The oil:dispersant ratio was 5:1.

a b

FIGURE 1. Photographs of experimental site at Steart Saltmarsh showing a) application of viscous water-in-oil emulsion (mousse), and b) spraying of lightly weathered crude oil.

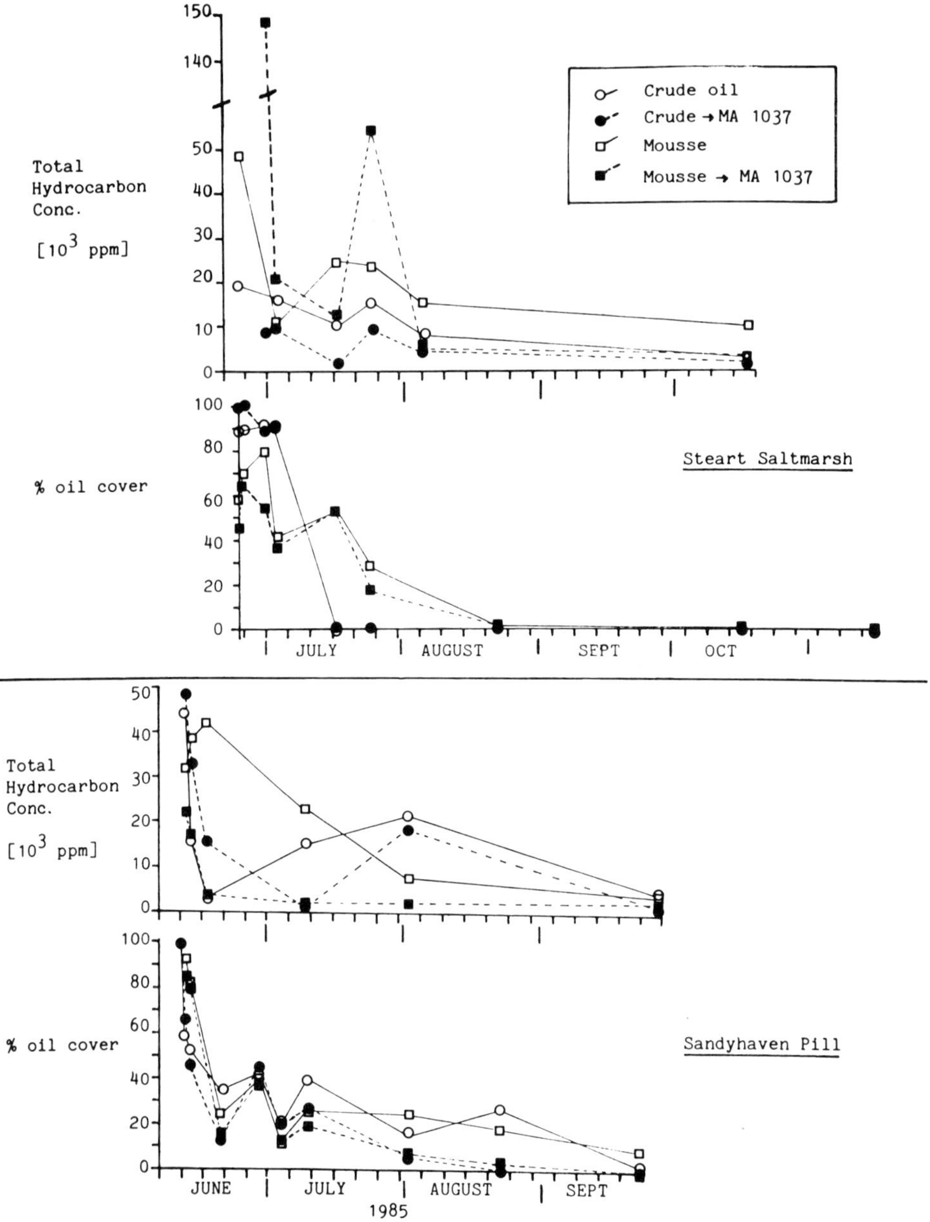

FIGURE 2. Total hydrocarbon concentrations (ppm dry mass) and % surface oil cover plotted against time at (top) Steart Saltmarsh and (bottom) Sandyhaven Pill. In spite of variability due to sediment and dead organic matter fluxes, dispersant use was effective on both types of oil.

On Steart saltmarsh, dispersant-treated oils were usually detected at lower sediment concentrations than were the untreated oils. Clear trends were not, however, visible until heavy rain removed the oiled, dead vegetation. Subsequent hydrocarbon concentrations in the crude-treated plots were reduced by half, and stable, lower concentrations were measured in the mousse plots (Figure 2). Although the dispersant-treated oils were both more damaging in the short term to the saltmarsh plant community, the longer term effects of untreated oil may prove more destabilising to the saltmarsh habitat as a whole. Untreated oil may be resuspended after erosion to form oil slicks, when public interest in the original spill and the sense of responsibility for its cleanup have all but disappeared.

On the sandflat at Sandyhaven Pill, the dispersant-treated mousse was detected at consistently lower concentrations than was the untreated mousse. The crude oil-treated plots showed more temporal variability than the mousse-treated plots. Dispersant-treated crude was ultimately detected at half the concentration found in the untreated crude plots (Figure 2). With the slightly higher energy conditions on the sandflat, dispersant use did not dramatically alter the fate of most of the oil. The ca. 10% which remained in the beach was visually less obvious and faster to disappear than was the untreated oil. This implies that if the cleanup authorities can resist amenity and conservationist pressures to disperse the remaining oil then such tidal flats will eventually self-clean.

Several explanations may be advanced for the effectiveness of the dispersant on both oil types. Firstly, it helped prevent the burial of oil layers and the formation of surface 'asphalt pavements'. Secondly, the contact time of oils and dispersant (>2 hours pre-tidal coverage) may have contributed to the reductions in oil concentrations. Thirdly, at both sites there was sufficient energy input to assist dispersion. In conclusion, whilst earlier formulations of chemical dispersant have often been ineffective on low energy sediment shores, Enersperse 1037 was effective on both shores. It remains to be seen whether dose rates and application techniques will be developed to retain this effectiveness and mitigate some of the toxicity.

IMPACT OF OIL ON COASTAL INDUSTRIES

H.D. Parker, I.C. White and T.H. Moller

International Tanker Owners Pollution Federation Limited

Introduction

The impact of oil pollution on the natural environment and seaside amenities is widely recognised but the threat of oil spills to coastal industries is probably less well known. Sheltered waters and estuaries not only provide safe locations for oil terminals but also provide favourable conditions for mariculture. Refineries, petrochemical plants and electricity generating stations using crude oil and its products as a feedstock or fuel, are also very often sited immediately adjacent to oil terminals. Both mariculture and many industrial installations need a continuous supply of uncontaminated sea water. This, taken together with the evidence of tanker spill statistics which indicates that most spills occur in port or port approaches, leads to the conclusion that there is a high potential for pollution damage close to oil ports. In this presentation, examples of some of the possible effects of oil spills are briefly described.

Mariculture

The effects of oil on marine life may be caused by its physical nature (physical contamination and smothering) or by chemical components of the oil (tainting and toxic effects). Organisms cultivated on the sea surface or in the intertidal zone are most at risk from physical contamination by floating oil. In other cases, plants and animals are suspended beneath floating structures and although protected from floating oil, may suffer tainting or toxic effects due to the uptake of soluble components or dispersed oil. The following examples illustrate these two aspects of the problem.

Nori production. Nori (Porphyra) is a seaweed with thin sheet-like fronds and whereas most seaweeds are protected by a mucous film that prevents oil sticking to the plant tissue, Nori appears to be oleophilic and even small quantities of oil are attracted to it. It is grown extensively in temperate regions of the Far East where it forms part of the staple diet and consequently its purity is carefully safeguarded. It is cultivated in shallow inshore waters on open mesh nets, typically 18 x 1.2 metres, held horizontal in the water by floats or bamboo stakes. The nets are artificially seeded and are set out in September/October. Harvesting begins in November/December and continues every 10-15 days during the winter months until the end of the growing season in March/April. As many as 4000 nets may be set along each kilometre of coastline and every net can produce between 50 and 200 kg Nori in a growing season. The harvested weed is washed, shredded and then compressed into thin sheets and dried with roughly 100 sheets, 200 x 220 mm, to a kilogram. At the end of the season the nets are removed from the floats and poles and stored until the following autumn.

Kuiper, J., Van den Brink, W.J. (eds), Fate and Effects of Oil in Marine Ecosystems. ISBN 90 247 3489 4

The location and intensive nature of Nori cultivation means that it is highly vulnerable to contamination by floating oil. Even if the plants themselves are not directly oiled, the structures on which they are grown may be affected and would have to be removed or replaced in order to avoid any chance of contamination from this source. In addition, because of the high quality demanded of this product, evidence of only a few small spots of oil on isolated fronds may be sufficient to render the crop from an affected area unmarketable. It is worth noting, however, that this threat does not exist all year; during the summer the equipment is in store.

Mussel cultivation. One method of mussel (*Mytilus*, *Perna perna*) cultivation commonly practised in many parts of the world is to grow them on vertical ropes suspended from floating structures. The mere presence of the floats and associated rope work encourages the natural settlement of larvae or spat. When the juveniles have grown to 100-200 mm they are collected from the ropes and floats and transferred to the vertical ropes where they are held in place by a sock of netting until re-established. They are harvested at 8-10 months and a yield of about 10 kg/metre of rope can be anticipated.

Although mussels and other organisms cultivated in this way are not directly exposed to floating oil, they may be contaminated by dispersed oil. Mussels are filter feeders and process large volumes of water from which food particles are extracted. In so doing, dispersed oil is also ingested and soluble components can accumulate in the flesh of the animal giving it an oily taste. If shellfish are to be served in the shell, clearly any trace of oil on the shell would be unacceptable and since several mussels are usually cooked together, one oiled shell can spoil all the others. Soluble components leaching from contaminated surface gear may also lead to tainting over prolonged periods.

Similar problems exist for produce stored in floating cages awaiting optimal market conditions and for salmonid rearing projects which are now widespread in Northern Europe, where the fish are maintained in enclosed nets below floating rafts.

Industrial water intakes

Water drawn from the sea for industrial purposes is usually passed through screens, filters and sediment traps to remove fish, weed and debris. The problem of oil contamination is rarely considered in the design of such intakes which are frequently only just below the low water mark so providing little protection from floating oil. As a result, in some designs fluid oils can become entrained in the flow, especially where flow rates are high. In addition, neutrally buoyant oils, such as oils which have incorporated sediment or heavy fuel oils in brackish estuarine waters, can be drawn directly into water intakes. The effect of such contamination is likely to vary but electricity generating stations and desalination plants are believed to be amongst the most vulnerable.

Electricity generating stations. The steam turbines used to drive electrical generators demand water flows of several thousand cubic metres per hour to condense steam exhausting from the turbine. Although this water is filtered, oil can be drawn into the condenser. Concern is often expressed that the heat transfer efficiency could be impaired or that the narrow bore tubes of the heat exchanger might become blocked by oil. A reduction in output or total shutdown might then be required while cleaning was carried out.

In a number of incidents, small quantities of oil have been drawn into power station cooling water intakes and although some oil was removed by the filters, the remainder passed through the heat exchangers without observable effect. It is assumed that due to the high temperatures experienced in the condenser, the oil was simply flushed through the system.

Desalination plants. In places where fresh water is scarce, desalination plants are often run in conjunction with electricity generation. While most of the water drawn from the sea is used for cooling purposes some is used to produce desalinated water for steam production and potable water supplies. Although little actual experience is available, there is some evidence to suggest that hydrocarbons carried over into the distillate impart an unpleasant taste to the water rendering it unfit to drink.

Conclusions

The effect of oil on marine organisms has been the subject of considerable research over the last twenty years and it is clear from the examples given here, that even low levels of contamination can have far reaching consequences for mariculture. The situation for industrial intakes is far less clear and concerns for the impact of oil remain largely conjectured. In view of the potential economic consequences, an understanding of the nature of such damage and identification of practical preventative and remedial measures are fruitful areas for further investigation.

THE EFFECTS OF OIL POLLUTION ON INTERACTING SALT MARSH SPECIES

M. SCHOLTEN (*) , P. LEENDERTSE and P.A. BLAAUW

Department of Ecology and Ecotoxicology, Biological Laboratory, Free University, P.O. Box 7161, 1007 MC Amsterdam, The Netherlands

(*)Present adress:Department of Biology Division of Technology for Society-TNO, Laboratory for Applied Marine Research,P.O. Box 57,1780 AB Den Helder, The Netherlands

1.INTRODUCTION

Salt marshes run a high risk of becoming polluted in case of an oil spill on the sea (e.g. 1,3,4,5). The effects of oil on the salt marsh vegetation can be grouped into:
a. Direct effects on plant growth, either through a reduction in gasexchange and light reception due to oil deposition on the leaves, or to physiological disturbances, effected by oil compounds taken up through stomata or roots.
b. Indirect effects on plant growth, either due to oil-mediated environmental changes,such as a stimulated N-fixation (6), soil anaerobiosis, shifts in interspecific competitive relationships, or to oil-induced changes in the level of tolerance of environmental stress factors.

2. MATERIALS AND METHODS

The effects of a topped North Sea crude oil (Forties) on salt marsh vegetation were tested in a greenhouse, using two experiments.
In both experiments the low salt marsh species <u>Halimione portulacoides</u>, <u>Puccinellia maritima</u>, <u>Spartina anglica</u> and <u>Triglochin maritima</u> were grown together in PVC-containers (ø 30 cm, depth 13.5 cm) on salt marsh soil with a salinity of 250 mmol NaCl . l^{-1} .
In order to discriminate between direct and indirect effects the oil was supplied in different ways:
a. In the first experiment (only mixed cultures) directly on the plant (P), on the soil surface (S), interstitially within the 5 cm top layer of the soil (I), or combined (ISP) . Oil supplies varied from 0 to 382 ml/m^2 soil surface.
b.In the second experiment directly on the plant (P), interstitially (I) or combined (IP) in mixed cultures. In mixed cultures the interstitial supply was performed at three nitrogen levels(low nitrogen (LN): 2.5 mg NO_3-N/week/container; medium nitrogen (MN) 13.4 mg NO_3-N/week/container and high nitrogen (HN): 80.0 mg NO_3-N/week/container) in order to detect any nutritive effect.
In order to detect the effects of oil on competitive interactions monocultures of the four species were tested as well, using interstitial oil (I) as well as a combined supply (IP). Oil supplies varied from 0 to 336 ml/m^2 soil surface.

3. RESULTS

3.1. <u>Effects on mixed cultures</u>

The first objective is to assess the effects of oil on the marsh vegetation. Oil causes a dose-dependent decrease in biomass production of the mixed cultures (fig. 1). This effect is mainly due to oil deposition on the plants themselves (fig. 2), indicating the relative importance of direct effects. Interstitial oil, or oil on the soil surface does not affect, or even stimulates the biomass production.
Growth stimulation by interstitial oil is probably due to a nutritive effect, indicated by its absence at a high soil nitrogen level (fig. 3). The nitrogen levels in the plant tissues, however, are not significantly increased by interstitial oil.

Kuiper, J., Van den Brink, W.J. (eds), Fate and Effects of Oil in Marine Ecosystems. ISBN 90-247-3489-4

3.2. Effects on separate species in mixed cultures

The second objective is to investigate the sensitivity to oil of the seperate species within the entire vegetation. When growing in a mixed culture, all the species suffer from a combined oil supply (ISP or IP), usually in a dose-dependent way; Puccinellia (1st exp.) and Triglochin (2nd exp.), however, don't respond at the lowest dose applied (fig.4).

Oil deposition on the plants themselves causes a decreased biomass production in all the species. Interstitial oil causes growth stimulation of Puccinellia (both experiments), Halimione and Spartina (1st exp.). Oil on the soil surface stimulates the growth of Puccinellia and Halimione in the 1st experiment (table1).

3.3 Effects on monocultures

When growing in monoculture, each species suffers from oil on the shoots as well as from interstitial oil, even at low levels, apart from Halimione , which shows a slightly stimulated growth in the latter case (table 2). The growth reduction from interstitial oil is not associated with decreased photosynthetic rates (table3). The differences in the reaction to oil between monospecific and mixed cultures indicate that interspecific interactions may influence oil effects on species.

3.4. Effects on competitive interactions

The effect of oil on the production of a particular species obviously depends on the presence of other species, especially in the case of Halimione and Puccinellia. A consistently negative correlation between the biomass production of these two species in the mixed cultures over all treatments provides evidence for a strong competitive interaction between them. Without oil Halimione seems to have a competitive advantage over Puccinellia . This advantage progressively declines with increasing oil supply and is even reversed in the case of interstitial oil only (fig.5).

Similar experiments with high marsh plants suggest comparable shifts in competitive relationships due to oil pollution giving Festuca rubra an advantage over Juncus gerardii and Artemisia maritima (Scholten & Leendertse unpublished).

As a consequence of the importance of plant interactions (Scholten & Blaauw in prep) in salt marsh vegetations, the responses to oil of plant species within a multispecies vegetation can not be predicted from monospecies experiments.

Reported vigorous growth of Puccinellia and Festuca following oil deposition (2,5) may in fact be caused by changed competitive relations.

Table 1: Biomass production of oil-treated plants compared to non-treated plants,(+) or (++) indicate slight resp. strong(more than 50%) stimulation, (-) or (--) slight resp. strong inhibition.

exp.	treatment	dose(ml/m^2)	Halimione	Puccinellia	Spartina	Triglochin
1	I	126	+	+	+	-
	S	126	++	++	-	--
	P	126	--	--	--	-
2	I	42	-	++	--	0
	I	84	-	+	-	-
	I	168	-	+	-	0
	I	336	--	0	-	-
	P	28	--	-	--	+

Table 2: As table 1, for monocultures 2nd exp.

treatment	dose(ml/m^2)	Halimione	Puccinellia	Spartina	Triglochin
IP	56	-	-	--	--
IP	112	--	--	--	--
I	42	+	-	-	-

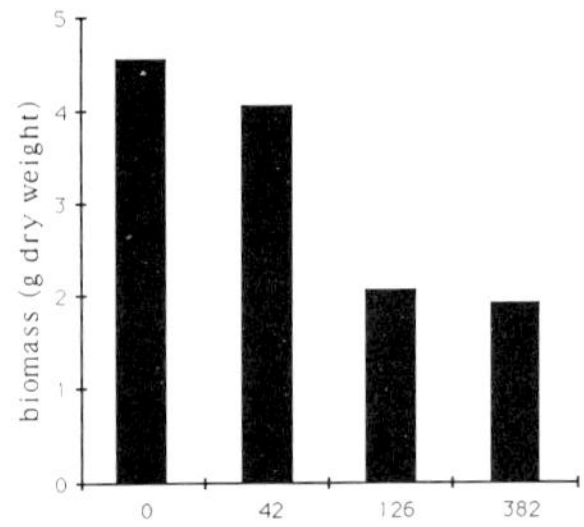

Fig.1a

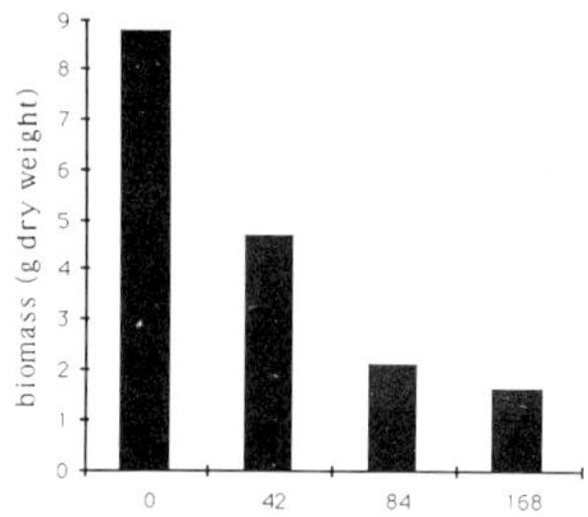

Fig.1b

Fig.1:The effect of combined oil application on total biomass production of a mixed vegetation in the 1st (a) and 2nd (b) experiment.

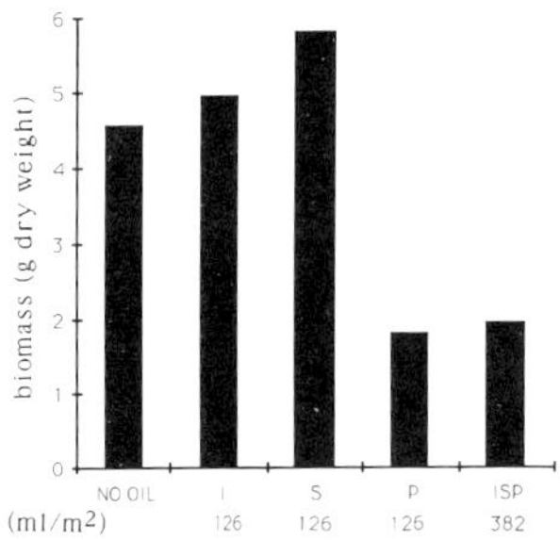

Fig.2a

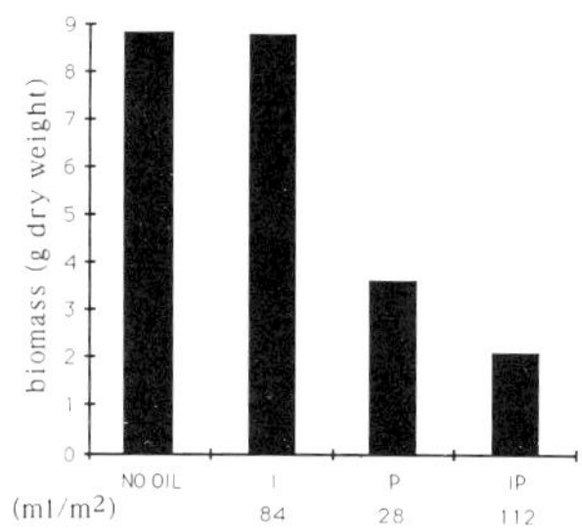

Fig.2b

Fig.2:The effect of different modes of oil application on the total biomass of a mixed vegetation in the 1st (a) and 2nd (b) experiment.

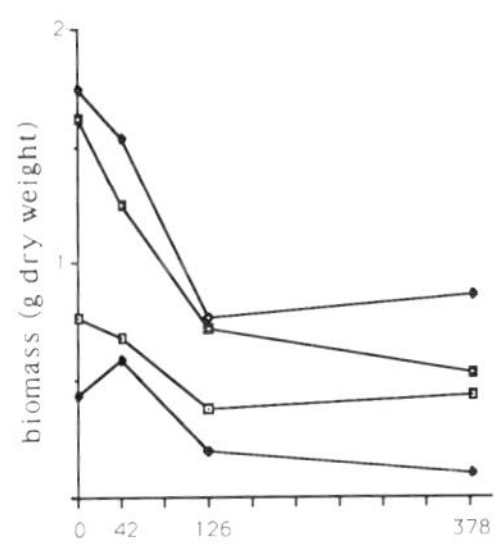

Fig.4a

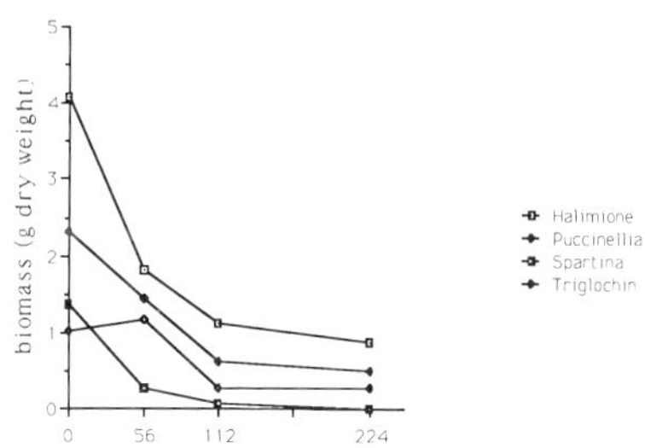

Fig.4b

Fig.4:The effect of combined oil application on the biomass of seperate species in a mixed vegetation in the 1st (a) and 2nd (b) experiment.

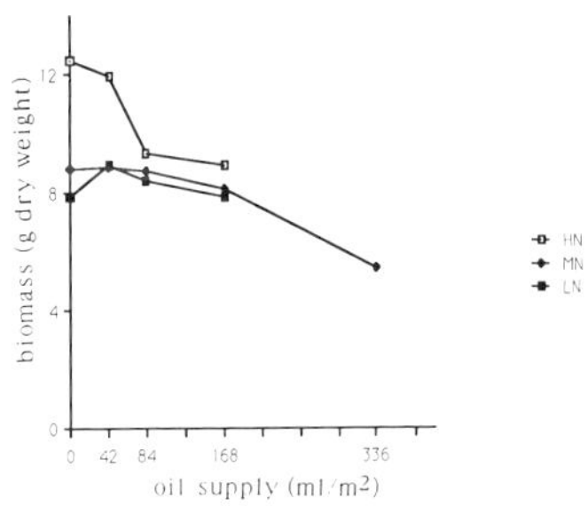

Fig.3:The effect of interstitial oil on a mixed vegetation at different nitrogen levels. LN: low nitrogen; MN: medium nitrogen; HN: high nitrogen.

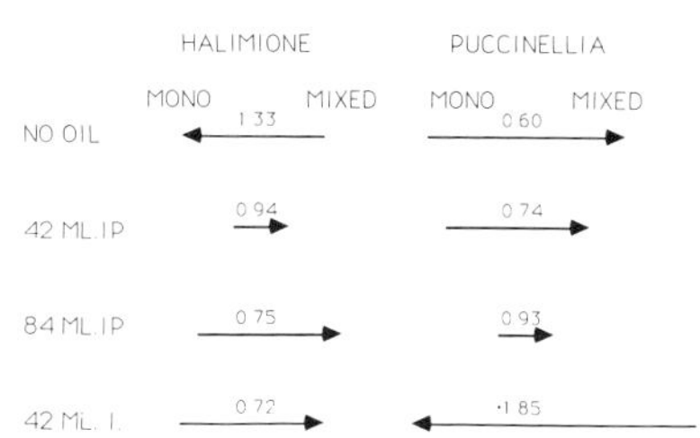

Fig.5:The differences between mono and mixed cultures in different treatments. The arrow indicates the direction of biomass change (to the left means mixed higher than mono, to the right mono higher than mixed). The figures above the arrow give the ratio mixed/mono.

Table 3: Net photynthesis rates (μmol CO_2/s/m^2 leaf area, n=4) of 3 species. Oil was applicated two weeks before measuring photosynthesis rates.

treatment	Puccinellia	Spartina	Triglochin
no oil	13.5 ± 2.6	13.5 ± 2.2	20.8 ± 3.0
56 ml/m^2 I	13.0 ± 2.8	14.7 ± 2.4	24.3 ± 4.4

4. CONCLUSIONS

-Oil pollution causes a dose-dependent negative effect on the production of a low salt marsh vegetation.
-Negative effects on growth are mainly due to deposition on the shoots.
-Growth reduction induced by interstitial oil is not associated with reduced rates of net photosynthesis.
-Lower concentrations of oil in the soil can stimulate the growth at low soil nitrogen levels.
-Interspecific interactions affect the reaction of plantspecies to oil pollution.
-Increased growth of Puccinellia following oil pollution may be due to a decreased competitive pressure from Halimione.

Acknowledgments: Thanks are due to our colleagues H. Schat and M. Stroetenga for comments and assistance.

5. REFERENCES

1. Baker JM: The effects of a single oil spillage. In: Cowell EB(ed): Ecological effects of oil pollution on littoral communities. Barking, UK. Applied Science Publishers, 1971.
2. Baker JM: Growth stimulation following oil pollution. In:Cowell EB(ed): ibid, 1971.
3. Géhu JM and Géhu-Franck J: Evolution des prés-salés nord amoricains sous l'impact de la marée noire. In: Conan G, Laubier L, Marchand M and d'O Zouville L(ed): Amoco Cadiz. Proc. Int. Symp. Centre Oceanologique de Bretagne, Brest, France,1979.
4. Hampson GR and Moul ET: No.2 fuel oil spill in Bourne, Massachusetts: immediate assesment of the effects on marine invertebrates and a 3-year study of growth and recovery of a salt marsh. J. Fish. Res. Board. Can 35:731-744,1978.
5. Stebbings RE: Recovery of salt marsh in Brittany sixteen months after heavy pollution by oil. Environ. Pollut. 1:163-167, 1970.
6. Thomson AD and Webb KL: The effects of chronic oil pollution on salt marsh nitrogen fixation. Estuaries 7: 2-11, 1984.

SECTION III

MONITORING OF OIL IN ABIOTIC COMPARTMENTS OF MARINE ECOSYSTEMS AND MONITORING OF BIOLOGICAL EFFECTS OF OIL SPILLS BOTH ON THE SPECIES AND THE ECOSYSTEM LEVEL

THE MIGRATION OF STRANDED OIL FROM A SHELTERED BEACH INTO ADJACENT NEARSHORE SUBTIDAL SEDIMENTS

E.H.OWENS, Geoscience Services, Aberdeen, Scotland, AB2 9JX, UK.
B.HUMPHREY, Seakem Oceanography, Sidney, British Columbia, V8L 3S1, Canada.
G.A.SERGY, Environment Canada, Edmonton, Alberta, T5K 2J5, Canada.

ABSTRACT

An experimental release of oil onto the water surface in a remote nearshore environment was allowed to strand on the adjacent beach. The intertidal and subtidal sediments were sampled over the following 4-year period to monitor the fate of the oil. Over this period the hydrocarbon concentrations in the nearshore subtidal sediments progressively increased as the volume of oil in the intertidal zone decreased. $15m^3$ of oil was initially released and $5.5m^3$ was recovered intentionally by skimming and an estimated $4.2m^3$ was lost immediately following the release to the atmosphere by evaporation and to the ocean by dispersion and dissolution. At the time of the last field survey an estimated $1.3m^3$ remained on the beach and $0.028m^3$ had accumulated in the subtidal sediments, with $4.0m^3$ having been removed from the shoreline and transported from the study area.

1. INTRODUCTION

The Baffin Island Oil Spill project (BIOS) involved a series of related studies that investigated the fate and effects of untreated and treated oil spills in a northern marine environment (17). One component of this multidisciplinary experiment had, as its objective, the short- and long-term monitoring of a nearshore release of 15 m^3 of aged medium crude (Lagomedio) oil. The release took place on 19 August 1981 and the oil was allowed to drift onto an adjacent gravel-cobble beach. In subsequent years, up to August 1985, the shoreline oil cover was measured and the beach and subtidal sediments were sampled and analysed for hydrocarbon composition and concentrations. This data set was used to determine quantitative changes in (a) the distribution of the stranded oil, and (b) the budget of oil on the shore (16).

Other studies associated with the BIOS experimental oilspill are reported elsewhere. The fate of oil, in terms of concentrations and composition changes, was monitored in four major environmental components: the water column (10); the intertidal beach sediments (13); the subtidal sediments (1); and the tissue of selected benthic invertebrates (9). Biodegradation of oil was monitored in the intertidal and subtidal sediments (7)(3). The effects of oil on bacterial numbers and micro-heterotrophic activity were monitored in the water column and subtidal sediments (2). Macrobiological studies monitored oil effects on the macrophytic algae (6), the benthic infauna (4) and the epibenthos (5).

The objectives of this paper are to present results from the experiment that demonstrate (i) the character of the chemical and physical changes in the stranded oil, and (ii) that oil was removed from the intertidal zone and became deposited in the adjacent nearshore subtidal sediments. A review of published and unpublished literature has not identified a consistent data set on the migration of oil from an intertidal to a subtidal environment. Most previous studies have

Kuiper, J., Van den Brink, W.J. (eds), Fate and Effects of Oil in Marine Ecosystems. ISBN 90-247-3489-4

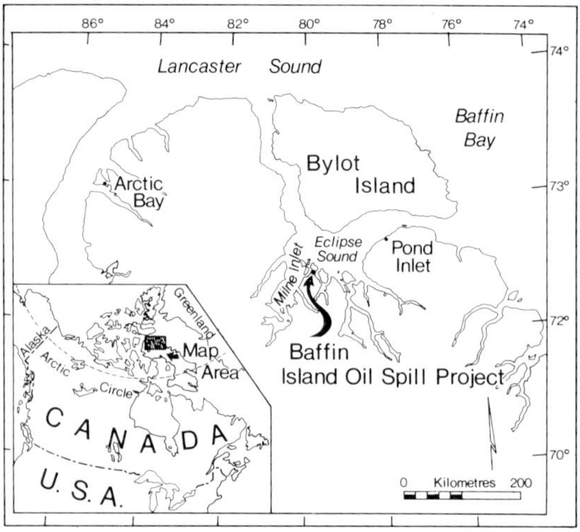

FIGURE 1 Location of Study Area

involved ad hoc sample collection following a spill and have not produced long-term sampling programs or have suffered from cross-contamination.

2. STUDY AREA

The study site is located on the eastern shore of Ragged Channel, Baffin Island in the Canadian Arctic (Fig. 1). The beach is exposed to waves with a fetch of less than 10 km and is subject to low wave-energy levels during the open-water season, which is an average of 65 days per year (late July - early October). Tides are semi-diurnal and range between 1.0m at neap and 2.0m at spring tide. The beach is defined by two bedrock outcrops about 400m apart and the intertidal width varies to a maximum of 50m. A gravel-cobble ridge characterizes the lower intertidal and gives way landward to a low trough of silt and sand. The upper intertidal zone has a sand-gravel beach face with a low cobble berm at the high-water mark. From the beach, the nearshore seabed slopes at about 4 degrees and the poorly sorted fine-grained sediments have a mean size of 3.4ϕ (approx. 0.075mm).

The intertidal areas are biologically barren, as is common in the Arctic. Conversely, the benthic subtidal productivity becomes significant in water depths greater than a few meters.

3. EXPERIMENTAL DESIGN

The BIOS Project was a large multi-disciplinary study conducted between 1980-83. The results in this paper are based upon an integration of data subsets from various BIOS component studies as well as additional studies at the site in 1985 (8)(15). The experimental design is therefore a composite extraction from those other studies.

Intertidal oil surface cover was determined by a series of ground mapping surveys carried out in 1981, 1982, 1983 and 1985. Changes in the composition and concentration of intertidal oil were determined from samples collected for hydrocarbon analysis systematically from the surface and subsurface of the upper, middle and lower one-third of the intertidal zone along pre-established profiles. Sampling took place in 1981, 1982, 1983 and 1985. In 1983 and 1985 samples were collected at additional

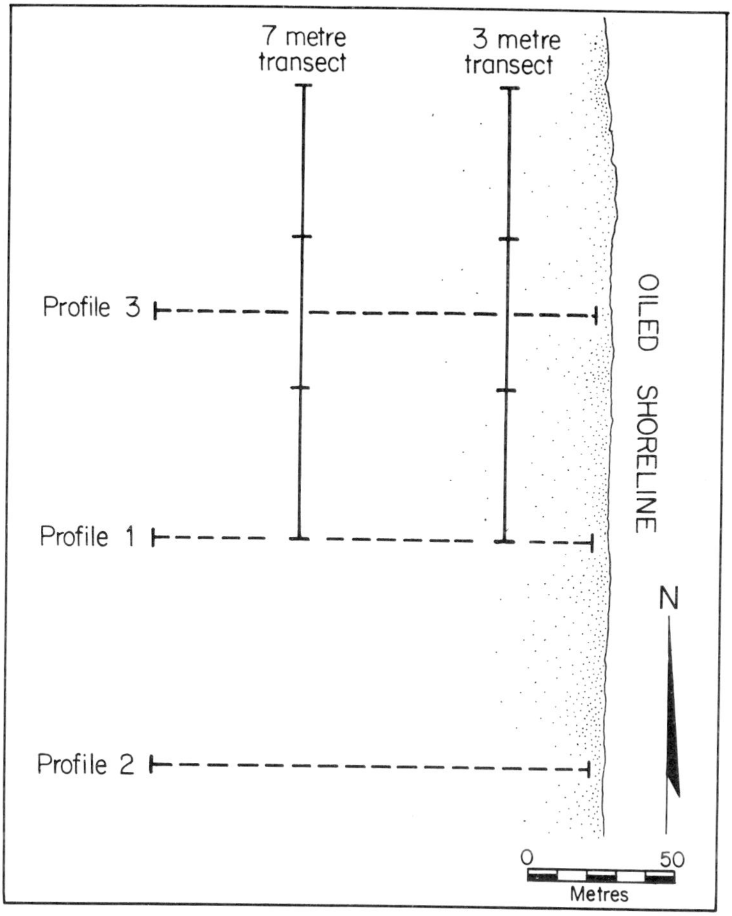

FIGURE 2 Location of Subtidal Sampling Transects and Profiles

locations identified as providing meaningful information, e.g. the asphalt pavement. Samples were analyzed by infra red spectroscopy (IR), gas chromatography with flame ionization detection (GC/FID), and with mass spectrometric detection (GC/MS)(8).

Subtidal sediments were sampled by SCUBA divers at depths up to 12m. The top 2cm were collected by gently scraping the jar along the substrate. In 1981, 1982, 1983 and 1985 samples were taken at predetermined random locations on two permanently marked 150m transects set on the 3m and the 7m isobaths (Fig.2). These transects were parallel to the shoreline at distances of about 50 and 100m respectively. Two single stations at 10m were also sampled during this period. Transects perpendicular to the shore were sampled systematically in 1983 and 1985 at depths from 0 to 12m. Subtidal profile 1 (Fig. 2) is located midway between beach profiles 4 and 5 (Fig. 3) and profile 3 is opposite beach profile 2. Sediments further offshore in deeper water were collected at six locations using a dredge. Subtidal sediment samples were analyzed by synchronous scan ultra-violet fluorescence (UV/F), by GC/FID, and GC/MS(8).

The surface slick was released on a rising tide and, under the influence of an onshore wind, the oil was naturally deposited on the beach as the tide receded. Control of the slick and the prevention of cross-contamination were accomplished by booming off the test area. To model the normal removal by tide, wind and currents, the oil remaining on the water surface after completion of the full tidal cycle was collected with skimmers. In terms of initial oil retention, it appeared that relatively stable conditions were reached after about 48 hours (four complete tidal cycles) when oil on the beach was no longer being refloated in large quantities. During this 48-hour period, $5.5m^3$ of oil were recovered from the water surface by skimming and $5.3m^3$ (about one-third of the original total) were left stranded on the beach. Evaporation and dissolution in seawater accounted for the balance.

4. INTERTIDAL STRANDED OIL

4.1 Surface Oil Cover

A set of 19 oil survey transects established in 1981 were resurveyed in 1982, 1983 and 1985 to provide data on changes in the surface oil cover (12). The results show that, with a total area of about $15,000m^2$ on this beach, initially $8,570m^2$ were contaminated and that almost half of the oiled area had a 75 to 100% oil cover (Table 1: Fig.3).

By mid-August 1982, a period equivalent to 3 open-water months, the total contaminated area had increased slightly due to spreading. The area with a 75 to 100% cover decreased to 1/3 of the original area in that category and the area with a 0.1 to 24% cover increased by a factor of 2.6 (Table 1). The action of physical shoreline processes over a further 3 month open-water period (to August 1983) reduced the total oiled area to $3925m^2$, which represents a reduction by more than half of the August 1982 area. Each of the oil cover categories shows a similar reduction over that time period.

The next survey, in August 1985, represents 12 open-water months after the release of the oil. The results show virtually no change in the oil cover from the previous survey in 1983. Over the entire study period the data show an initial small increase in the total oil cover over the first 3 month open-water period, followed by a major decrease over the next 3 months (a reduction of more than 50%), and by a period of relative stability over the next 6 open-water months.

A more accurate description of these changes is obtained if weighted values are used that integrate the per cent oil cover values with the area. Each oil distribution observation recorded the estimated surface oil cover (to the nearest 5%) over an area of $40m^2$: thus 9 observations of a 10% cover equates to an Equivalent Area of 100% Oil Cover of $36m^2$, and 5 observations of 80% would give $160m^2$. The Equivalent Area of 100% Cover for each year (Table 2) is obtained by summing the 21 individually calculated values. This figure changes from $4850m^2$ in 1981 to $1200m^2$ in 1985, a

TABLE 1 Intertidal Surface Oil Distribution.

	TOTAL AREA OILED	OIL DISTRIBUTION BY CLASS			
		0	0.1 to 24%	25 to 74%	75 to 100%
AUGUST 1981	$8,570m^3$	$6,200m^2$	$2,015m^2$	$2,845m^2$	$3,710m^2$
AUGUST 1982	9,600	5,170	5,200	3,095	1,305
AUGUST 1983	3,925	10,845	2,120	1,190	615
AUGUST 1985	4,440	10,330	1,830	2,100	510

TABLE 2 Changes in Equivalent Area of 100% Cover and Average Oil Cover

	Equivalent Area of 100% Oil Cover	Average Surface Oil Cover
AUGUST 1981	4,850m^2	57%
AUGUST 1982	3,282	34
AUGUST 1983	1,337	34
AUGUST 1985	1,200	27

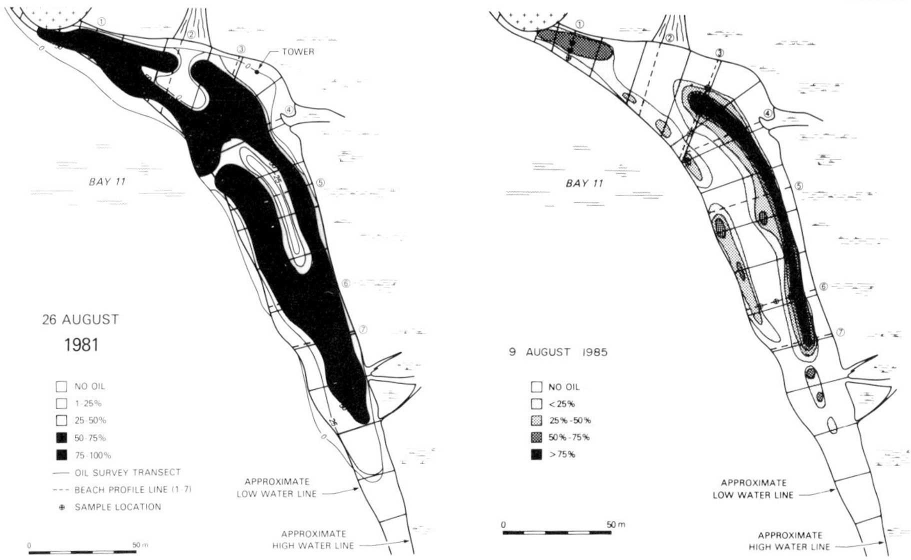

FIGURE 3 Surface Oil Distribution Maps (from (14)).

reduction to one quarter of the initial value. This parameter also shows a decrease to two-thirds the initial value between 1981 and 1982, whereas the total oiled area actually increased slightly over that period. The average surface oil cover (the Equivalent Area of 100% cover divided by the Total Area Oiled) also decreased from a value of 57% in 1981 to 34% in 1982 (Table 2). Subsequent values show no change in the average oil cover but a reduction by more than 50% in the Equivalent Area of 100% cover between 1982 and 1983, and only a relatively small reduction in both values between 1983 and 1985.

This data set indicates that approximately 3/4 of the oil that was stranded and initially remained on the shoreline was removed by natural shoreline processes over a period of about 12 open-water months.

4.2 Extractable Hydrocarbons

A series of surface and subsurface sediment samples were collected systematically on profiles 1,3, and 6 from the upper, middle and lower sections of the intertidal zone. These 9 samples were analysed for total extractable hydrocarbons (th) and the mean values for each sample set (Table 3) show that the initial high oil-in-sediment concentrations decreased but were nevertheless still relatively high at the time of the last sample collection in 1985. The increase in the mean values in the 1983 data set is attributed to the formation of an asphalt pavement between the 1982 and 1983 surveys (15) and to the fact that two of these systematic samples with high oil concentrations were collected on profiles 3 and 6

TABLE 3 Total Extractable Hydrocarbon Concentrations and Computed Oil Volumes. Mean value is based on number of samples in bracket (a = systematic sample set: b = "a" plus additional samples).

			Total Hydrocarbons Surface (mg/kg)		Subsurface (mg/kg)		Computed Surface Oil Volume
19 August	1981		17,400	(6)	-		$5.3m^3$
20 August	1981		7,100	(9)	170	(9)	2.1
28 August	1981		6,700	(9)	900	(9)	2.0
15 September	1981		5,900	(9)	200	(9)	1.8
10 August	1982		4,400	(9)	1,000	(9)	1.5
16 August	1983	a	11,870	(9)	1,300	(9)	
		b	8,150	(19)			1.1
9 August	1985	a	2,601	(9)	630	(9)	
		b	8,400	(23)			1.3

from the asphalt pavement. These two samples had concentrations of 27,000 and 58,000mg/kg respectively and thus biased the average values. Additional samples collected from the asphalt pavement had mean concentrations of 19,800 and 19,000mg/kg in 1983 and 1985 respectively (i.e. 2% oil-in-sediment by weight)(15).

Subsurface values (samples collected from the 5 to 10cm depth interval) were at least an order of magnitude lower than the surface samples (collected from the upper 2cm). The subsurface concentrations indicate that there has been an increased penetration of the oil into the beach sediments through time.

The most significant point, for this discussion, is that the oil-in-sediment concentrations remained high throughout the study period, and that concentrations were highest in the asphalt pavement.

4.3 Estimated Shoreline Surface Oil Volume

Several methods were developed to estimate the volume of oil and of contaminated sediments on the shore (13). The most accurate computation integrates the total hydrocarbon concentrations obtained from the sediment sample analyses with the surface oil cover data and with the observed depth of oil penetration. This method provides values of $5.3m^3$ of oil on the beach surface (Table 3) and $1.1m^3$ in the subsurface sediments on 19/20 August 1981. This value corresponds closely with deductive calculations that indicate a volume of $6.8m^3$ on 21 August 1981(13).

The computed volume of oil on the beach surface decreased rapidly over the initial one-month period up to 15 September 1981 (Table 3) and thereafter the total volume remained relatively constant. In 1985 an estimated $1.3m^3$ of oil remained in the intertidal zone. The small increase in the computed oil volume between 1983 and 1985 is considered to be an artifact of the method used to compute the volume, which was influenced by the slight increase in the total oiled area over that period. A significant proportion of the oil was contained in the asphalt pavement: $0.6m^3$ (55%) of the 1983 surface oil volume, and $0.35m^3$ (27%) of the volume in 1985 (15).

4.4 Weathering of the Stranded Oil

The analysis by GC/MS of samples collected in 1985 from the intertidal zone shows a distinction between those with high oil-in-sediment concentrations (>5000mg/kg) versus those with lower concentrations (Table 4). The high oil concentration samples, and particularly those from the asphalt pavement, are less weathered, independent of whether the sample was

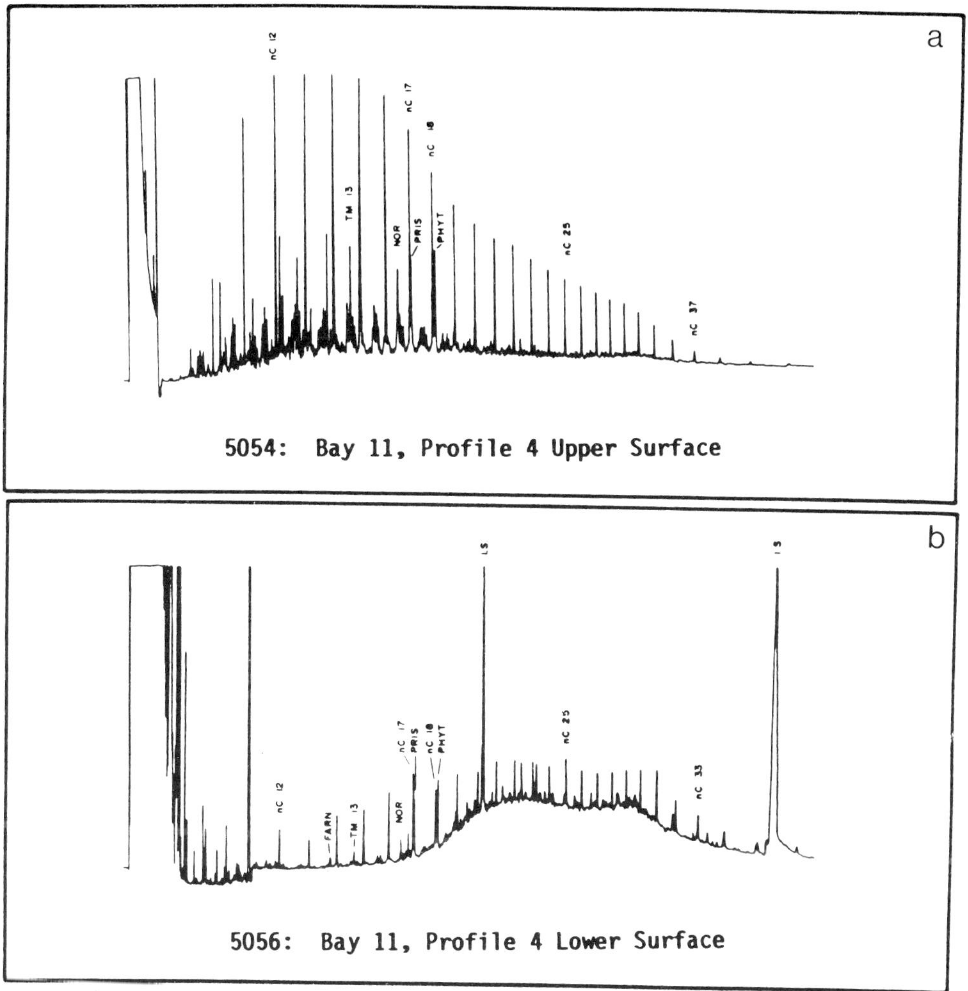

FIGURE 4 GC traces of beach sediment samples: (a) upper intertidal zone asphalt pavement (total hydrocarbon concentration - 6100 mg/kg); (b) lower intertidal zone (th 47 mg/kg). Oil was stranded on 19 August 1981, samples were collected on 9 August 1985.

TABLE 4 Weathering Indices: (a) 1985 Intertidal Sediment Samples - mean value of indices from low and high concentration samples; and (b) Original Lagomedio Crude Oil (indices are defined in (8))

th (mg/kg)	n	SHWR	ALK/ISO	$\frac{C_{18}}{PHYT}$	$\frac{PRIS}{PHYT}$	AWR
(a) < 5000	10	1.14	0.6	0.45	0.5	1.64
> 5000	6	1.58	0.7	1.22	0.65	2.18
(b) Aged Crude	-	2.28	2.50	1.61	0.74	3.47

taken from the surface or subsurface sediments. This feature is evident in all of the diagnostic weathering ratios and also in the chromatograms (Fig.4a), which show distinct, well-resolved alkanes from C_{10} up, with a low UCM (unresolved complex mixture). The lower concentration samples are more weathered (Table 4) and the chromatograms (Fig. 4b) are characterized by a significant UCM and the loss of alkanes relative to isoprenoids, although some alkanes can be resolved.

5. SUBTIDAL SEDIMENTS AND OIL

The macrobenthos investigations conducted over the five-year study period included the collection and analysis of subtidal sediment samples to provide data on sediment hydrocarbon concentrations for use as covariants in the biological statistical analyses. After the results from the second sampling of 1981 and from the 1982 sampling were available, the apparent increase in subtidal sediment hydrocarbon concentrations was sufficiently interesting that additional profiles perpendicular to the shore were sampled and analyzed in 1983 and 1985. The results of all four sample sets provide a picture of the movement of oil components from the oiled intertidal beach to the adjacent subtidal sediments.

5.1 Depth Transects

The transects used for the macrobenthic fauna studies were set parallel to the shoreline at 3 and 7m depths. Samples were collected on the seaward and landward sides of each of these two 150m-long transects for comparison to population studies and tissue hydrocarbon determinations. The hydrocarbons were determined primarily by synchronous scan fluorescence (8), an analytical technique which is sensitive to the aromatic nucleus of the components, but is not restricted to parent polycyclic aromatic hydrocarbons (PAH)(11).

As the variance of the analytical results was proportional to the mean value for the transect the statistical analyses were performed on log transformed data. Changes in sediment hydrocarbon concentration, with 95% confidence, occurred at both depths between August 1981 and September 1981, September 1981 and August 1982, and August 1982 and August 1983. At the 7m depth there was no significant change between 1983 and 1985 (Table 5).

TABLE 5 Subtidal sediment mean hydrocarbon concentrations (mg/kg) (figure in brackets = number of samples: n/s = no sample)

Date	Depth Transect 3 m	Depth Transect 7 m
Pre-spill 1981	0.2 (4)	0.5 (4)
AUGUST 1981	0.1 (7)	0.2 (7)
SEPTEMBER 1981	0.6 (9)	2.1 (20)
AUGUST 1982	3.2 (17)	5.1 (18)
AUGUST 1983	27.0 (18)	15.0 (18)
AUGUST 1985	n/s (-)	15.0 (34)

5.2 Perpendicular Profiles

The results of the 1982 analyses prompted the collection and analysis of additional subtidal sediments in 1983 (a) perpendicular to the beach at the south end of the experimental area (Fig. 2) and (b) from the deepest part of the bay. Some sediment cores from within the experimental area were also analyzed. The results of these analyses then prompted an expanded sampling program for 1985 in an attempt to determine the areal extent and the concentration contours of the contaminated sediment.

The central profile (Profile 1 in Fig.2) was sampled in 1983 and 1985, whereas Profiles 2 and 3 were sampled in 1985 only. The hydrocarbon concentrations of samples collected along these profiles indicate that the sediment had been contaminated by hydrocarbons from the beach, with oil-in-sediment concentrations up to 600mg/kg in the sediments immediately adjacent to the oiled intertidal area. The greatest impact was observed along Profile 1, at the south end of the oiled area, with a lesser impact at the center (Profile 2) and almost no impact 75m south of the oiled area. The sediment hydrocarbon concentrations decrease with distance from the low-tide line. As the permanent profile markers along Profile 1 were damaged between 1983 and 1985 no direct comparison is possible between these data sets.

5.3 Estimated Subtidal Oil Volume

Assuming an average hydrocarbon concentration of 15 mg/kg over the entire experimental area, and an impacted area of 37,500m^2 (based on the decrease of contamination with depth) the oil budget for the top 2cm of subtidal sediments accounts for only 0.028 m^3 of oil.

5.4 Composition of Subtidal Oil

The compositional analysis of the extracted hydrocarbons from the subtidal sediments, using GC/FID and GC/MS, indicates that the subtidal oil is very weathered. Indeed, very few of the compounds used to identify the oil remain, and many of the sediment hydrocarbon compositions resemble those of natural hydrocarbons. The UV/F patterns, however, remain similar to Lagomedio crude oil. The three-dimensional Total Luminescence pattern indicates that, although the parent PAH compounds may no longer be present, oxidation or metabolic products from them are there. These PAH derivatives are usually more polar, and may be more mobile, than the parent.

6. DISCUSSION

Data collected from the intertidal zone indicate that 4 years after the release of 15m^3 of oil onto the nearshore waters of the study site approximately 1.3m^3 remained on the adjacent shoreline. A significant proportion of this oil was contained in an asphalt pavement in which (a) oil concentrations were high (mean 2% by weight), and (b) weathering was considerably less advanced as compared to non-pavement oil samples.

Over the study period, which equates to approximately 12 months of open-water conditions on the shoreline, there was an initial rapid physical removal of about 3.8m^3 from the shoreline, between 1981 and 1982, that was accompanied by a spreading of the oil over previously uncontaminated sediments. This was followed, between 1982 and 1983, by a reduction in the oiled area and by a decrease in the volume of oil on the shore by about 0.4m^3. Apparently little oil was removed over the 6-month open-water period between 1983 and 1985.

The increase in the hydrocarbon concentration in the subtidal sediments is significant, but a subtidal oil budget estimate accounts for less than one percent of the oil which has been naturally removed from the beach during four open-water seasons (4.0m^3). The final disposition of the unaccounted 99% is probably in the water column. Thin sheens of oil have been observed coming off the beach, especially during periods of rain. The hydrocarbons in these sheens would be photolysed on the surface and also would be naturally dispersed into the water column.

The hydrocarbons found in the subtidal sediments are very weathered, retaining few characteristics of the original oil or the stranded oil. It is likely that oxidized and degraded components are preferentially removed from the stranded oil.

Although it is evident that oil has been removed from the beach and deposited in the adjacent nearshore subtidal sediments, the actual amounts found in the sediments account for only a very small fraction of the stranded oil that was removed. Evidently the majority of the 4.0m^3 of oil that was removed from the intertidal beach between 1981 and 1985 was transported away from the study area to more remote locations. The oil that has accumulated in the subtidal sediments of the study area is likely to persist for a considerable length of time as this is a low-energy subtidal environment not subject to rapid rates of sediment transport and redistribution.

The increases in the subtidal oil concentrations were greatest after September 1981 (Table 5). This indicates that the amounts of hydrocarbons deposited on the seabed immediately following the release of oil onto the water surface in August 1981 were relatively small compared to the subsequent, secondary, deposition of oil removed from the beach. The largest increase in concentrations on both depth transects occurred between August 1982 and August 1983. Between the 1983 and 1985 sample intervals the intertidal oil budget remained constant (Table 3), so that apparently little or no oil was removed from the beach, and over the same time interval no increase in subtidal oil-in-sediment concentrations was recorded on the 7m transect (Table 5).

7. CONCLUSIONS

(a) Three quarters of the oil that was stranded on the shoreline from a controlled nearshore spill in 1981 was removed by natural processes over the 12-month open-water study period.

(b) By 1985, approximately 4.0m^3 of oil had been removed from the beach. Of the 1.3m^3 that remained in the intertidal zone, 0.35m^3 was contained in an asphalt pavement in which the mean total hydrocarbon concentration was 19,000 mg/kg (2%) oil-in-sediment by weight.

(c) On intertidal sections where oil concentrations remained high in 1985 (greater than 5,000 mg/kg), the oil remained relatively unweathered.

(d) Oil accumulated in the nearshore subtidal sediments, with progressively higher concentrations recorded at each sample interval up to and including the 1983 sample set.

(e) Maximum subtidal oil-in-sediment concentrations were 600 mg/kg, adjacent to the oiled shoreline, and decreased seawards (with depth) and alongshore away from the contaminated beach section.

(f) The volume of oil in the subtidal sediments represents, (i) less than 0.2% of the original total of spilled oil, or (ii) less than 1% of the oil that was removed from the beach. The majority of this oil in the subtidal sediments is a result of secondary deposition following removal of oil from the beach, rather than from primary deposition immediately following the release of oil on the nearshore water surface.

(g) Compositional analyses indicate that the oil in the subtidal sediments is derived from the original batch of spilled oil.

(h) The contaminated beach remains a potential source for the release of further relatively unweathered hydrocarbons into the adjacent nearshore environment.

REFERENCES

1. Boehm, PD, Steinhauer, MS, Green, DR, Fowler, B, Humphrey, B, Feist, DL and Cretney, WS: Comparative Fate of Chemically Dispersed and Beached Crude Oil in Subtidal Sediments of the Arctic Nearshore. Arctic 40 (Supplement), 1987.

2. Bunch, JN, Effects of Petroleum Releases on Bacterial Numbers and Microheterotrophic Activity in the Water and Sediment of an Arctic Marine Ecosystem. Arctic, 40 (Supplement), 1987.
3. Bunch JN and Cartier, T: Microbiology 1: Effects of Petroleum Releases on the Microheterotrophic Flora of Arctic Sediments - 1983 Study Results. Baffin Island Oil Spill (BIOS) Project Working Report 83-5; Environmental Protection Service, Environment Canada, Ottawa. 44p, 1984.
4. Cross WE and Thomson, DH: Effects of Experimental Releases of Oil and Dispersed Oil on Arctic Nearshore Macrobenthos: I. Infauna. Arctic, 40 (Supplement), 1987.
5. Cross WE, Martin, CM and Thomson, DH: Effects of Experimental Releases of Oil and Dispersed Oil on Arctic Nearshore Macrobenthos: II Epibenthos. Arctic, 40 (Supplement), 1987
6. Cross WE, Wilce, RT and Fabijan, MF: Effects of Experimental Releases of Oil and Dispersed Oil on Arctic Nearshore Macrobenthos: III Macroalgae. Arctic, 40 (Supplement), 1987.
7. Eimhjellen, K and Josefson, K: Microbiology 2: Biodegredation of Stranded Oil - 1983 Results. Baffin Island Oil Spill (BIOS) Project Working Report 83-6; Environmental Protection Service, Environment Canada, Ottawa, 58p, 1984
8. Hope, D and Humphrey, B: Long Term Fate and Persistence of Stranded Oil at Cape Hatt, N.W.T. and from the "Amoco Cadiz" Spill - Part 2, Cape Hatt Sub-Tidal Sediments. Environment Canada, Unpub. Rept. No. EE-84, River Road Environmental Technology Centre, Ottawa, 35p, 1986.
9. Humphrey, B and Boehm, PD, Hamilton, MC and Norstrom, RJ: The Fate of Chemically Dispersed and Untreated Crude Oil in Arctic Benthic Biota. Arctic, 40 (Supplement), 1987.
10. Humphrey, B, Green, DR, Fowler, BR, Hope, D and Boehm, PD: The Fate of Oil in the Water Column Following Experimental Oil Spills in the Arctic Marine Nearshore. Arctic, 40 (Supplement), 1987.
11. Humphrey, B and Vandermeulen, JH: Characterization of Fifteen-Year-Old Stranded Oil. Proc. 9th (AMOP) Arctic Marine Oilspill Tech. Seminar, Edmonton, Conservation and Protection, Environment Canada, 29-38, 1986.
12. Owens, EH: Estimating and Quantifying Oil Contamination on the Shoreline. Marine Pollution Bulletin, (in press).
13. Owens, EH, Harper, JR, Robson, W and Boehm, PD: Fate and Persistence of Crude Oil Stranded on a Sheltered Beach. Arctic, 40 (Supplement), 1987.
14. Owens, EH, Hope, D and Humphrey B: Long Term Fate and Persistence of Stranded Oil at Cape Hatt, N.W.T. and from the "Amoco Cadiz" Spill - Part 1, Cape Hatt Shoreline Study. Environment Canada, Unpub. Rept. No.EE-84, River Road Environmental Technology Centre, Ottawa, 98p, 1986.
15. Owens, EH, Robson, W and Humphrey, B: Data on the Character of Asphalt Pavements. Proc. 9th (AMOP) Arctic Marine Oilspill Program) Technical Seminar, June 10-12, Edmonton, Alberta, 1-17, 1986.
16. Owens, EH, Robson, W, Humphrey, B, Hope, D, and Harper, JR: The Fate of Stranded Oil Four Years After an Experimental Spill on a Gravel Beach. Proc. 1987 Oil Spill Conf., Baltimore, Maryland, Amer. Petr. Instit., Washington, D.C., (in press), 1987.
17. Sergy, GA and Blackall, PJ: Design and Conclusions of the Baffin Island Oil Spill Project. Arctic, 40 (Supplement), 1987.

KINETICS OF AROMATIC HYDROCARBON DEPURATION BY OYSTERS IMPACTED BY THE *AMOCO CADIZ* OIL SPILL

D.S. PAGE*, J.C. FOSTER* AND E.S. GILFILLAN**

HYDROCARBON RESEARCH CENTER* AND MARINE RESEARCH LABORATORY**
BOWDOIN COLLEGE, BRUNSWICK, MAINE 04011 USA

1. INTRODUCTION

After oil spills, various species of filter-feeding bivalve molluscs have been shown to absorb petroleum hydrocarbons from the water column and suspended particulate matter. There have been numerous published studies in which hydrocarbon concentrations in oil exposed bivalves have been measured both in the field and in the laboratory (for reviews see 1,2,3). Given the commercial value of such species as clams, oysters, and mussels, it is important to understand the kinetics of oil depuration from bivalve tissues with the goal of possible salvage of impacted animals and to understand the basic physiological processes involved.

During the *Amoco Cadiz* oil spill of March, 1978, in Brittany, France, large quantities of commercial oyster stocks were impacted by crude oil (4). Measurements of hydrocarbons in the tissues of oysters sampled from the heavily impacted estuaries, Aber Benoit and Aber Wrac'h, sampled between 3 and 16 months post-spill demonstrated that the predominant tissue hydrocarbon fractions underwent changes that reflected the overall oil weathering in the sediments (5,6). The changes in the hydrocarbon distributions in sediment samples from the spill zone due to biodegradation and other weathering factors were documented by Boehm, et al., (7). These workers observed an initial rapid loss of linear alkanes and light aromatics followed by the slower breakdown of branched alkanes, unresolved complex material, and alkylated 3-ring and higher polynuclear aromatics and dibenzothiophenes. Because of the possibility that the dibenzothiophenes depurated more slowly than other types of aromatic hydrocarbons (5), the present study was undertaken to measure the kinetics of hydrocarbon depuration in oysters from the spill zone that had reached a steady state with respect to environmental hydrocarbon exposure.

2. PROCEDURE

2.1. *Study plan*

A sample of approximately 500 live Japanese oysters (*Crassostrea gigas*) were sampled in June, 1979 from an oil-impacted location in the Aber Wrac'h, Finistere, France. These oysters had been placed in this intertidal area prior to the *Amoco Cadiz* oil spill of 1978. In addition, a surface sediment sample was taken from the same area. Reference oyster samples for hydrocarbon analysis only were taken at the following locations outside the spill zone: Roscanvel (Rade de Brest); Lezardrieux (Trieux River); Cancale (Baie du Mont St. Michel). The Aber Wrac'h oyster samples were returned live to the Bowdoin Marine Research Laboratory in Cundy's Harbor, Maine, USA, and allowed to depurate any hydrocarbons in the tissues for a total of 96 days. Duplicate samples of oysters were

Kuiper, J., Van den Brink, W.J. (eds), Fate and Effects of Oil in Marine Ecosystems. ISBN 90-247-3489-4

taken on a random basis for hydrocarbon analysis by liquid and gas chromotography at 0, 15, 26, 40, 54, 77, and 96 days after the onset of depuration. Analysis of the tissues for aliphatic and aromatic hydrocarbons showed progressive loss of hydrocarbons to background levels with a half-life of about 23 days (Page, unpublished data).

In 1982, the archived hydrocarbon fractions and total extractable residues from the depuration experiment, the surface sediment, and the reference oyster samples were re-analyzed for specific aromatic hydrocarbons by gas chromatography/mass spectrometry (GC/MS) in the specific ion mode. The concentrations of the monomethyl (C1-), dimethyl (C2-), and trimethyl (C3-) dibenzothiophene isomers and the C1-, C2-, and C3- anthracene/phenanthrene isomers were determined.

2.2. Analytical Methods

2.2.1. Extraction and liquid chromatography. Oyster tissue samples were extracted by refluxing a homogenized sample of 60-100 grams of shucked meats with 2 M KOH in methanol for three hours to extract and saponify the tissues. The cooled homogenate was centrifuged to remove solid matter and the supernatant was extracted with three portions of pentane. The solid residue was extracted with methylene chloride. The lipid extracts were combined, concentrated to a residue. All or part of the total extractable residue was separated into fractions by liquid chromotography using a silica/alumina column. The hexane eluate contianed the aliphatic fraction of hydrocarbons in the sample. Elution with 50/50 hexane/methylene chloride yielded an aromatic fraction of hydrocarbons containing any aromatic hydrocarbons present plus other substances that behave as aromatics in the liquid chromotography procedure. Each fraction was quantitated by weighing in tared containers on a microbalance. The sediment sample was extracted and analyzed using the procedures reported by Page, et al. (8).

2.2.2. Gas chromatographic methods. Gas chromatographic analysis of the aromatic fraction was performed by injecting a known aliquot of the sample on a Perkin-Elmer Sigma 3B GC (FID), equipped with an MS-41 injection accessory, using a 50 m x 0.32 mm ID fused silica capillary column coated with SE-52. The carrier gas was helium with a linear flow velocity of approximately 40 cm/sec at 80 C. The temperature program was 80 C for 1 min., then 4 C/min to 280 C/min, with a final hold at 280 C for 16 min. The output signal was integrated for each run using a Hewlett-Packard 3390A computing integrator.

Gas chromatographic-mass spectrometric (GC/MS) analyses of aromatic fractions were performed on a Hewlett-Packard 5992B operated in the selected ion monitoring (SIM) mode. A 15 m x 0.32 mm ID SE-52 fused silica capillary column was used with a SGE Universal Injector operated in the split mode. The gas chromatographic conditions were as above. The electron multiplier was operated at 2000 volts. Prior to analysis each sample was spiked with 3800 ng of D_{10}-anthracene per mg of aromatic residue weight. A known aliquot of the sample was then injected. During each run, the following selected ions were monitored, each one representing the molecular ion of one of the groups of compounds of interest: D_{10}-anthracene m/e 188; C_1-anthracenes & phenanthrenes m/e 192, C_1-dibenzothiophenes m/e 198; C_2-anthracenes & phenanthrenes m/e 206; C_2-anthracenes m/e 212, C_3-anthracenes & phenanthrenes m/e 220, C_3-dibenzothiophenes m/e 226. Dwell time for each ion monitored was 50 msec, for a total cycle time of .35 sec. Resolvable components were computer-integrated and the areas stored on disk. Retention times for the compounds of interest had been previously established by mass spectrometric peak identification of the aromatic fraction of a crude oil and by extracted ion current profiling (EICP) with the MS operated in the scanning mode (50-350 m/e, cycle time approximately .4 sec). Areas from the samples with retention times corresponding to those of the compounds of interest were summed and corrected for aliquot size. Parts per billion values are reported as D_{10}-anthracene equivalents by assigning to all summed peaks a response factor equal to that for D_{10}-anthracene.

3. RESULTS

3.1. GC/MS results. The results of the GC/MS analysis of the depuration samples and surface sediment sample are given in Table 1. The results of the analyses of the reference oyster samples are given in Table 2. Also given in both tables are the gravimetric results for the overall aromatic hydrocarbon fractions in the samples. Figure 1 shows the capillary gas chromatograms in an overlay fashion for the total aromatic hydrocarbon fractions in the oyster sampled at 0 days and 96 days.

TABLE 1. Results of GC/MS analyses of aromatic hydrocarbons in oysters from the *Amoco Cadiz* spill zone depurated in clean water. Results for the surface sediment sample are also given. All quantities are given on a dry sample weight basis. In the table, DBT=dibenzothiophene isomers, AP=anthracenes & phenanthrene isomers and nd=not detected.

Days Depurated	Total ppm Aromatics	Parts per billion relative to D10 Anthracene C1DBT	C2DBT	C3DBT	C1AP	C2AP	C3AP
0	725	23	1580	2860	nd	539	270
15	599	14	1060	4160	nd	125	1460
26	381	12	143	777	nd	31	253
40	318	8	43	303	nd	13	117
54	186	nd	21	122	nd	nd	140
77	71	4	11	11	14	25	5
96	46	3	8	6	7	4	7
Surface Sediment	1930	85	1930	3470	249	1040	606

TABLE 2. Results of GC/MS analyses of aromatic hydrocarbons in oysters from reference areas outside the *Amoco Cadiz* spill zone. All quantities are given on a dry tissue weight basis. In the table, DBT=dibenzothiophene isomers and AP=anthracenes & phenanthrene isomers.

Location	Total ppm Aromatics	Parts per billion relative to D10 Anthracene C1DBT	C2DBT	C3DBT	C1AP	C2AP	C3AP
Roscanvel*	33	294	120	91	39	39	22
Lezardrieux**	62	33	195	265	44	102	118
Cancale***	20	15	97	93	31	94	19

*Rade de Brest ** Trieux River *** Baie du Mont St. Michel

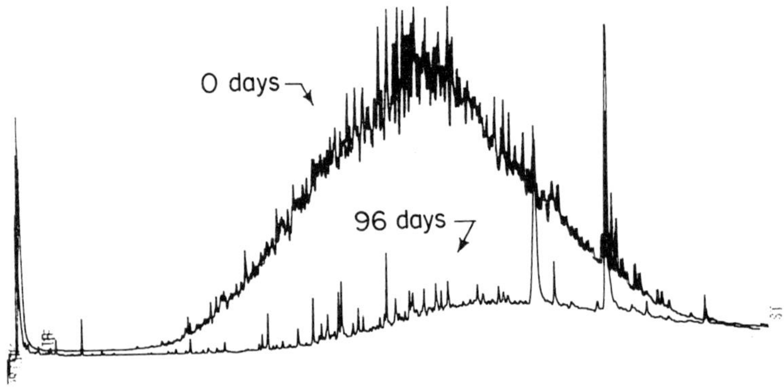

FIGURE 1. Capillary gas chromatograms of the aromatic hydrocarbon fraction from oysters sampled at the onset of depuration (0 days) and at the end of depuration (96 days)

3.2 Data analysis. The results in Table 1 were analyzed using two pharmacokinetic models: a one compartment single exponential release model and a two compartment double exponential release model in which depuration from one of the two compartments is slower than the other (9). The time series data were fitted to the exponential expressions using a non-linear least squares program running on an Apple IIe microcomputer (NLLSQ, CET Research). The two compartment model described the data better than the single compartment model, particularly for the rapid loss of hydrocarbon material in the initial period of depuration. The results of the computer fits are given in Table 3 for this study along with published depuration data treated in the same way for the sake of comparison.

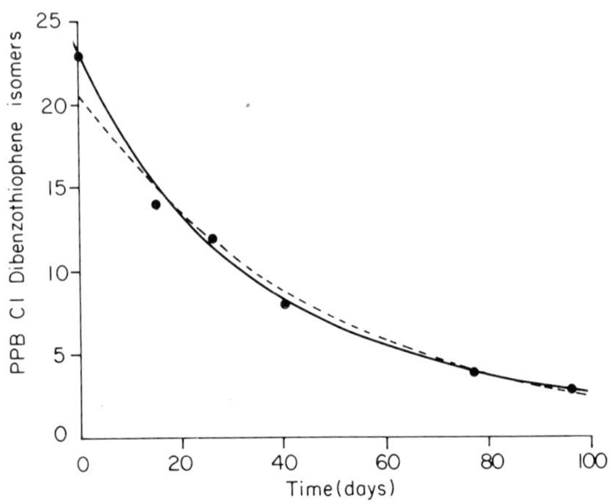

FIGURE 2. Computer generated graphs of best fits to the C1-dibenzothiophene depuration data given as the solid circles on the graph. The solid line is the double exponential fit and the dashed line the single exponential fit.

TABLE 3. Estimates of the parameters for the computer fit of depuration data to the expression:

$$\text{tissue concentration} = Ae^{-k1t} + Be^{-k2t}$$

where the values of A and B represent the initial concentration of hydrocarbon in the fast and slow depurating compartments respectively, where k1 and k2 are rate constants with the units $days^{-1}$ and where half-life = 0.693/k.

Reference	Substance	A	k1	Half Life (days)	B	k2	Half Life (Days)
This study	Total Aromatics	869 ppm	.030	23	-	-	-
	C1DBT	10 ppb	.050	14	13ppb	.016	43
	C2DBT	2000	.11	6.3	70	.012	30
	C3DBT	3800	.069	10	-	-	-
	C2AP	480	.14	5	65	.025	28
Michel (10) (Station 1)	Total Aromatics	281 ppm	.071	9.8	224 ppm	.011	63
Farrington, et al. (11)	C1-phenanthrene	1100 ppb	.17	4.1	890 ppb	.035	20

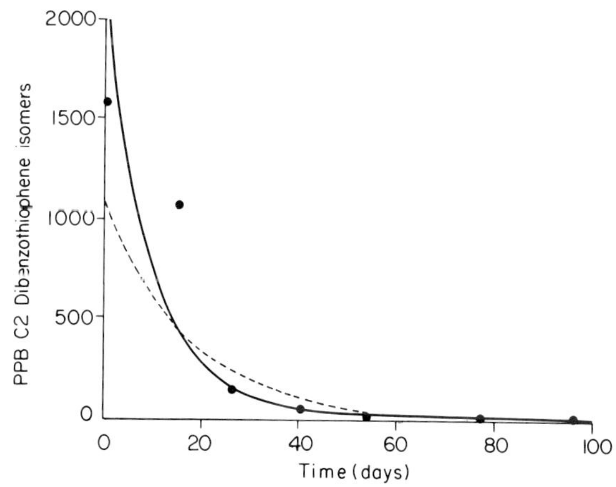

FIGURE 3. Computer generated graphs of best fits to the C2-dibenzothiophene depuration data given as the solid circles on the graph. The solid line is the double exponential fit and the dashed line the single exponential fit.

For the two sets of parameters in Table 3 for which no values were given, the amount of material in the slow depurating compartment was too small to allow estimation of the parameters. Figures 2, 3 and 4 show computer generated plots of the single exponential and double exponential fits of the C1DBT, C2DBT and C2AP data respectively.

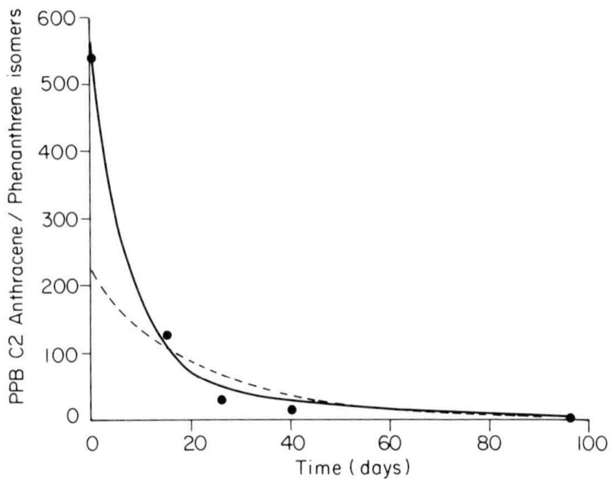

FIGURE 4. Computer generated graphs of best fits to the C2-anthracene/ phenanthrene depuration data given as the solid circles on the graph. The solid line is the double exponential fit and the dashed line the single exponential fit.

4. DISCUSSION

Figure 1 shows that the oysters progressed from an initial aromatic hydrocarbon distribution in which unresolved petrogenic material predominated to a distribution after 96 days of depuration in which biogenic resolved material was the predominant feature of the gas chromatogram. The results in Table 2 show that the oysters depurated in Maine had concentrations of the substances measured that were considerably below those found in oysters from reference areas in Brittany outside the spill zone. The presence of hydrocarbon pollutants in even small amounts in the water column and suspended particulate matter can limit the ability of these animals to depurate hydrocarbons from their tissues. The fact that the oysters in this study depurated to very low concentrations after a relatively short period of time is probably a function of the comparative absence of hydrocarbon pollutants in the water at the lab in Cundy's Harbor, Maine, and illustrates the importance of water quality in conducting these types of experiments.

Because of differences in the length of oil exposure, type of oil product, and environmental variables in the field or the laboratory, there are considerable differences in measured half-life values for the loss of hydrocarbons from animal tissues. This is illustrated in Table 4 which summarizes some depuration half-life values.

The kinetics of hydrocarbon depuration in filter feeding bivalves are dependent on a number of factors including length of oil exposure (1,5) and physiological well-being of the animal (22). This latter factor reflects the degree of stress imparted to the animal by the hydrocarbons present as well as other stressors, such as metals and other pollutants present in the depuration medium. The graphs shown in Figures 2, 3 and 4 clearly show that

TABLE 4. Some measured biological half-life values for the depuration of hydrocarbons in filter feeding bivalves.

Organism	Hydrocarbon Measured	Half-life (days)	Reference
Mussel	Benzo(a)pyrene	1.3	(12)
Oyster	Total hydrocarbons	5	(13)
Clam	Benzo(a)pyrene	2-5	(14)
Mussel	Total hydrocarbons	3.5	(15)
Mussel	Total aromatics	28-35	(16)
Mussel	Total hydrocarbons	2.7-3.5	(17)
Mussel	Benzo(a)pyrene	16	(18)
Clam	Total hydrocarbons	120	(19)
Clam	Total hydrocarbons	11-55	(20)
Mussel	Specific aromatics	6-12	(11)
Clam	Specific aromatics	5-7	(21)

the two compartment model fits the experimental hydrocarbon data much more closely than the single compartment model. The graphs also show the importance of early time period data in depuration experiments. Fitting such data to single exponential rate functions can lead to erroneous estimates of the actual rate parameters.

The biphasic nature of the kinetics of hydrocarbon depuration in filter feeding bivalves has been noted by a number of workers (10,11,13,15,17,19, 23). While the nature of the two compartments in these animals has not been clearly identified with actual organis within the animal, it has recently been suggested by Widdows, et al., (23) that the slower depuration process corresponds to the release of hydrocarbon from lipid-rich cellular structures within the animal. The extent to which the hydrocarbon material in the animal is distributed between the fast and slow depurating compartments is probably a function of exposure time, of the nature of the petroleum product absorbed, and of the physiological state of the organism. For the results igven in Table 3, the data of Michel (10) and Farrington, et al., (11) both pertain to in-situ depuration studies of animals impacted by #2 fuel oil (gas-oil). In both cases, the lighter refined product appeared to distribute itself more equally between the fast and slow depurating compartments in the mussels than the hydrocarbons derived from the heavily weathered Amoco Cadiz light Middle Eastern crude oil in the oysters sampled in this study. The smaller fraction of material in the slower depurating compartment in the Aber Wrac'h oysters may also reflect the poor physiological state of the animals where lower amounts of total body lipid would result in a smaller lipid compartment for absorbing hydrocarbons.

The data in Table 1 show that the distribution of aromatic hydrocarbons measured in the oyster tissues closely reflects the distribution of hydrocarbons in the surface sediment. Of the specific compounds analyzed, those present in the greatest concentration at the onset of depuration were the di- and trimethyl dibenzothiophene isomers (C2DBT and C3DBT). For these two groups of compounds, the results in Table 3 shows that the greatest proportion of the material present was in the rapidly depurating compartment. Moreover, there does not appear to be large differences between the rate of loss of the sulfur heteroaromatics and of the other polynuclear aromatics measured.

The half-life values for the depuration of the substances analyzed from the rapidly depurating compartment and the slowly depurating compartment in the present study are comparable with the values obtained from the analysis of the data of Michel (10) and Farrington, et al., (11). This suggests that the values for the rate of hydrocarbon depuration in different species of filter feeding bivalves may be similar for a given family of substances. The wide range of half-life values in Table 4 may therefore reflect the measurement of either the rapid process or the slow process. Comparison of the data in Table 3 with the half-life values in Table 4 indicates that shorter half-life values in Table 4 indicates that shorter half-life values in Table 4 probably pertain to depuration in animals with little material in the slowly depuration compartment. The large half-life values in Table 4 most likely pertain to animals in which the slowly depurating compartment contained the major fraction of the hydrocarbon body burden.

5. CONCLUSIONS

The major conclusions of this study may be summarized as follows:

- A population of Japanese oysters (Crassostrea gigas) from an oil-impacted estuary in the Amoco Cadiz oil spill zone that had reached a steady state with respect to environmental exposure to weathered crude oil was able to reach background levels of hydrocarbons during a depuration period of 96 days.
- The kinetics of this depuration process was biphasic in nature for certain specific aromatic compounds analyzed indicating a 2 compartment hydrocarbon release model. The rate parameters that were estimated for the data in this study were comparable to rate parameters estimated from published depuration studies.
- The animals depurated in Maine reached concentrations of specific aromatic compounds that were considerably below those measured in the tissues of oysters sampled from reference areas outside the Amoco Cadiz oil spill zone.

REFERENCES

1. Lee, R.F. Accumulation and Turnover of Petroleum Hydrocarbons in Marine Organisms. In D. Wolfe(ed) Fate and Effects of Petroleum Hydrocarbons in Marine Organisms and Ecosystems, pp. 60-70. Pergamon, New York, 1977.

2. Marchand, M. and Cabane, F. Hydrocarbures dans les Moules et les Huitres. Rev. Int. Oceanogr. Med. LIX pp. 3-30, 1980.

3. National Academy of Sciences. Oil in the sea, Inputs Fates and Effects. pp. 297-302. National Academy Press, Washington, D.C. (USA), 1985.

4. Maurin, C. Consequences de L'Accident de l'Amoco Cadiz sur des Resources Marines Vivantes Exploitables. pp. 667-677. In Amoco Cadiz; Consequences d'une pollution accidentelle par les hydrocarbures; Fates and Effects of the Oil Spill. Centre National Pour L-Exploitation des Oceans. Paris, 1981.

5. Friocourt, M.P., Gourmelin, Y., Berthou, F., Coosn, R., and Marchand, M. Effets de la Pollution de l'Amoco Cadiz sur l'Ostreiculture en Bretagne Nord: Suivi Chimique de la Pollution, de l'Epuration et de l'Adaptation. pp. 617-631. In Amoco Cadiz; Consequences d'une pollution accidentelle par les hydrocarbures, Fates and Effects of the Oil Spill, Centre National Pour L'Exploitation des Oceans. Paris, 1981.

6. Laseter, J.L., Lawler, G.C., Overton, E.B., Patel, J.R., Holmes, J.P., Shields, M.I., and Mayberry, M. Characterization of Aliphatic and Aromatic Hydrocarbons in Flat and Japanese Oysters and Adjacent Sediments Collected from l'Aber Wrac'h Following the Amoco Cadiz Oil Spill. pp. 633-644. In Amoco Cadiz, Consequences d'une pollution accidentelle par les hydrocarbures; Fates and Effects of the Oil Spill. Centre National Pour L'Exploitation des Oceans. Paris, 1981.

7. Boehm, P.D., Fiest, D.L., and Elskus, A. Comparative Weathering Patterns of Hydrocarbons from the Amoco Cadiz Oil Spill Observed at a Variety of Coastal Environments. pp. 159-173. In Amoco Cadiz, Consequences d'une pollution accidentelle par les hydrocarbures, Fates and Effects of the Oil Spill. Centre National Pour L'Exploitation des Oceans. Paris, 1981.

8. Page, D.S., Gilfillan, E.S., Foster, J.C., Hotham, J.R., Hanson, S.A., Vallas, D., and Gerber, R.P. Tidal Area Dispersant Project: Fate of Dispersed and Non-Dispersed Oil in Two Nearshore Test Spills. pp. 280-298. In: T.E. Allen(ed). Oil Spill Chemical Dispersants: Research, Experience and Recommendations, American Society for Testing Materials STP 840, Philadelphia, Pennsylvania (USA), 1984.

9. O'Flaherty, E.J. Toxicants and Drugs: Kinetics and Dynamics. Wiley-Interscience, New York, New York (USA), 1981.

10. Michel, P. Cintetique d'Epuration "in Situ" de Moules Contaminees par une Gas-Oil. Science et Peche. Bull. Inst. Peches Marit. (Nantes, France) 259 pp. 1-7, 1976.

11. Farrington, J.W., Davis, A.C., Frew, N.M., and Rabin, K.S. No. 2 Fuel Oil Compounds in *Mytilus edulis*, Retention and Release after an Oil Spill. Mar. Biol. 66 pp. 15-26, 1982.

12. Lee, R.F., Saurheber, R. and Benson, A.A. Petroleum Hydrocarbons: Uptake and Discharge by the Marine Mussel *Mytilus edulis*. Science 177 pp. 344-346, 1972.

13. Stegeman, J.J. and Teal, J.M. Accumulation, Release and Retention of Petroleum Hydrocarbons by the Oyster *Crassostrea virginica*. Mar. Biol. 22 pp. 37-44, 1973.

14. Neff, J.M. and Anderson, J.W. Accumulation, Release, and Distribution of Benzo(a)pyrene-C^{14} in the Clam Rangia cuneata. In: 1975 Oil Spill Conference Proceedings pp. 469-472. American Petroleum Institute, Washington, D.C. (USA), 1975.

15. Fossato, V.U. Elimination of Hydrocarbons by Mussels. Mar. Pollut. Bull. 6 pp. 7-10, 1975.

16. DiSalvo, L.H., Guard, H.E., and Hunter, L. Tissue Hydrocarbon Burden of Mussels as a Potential Monitor of Environmental Hydrocarbon Insult. Environ. Sci. Technol. 9 247-251, 1975.

17. Fossato, V.U. and Canzonier, W.J. Hydrocarbon Uptake and Loss by the Mussel Mytilus edulis. Mar. Biol. 36 243-250, 1976.

18. Dunn, B.P. and Stich, H.F. Release of the Carcinogen Benzo(a)pyrene from Environmentally Contaminated Mussels. Bull. Envir. Contam. Toxicol. 15 pp. 398-401, 1976.

19. Boehm, P.D. and Quinn, J.G. The Persistence of Chronically Accumulated Hydrocarbons in the Hard Shell Clam Mercenaria mercenaria, Mar. Biol. 44 pp. 227-233, 1977.

20. Stainken, D.M. The Accumulation and Depuration of No. 2 Fuel Oil by the Soft Shell Clam, Mya arenaria. In D. Wolfe(ed) Fate and Effects of Petroleum Hydrocarbons in Marine Organisms and Ecosystems. pp. 313-322. Pergamon, New York, 1977.

21. Gilfillan, E.S., Page, D.S., and Foster, J.C. Tidal Area Dispersant Project; Fate and Effects of Chemically Dispersed Oil in the Nearshore Benthic Environment. p. 111. American Petroleum Institute Publication No. 4440, 1986.

22. Widdows, J., Donkin, P. and Evans, S.V. Recovery of Mytilus edulis L. from Chronic Oil Exposure. Mar. Environ. Res. 17 pp. 250-253, 1985.

23. Widdows, J., Joore, J.L., Clarke, K.R. and Donkin, P. Uptake, Tissue Distribution and Elimination of (1-^{14}C) Naphthalene in the Mussel Mytilus edulis. Mar. Biol. 76 pp. 109-114, 1983.

BIOLOGICAL AND CHEMICAL MONITORING OF SEABED SEDIMENTS AT THE SULLOM VOE NORTH SEA TERMINAL - A SYNTHESIS OF DATA FOR THE PERIOD 1978-1985

S.S.C. WESTWOOD, D.I. LITTLE, N.M. DODD and S.E. HOWELLS

Oil Pollution Research Unit, Field Studies Council,
Orielton Field Centre, Pembroke, Dyfed, U.K.

ABSTRACT: An integrated seabed macrobenthic and chemical monitoring programme has been carried out in Sullom Voe under the direction of the Shetland Oil Terminal Environmental Advisory Group (SOTEAG) to assess the impact of the development and operation of a North Sea crude oil handling terminal. The data have been collected annually since 1978 and have defined the variability within the system for macrobenthos and sediment parameters, including grain size, organic content and hydrocarbon contamination. Sullom Voe supports a diverse fauna but localised changes have occurred around areas of chronic input. However, the environmental impact of the Sullom Voe terminal which handles approximately 45% of North Sea oil has been minimal. At Station 1 in the inner basin biological differences have been recorded since baseline studies were undertaken. These differences are attributed to the partial deoxygenation of deep water during many summers. In 1985, inner Voe aliphatic concentrations were much higher than previous years and results suggest resuspended 'Esso Bernicia' fuel oil could be contributing to elevated concentrations although there are several other minor sources to be considered. Communities have shown little change around the main loading jetties with the exception of a single station at the mouth of Garths Voe. Disturbed communities have been recorded in Garths Voe near the tug jetty since 1979. Movement of sediment, elevated hydrocarbon concentrations near the jetty and sewage coming into the Voe via three burns have probably all contributed to this disturbance. In Orka Voe, east of the production and ballast water diffuser, high densities of opportunistic species were recorded in 1983 and 1984 coupled with elevated hydrocarbon concentrations in the sediments. There are many possible sources of this contamination. In 1985 there was a substantial reduction of indicator species suggesting the communities are under less stress than previous years but this can only be confirmed in future surveys.

1. INTRODUCTION

The Shetland Islands, the most northerly of the British Isles, lie over 150 kilometres off the north coast of Scotland. With the discovery of Shell's Brent field 200 km northeast of Shetland in 1971, closely followed by several other fields including the Cormorant, Thistle and Ninian fields, it became apparent that a pipeline would be the most economic means of linking oil fields to the closest land area. Sullom Voe in Shetland was identified as the best prospective site for a large North Sea terminal and the Brent System and Ninian pipelines were established for the transportation of crude oil ashore. During the following years a terminal was constructed capable of handling 1.41 million barrels per day (Auther, 1981) and the first oil arrived near the end of 1978. In 1984

Kuiper, J., Van den Brink, W.J. (eds), Fate and Effects of Oil in Marine Ecosystems. ISBN 90-247-3489-4

the Sullom Voe terminal was handling approximately 45% of the 125 million tonnes of oil produced in the North Sea (Department of Energy, 1985).

Several organisations were established to deal with the building and environmental impact of such a large project. The Sullom Voe Association Ltd. (SVA), composed of representatives from the Shetland Council and more than thirty oil companies acting through the majors BP and Shell, was formed to organise and supervise the design, construction and management of the terminal. The Sullom Voe Environmental Advisory Group (SVEAG) was initiated in 1974 to advise on environmental matters and undertake baseline surveys. SVEAG was succeeded in 1977 by the committee presently responsible for the monitoring programmes in Sullom Voe, the Shetland Oil Terminal Environmental Advisory Group (SOTEAG), comprised of a wide range of members with experience of environmental matters (Foxton, 1981).

On the initiative of SVEAG, baseline ecological data were gathered on macrobenthos (Addy, 1979 and 1981; Pearson and Eleftheriou, 1981), rocky shores (Hiscock, 1981), salt marshes (Dalby, 1981), soft shores (Jones and Jones, 1981), birds (Richardson _et al_., 1981), hydrocarbons (Douglas _et al_., 1981; Davies, 1981), heavy metals (Coackley _et al_., 1981), sediment chemistry (Stanley _et al_., 1981) etc. which formed the basis of a broad-based monitoring programme that could be modified as necessary in later years. The wisdom of this cautious approach was upheld when on 30 th December, 1978 Sullom Voe suffered its first and most serious oil spill to date. The tanker 'Esso Bernicia' struck a mooring dolphin and punctured a fuel tank, releasing 1174 tons of bunker oil which spread by wind and tide throughout Sullom Voe and Yell Sound.

However, as monitoring continued in Sullom Voe, the need arose to change the original programme to cope with permitted chronic inputs associated with oil industry activity. This paper is concerned specifically with the findings of the 1985 macrobenthic and chemical monitoring surveys but more generally with data over the 1978-1985 period and describes how these programmes have altered over the years to better suit monitoring requirements.

The programme initiated in 1978 consisted of 12 stations (Figure 1) placed at 1000-2000 m intervals along the length of Sullom Voe. Although this strategy was appropriate for large scale incidents, in terms of chronic inputs, a different approach was needed. To this end, three more intensive programmes were established, in addition to the original programme, to monitor potential point sources of contamination: (1) A grid of twenty stations was introduced around the four main loading jetties where small spillages occasionally occur; (2) A line of 7 stations was initiated in Garths Voe off the jetty where the tugs are moored and (3) A transect of five stations was established in Orka Voe, a partially reclaimed inlet to the east of the production and ballast water diffuser which also receives site run-off.

Until 1983 chemical monitoring was a separate programme undertaken by the Scottish Marine Biological Association (SMBA). The macrobenthic and chemical programmes were first coordinated with the establishment of the jetty grid in 1983 and were amalgamated into a single programme in 1985 to include all the previously separate stations sampled for both macrobenthos and sediment chemistry.

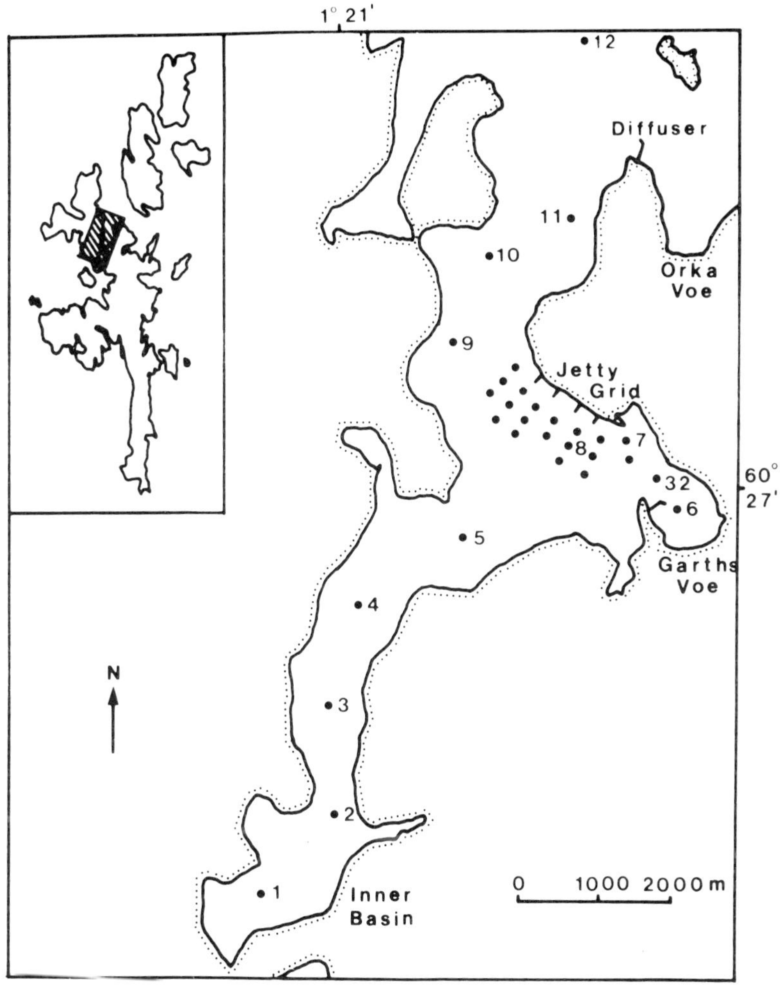

FIGURE 1. Location of orginal sampling stations and jetty sampling grid in Sullom Voe.

2. METHODS

Stations were located using lines of sight on fixed landmarks in conjunction with radar range and bearing to recognisable shore features. Quantitative macrofaunal samples were obtained using a Day grab. At all stations in the original sampling programme, five replicate biological grabs were taken. On the two transects in Garths Voe and Orka Voe, 3 replicate biological grabs were obtained at each station. On the jetty grid taking more than 2 grabs at each of the 20 stations was cost prohibitive. At all stations an extra grab sample was obtained for particle size analysis and percentage combustibles. By 1985, samples for sediment hydrocarbons, carbon and nitrogen and heavy metals were also obtained. In addition, oxygen, salinity and temperature profiles were recorded in the water column above all stations. Full details of field and laboratory methods are described in Westwood et al., 1986. All reports concerning monitoring in Sullom Voe, including full details of methods and the complete biological data, are available through the library of the Marine Biological Association of the United Kingdom at Plymouth.

3. RESULTS AND DISCUSSION

3.1. The original macrobenthic monitoring programme

During the years 1978-1983, the macrobenthic programme was composed of 12 stations throughout the Voe. Data from these years showed that Sullom Voe sustained a rich and varied fauna which remained largely unaffected by oil industry activity. Fluctuations in terms of numbers of species and individuals were noted from year to year and were partly attributed to the heterogeneity of the sediment. Sediments at Sullom Voe monitoring stations have ranged from fine silt to very coarse sand, moderately to poorly sorted. Also, due to the nature of the surrounding terrain, many of the sediments in the sheltered Voes were found to be heavily loaded with peat. However, even in areas of more uniform sediment, Buchanan et al., 1978 found that fluctuations in nearshore benthic communities was a common occurrence due to yearly differences in winter sea temperatures and differences in recruitment of larvae. Despite fluctuations in numbers of species and individuals, species composition at many stations in Sullom Voe has remained remarkably similar, although ranking of species may have changed from year to year (Woodman, 1984a). The regular occurrence of species would suggest that the basic communities are relatively stable. Data from the first few years revealed a natural range of variability that was a good basis against which to judge some subsequent changes. Because of the high degree of variation, trends were only analysed more closely for signs of pollution-induced changes if (1) they continued for more than two years, (2) changes occurring at a single station were not reflected at other stations (3) changes in faunal composition indicated that pollution stress may be having an effect on the community.

Station 1 in the inner basin of Sullom Voe has shown consistent biological differences from other stations since the beginning of the monitoring programme. Baseline studies (Dooley, 1981) and subsequent surveys have shown that the sediments and overlying water at this station are subjected to periods of partial deoxygenation due to the establishment of a thermocline at approximately 30 m depth in summer or autumn. Depletion of oxygen in the water column is accompanied by a rapid build up of sulphide in the water overlying the sediment surface (Stanley et al., 1981). Furthermore the inner basin is an area of high organic levels (Westwood et al., 1986) which encourages oxygen depletion (Rosenburg, 1980). Further stresses may be imposed on the inner basin environment by the enrichment of iron, chromium, copper, nickel, zinc and lead. At the junction of oxic and anoxic water masses, iron oxides may be precipitated by oxidation of soluble ferrous iron. These particles may adsorb humic substances, petrogenic hydrocarbons and trace metals (Tipping, 1984) and thus contribute to the sediment burden.

Fewer species have been found at this station in comparison to other areas of Sullom Voe and have been mainly limited to those found in estuaries and fjords where deoxygenation has occurred (e.g. Pectinaria koreni, Scalibregma inflatum and Abra alba, Rosenberg, 1980) or in the early successional phase when a marine system is subjected to high organic inputs (Pearson and Rosenberg, 1978; Pearson and Eleftheriou, 1981). These populations were considerably depleted or eliminated during summer months if a thermocline occurred.

Macrobenthic samples have not been taken at Station 1 since 1982. However, 1985 hydrocarbon data showed that aliphatic hydrocarbon concentrations were much higher than previous surveys (720 ppm in 1985, 549 ppm highest previous maximum). Although the highest concentration of petrogenic hydrocarbons were recorded at Station 1, it also showed the lowest proportion of petrogenic material relative to biogenic material. The source of the biogenic material is almost certainly the accumulation of peat from the surrounding land area. GC/MS analysis suggests that the petrogenic component is predominantly derived from Middle Eastern crude and may have resulted from resuspension of 'Esso Bernicia' fuel oil which contaminated shorelines in the area. However, other minor sources may include a diesel seep at Mavis Grind, spillages from the main loading jetties, road run-off or oils in domestic and industrial waste waters.

3.2. The jetty sample grid

Of the twenty stations sampled in 1983 (Figure 1), seven stations sampled in the 1976 baseline survey and the original sampling programme were incorporated into the jetty grid (full details in Woodman, 1984b). Faunal composition across the grid in 1983 showed reasonable uniformity, the two most dominant species, Mysella bidentata and Thyasira flexuosa, occurring at all stations. Sediment composition was responsible for the distribution of some species such as Turritella communis and Owenia fusiformis, which occurred in largest numbers in the slightly coarser sediment at the northwest end of the grid. Conversly Corbula gibba increased at the muddier southeast stations.

The most notable community changes around the jetties between 1976 and 1983 have been large increases in numbers of species and individuals. These trends have taken place at all stations except Station 32, and are attributed to natural variation which is characteristic of communities throughout the Voe. The community at Station 32, which is at the southeastern corner of the grid in the mouth of Garths Voe, had changed from one of the most diverse in 1976 to substantially less diverse than all other 19 stations in 1983. Sediment parameters remained similar between years and are unlikely to have been the cause. These changes may represent an extension of the disturbances recorded in Garths Voe (see below) into the jetty area.

Hydrocarbon analyses were carried out on sediments at all stations on the grid (see Stanley et al., 1984). Total alkanes and aromatics showed a significant correlation with percentange mud (Spearman rank, p = .001), i.e. both parameters increased with percentage mud towards the southeast corner of the jetty grid. Values for the unresolved complex mixture (ucm) followed the same distribution as total alkanes and aromatics. Studies in Milford Haven have shown that hydrocarbons are adsorbed more readily onto finer particles and tend to accumulate in muddy areas rather than in coarser sediment (Little et al., in press). This pattern was evident in the jetty area in Sullom Voe, suggesting that the distribution of hydrocarbons reflected sedimentological processes and the seabed topography of the area. Results also showed that n-alkanes were mainly of biogenic origin, predominantly from higher plant waxes. However, some weathered oil was also apparent as a ucm representing the high boiling petrogenic residues at most stations.

Biological data from the 1983 survey showed no link with hydrocarbon concentrations in terms of numbers of species, individuals or diversity except at Station 32. Reductions in numbers of species and diversity at Station 32 were noted compared with the large increase of those two parameters at all other stations. This station showed the highest total alkane (308 ppm) concentration of which 108 ppm was ucm. A linear regression carried out using arcsin-transformed percentage mud against total alkanes showed that most stations contained total alkane burdens consistent with the sediment type except at Station 32 where the total alkane value was anomalously high in relation to the percentage mud. Sources of contamination may arise from the main loading jetties as a result of sediment adsorption and subsequent transportation. An alternative explanation may be the position of Station 32 which is located at the mouth of Garths Voe, approximately 500 m northwest of Station 6. Station 6 has shown signs of contamination since 1979 and elevated hydrocarbon concentrations at Station 32 may be a result of disturbance and transportation of sediment from Garths Voe. Whatever the source of contamination there would appear to be some link between biological changes and elevated hydrocarbon concentrations.

3.3. Garths Voe

From the beginning of the monitoring programme in 1978, Garths Voe was traditionally monitored by sampling at one station (6) approximately 150 m away from the main tug jetty.

Sediment texture has changed considerably at this station from 1978 to 1985 (Table 1). Increases in percentage mud were apparent until 1981 and substantial decreases in this parameter have been recorded from 1982. Increased percentage mud in the early years may be a direct result of run-off due to building activity associated with the Sullom Voe terminal. In 1979, 'capitellids' (now identified as *Capitomastus minimus* and *Mediomastus fragilis*) and *Tubificoides benedeni*, which are indicators of pollution and disturbance (Pearson and Rosenburg, 1978; Dauvin, 1982) were first recorded in high abundances, although they had been absent in the 1978 samples. High numbers continued until 1983 when they virtually disappeared. However, sediment texture at this station had changed substantially in 1983, with a reduction in percentage mud from 58% to 34%.

TABLE 1. Percentage mud at Station 6 in Garths Voe from 1978–1985.

Year	1978	1979	1980	1981	1982	1983	1984	1985
Percentage mud	38.3	45.2	60.4	64.6	58.6	34.3	17.2	10.8

In order to establish whether this reflected a slight change in position rather than an actual decrease, a line of stations was established at 25 m intervals away from Station 6 in 1984 (Figure 2). Further changes to sediment were noted at Station 6 in 1984. Percentage mud was again reduced, from 34% to 17%. All the other stations on the transect showed fairly uniform mud levels at *ca.* 39%. Data showed that the three opportunistic species found at Station 6 from 1979–1982 were still present in large numbers in Garths Voe (ranging from 3180 m^{-2} to 980 m^{-2}) and that two showed a distinct gradient along the transect away from the tug jetty (Westwood, 1985). They were, however, still absent from Station 6 (Figure 3). One other species, *Apistobranchus tullbergi* was found in high abundances on the transect (*ca.* 3000 m^{-2}) but was patchily distributed. This species was the only one of the four dominant species to be recorded in substantial numbers at Station 6. *A. tullbergi* has not commonly been cited in the literature as indicative of enrichment, but Pearson (*pers. comm.*) has occasionally found it in enriched areas in the Clyde and Loch Eil. *Tubificoides benedeni*, *Capitomastus minimus* and *Mediomastus fragilis*, abundant along the transect but not found in any numbers at Station 6 are all deposit feeders and tend to be found in fine sediments (Pearson, 1971; Addy, 1979) which explains the absence of these species in high densities at this station. This, in combination with large numbers of *A. tullbergi*, is interpreted as a reflection of sediment type rather than an absence of organic enrichment at this station. The presence of a biological gradient away from the tug jetty suggests some form of disturbance or contamination in the vicinity of the jetty.

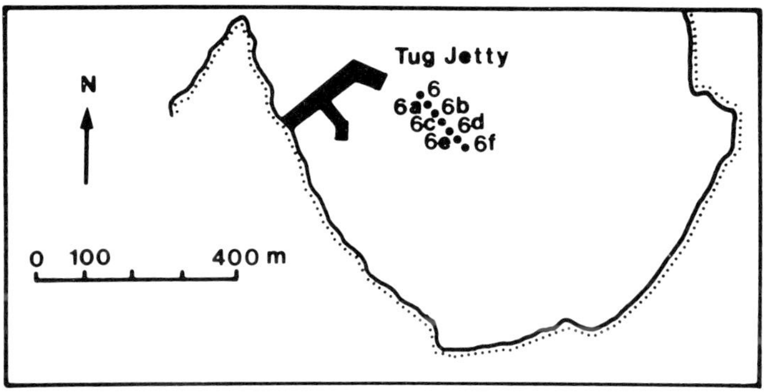

FIGURE 2. Location of sampling stations in Garths Voe.

Further changes were apparent on the transect in 1985 i.e. *Tubificoides benedeni*, and *Capitomastus minimus* were recorded in large numbers but *Mediomastus fragilis* had declined. The first order opportunist *Capitella capitata* (Pearson and Rosenburg, 1978) was recorded in high numbers in Garths Voe for the first time (1880 m^{-2}) A biological gradient was apparent on the transect but clearly the distribution of species had altered since 1984. In 1985 the peak of opportunists was recorded in the middle of the transect at Stations 6c and 6d (Figure 4). One of the major factors contributing to the change in distribution of species on the transect has undoubtedly been further change in sediment texture at these

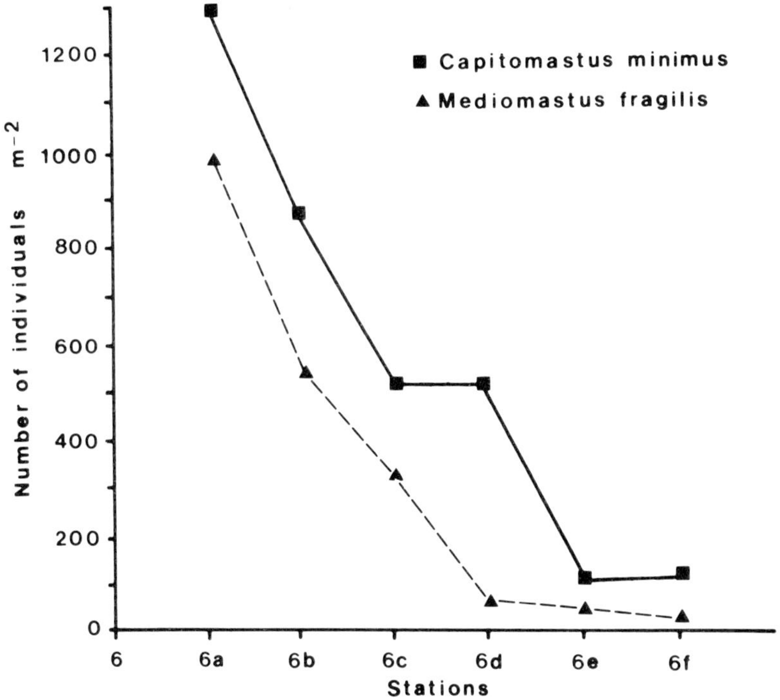

FIGURE 3. The distribution and abundance of opportunistic species in Garths Voe, 1984.

stations. A distinct sediment gradient could be discerned away from the tug jetty (Table 2). Percentage mud had again been reduced by half between 1984 and 1985 at Station 6 and by one third at 6a. Conversely percentage mud had increased at stations 6e and 6f.

Table 2. Percentage mud at stations on the transect in Garths Voe in 1985.

Station No.	6	6a	6b	6c	6d	6e	6f
Percentage mud	10.8	25.1	35.2	39.2	47.0	47.6	56.8

It is possible that tug movements in and out of Garths Voe have been contributing to some of the sediment changes, by redistributing sediments. Strong water movements from propellers could cause winnowing out of the fine material, and the observed losses of mud at some stations. However, large reductions in percentage mud only began in 1983 and the tug jetty has been in operation since 1978. Two new tugs, both Voith water tractors, went into service in 1983 which potentially have greater scope

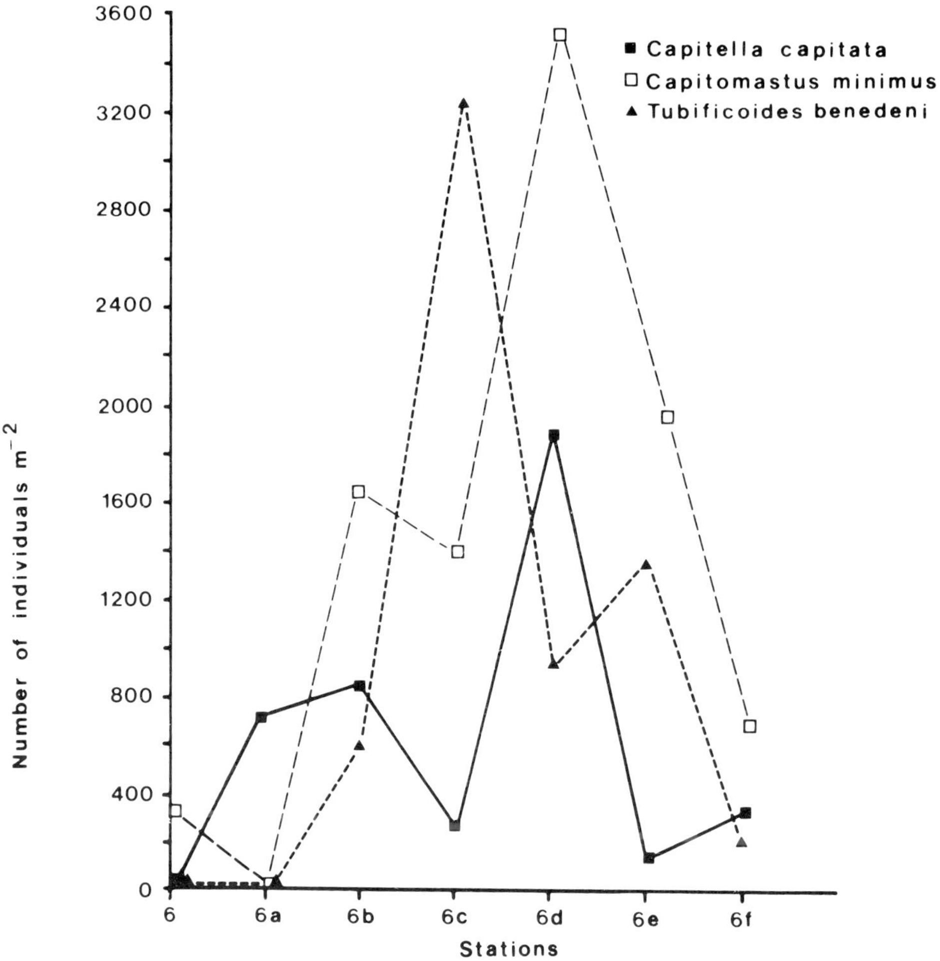

FIGURE 4. The distribution and abundance of opportunistic species in Garths Voe, 1985.

for sediment disturbance than the old twin-screw vessels (R. Gallagher, *pers. comm.*). However, these vessels went into service in July and September 1983 and the survey was carried out in June. The reduction in mud between 1982 and 1983 cannot therefore be explained in terms of disturbance by these vessels. Sediment disturbance by tug movement cannot be ruled out for 1984 and 1985 but data would tend to point to other causes, possibly exacerbated by tug movmement. It is possible that the jetties may have altered current patterns in the area and due to the distance of Station 6 from the jetty, the effects are only now being observed. A conclusive explanation is not available at this time, but whatever the cause, changes in species' distribution on the transect can nevertheless be linked to the changes in sediment texture. As

C. capitata, C. minimus and T. benedeni are all deposit-ingestors and tend to be found in fine sediments, the reduction of percentage mud over 3 years at stations nearest the jetty may be pushing opportunistic species further down the transect by limiting the habitat available for them to exploit.

Some studies (Eagle and Rees, 1973; Thistle, 1981) have shown that C. capitata has been found in high numbers as a response to purely physical disturbance. It is unlikely that this is the most influential factor in Garths Voe. Elevated numbers of 'capitellids' (C. minimus and M. fragilis) and T. benedeni have been recorded in Garths Voe since 1979 including a period of 3 years in which the sediment texture was very similar from year to year, suggesting limited physical disturbance. The appearance of C. capitella suggests continued and possibly increasing stress in the community. Petrogenic hydrocarbon concentrations in 1985 were elevated in Garths Voe, compared to other parts of Sullom Voe, in a pattern resembling the distribution of opportunistic species. The exception was at Station 6f where concentrations were anomalously high. This may have resulted from run-off from the burns at the head of the Voe (SOTEAG meeting, December 1985 pers. comm.). However, it must be pointed out that hydrocarbon samples are taken at one specific point and may be elevated by single events in time whereas the biological community data reflects the integrated conditions of both sediment composition and contamination over a much longer time span. Although a small proportion of the elevated hydrocarbon concentrations may be attributable to discharges from the main jetty area by adsorption of hydrocarbons to fine particles which are subsequently transported and deposited in Garths Voe, the pattern of petrogenic hydrocarbon concentrations on the transect suggests the source is more likely to be in the vicinity of the tug jetty. As with the distribution of the opportunistic species, hydrocarbon distribution is at least partially dependent on sediment distribution.

3.4. Orka Voe

Chemical monitoring in Orka Voe was first undertaken for SOTEAG by SMBA in 1980 at a single station which was found to be fairly heavily polluted by crude oil (Stanley and Pearson, 1981). In 1981 a series of 5 stations (Figure 5) were established, all of which showed signs of substantial oil pollution (Stanley and Pearson, 1982). Macrobenthic monitoring at these stations was initiated in 1983 and Scolelepis fuliginosa, Polydora sp. and Capitomastus minimus, species characteristic of organically enriched sediments (Pearson and Rosenburg, 1978), were recorded as dominant fauna at OV 2 and OV 5 (Stanley et al., 1984).

In 1984, Pearson et al. (1985) reported a gradient in species composition along the transect. Species associated with highly enriched sediments, Capitomastus minimus, Tubificoides benedeni and Polydora sp., were recorded at inner stations in conjunction with Abra nitida, a species known to predominate in moderately enriched areas (Pearson and Rosenburg, 1978; Pearson et al., 1985).

Enrichment opportunist species were recorded in low abundances on the transect in 1985 with the exception of Abra nitida, which occurred in substantial numbers at OV 1 (483 m^{-2}) and was also present on the list of numerically dominant fauna for OV 3 and OV 4 but in lower abundances. Large numbers of juvenile Abra spp. were also found at inner stations showing a similar pattern of distribution to that reported in 1984 (Westwood et al., 1986).

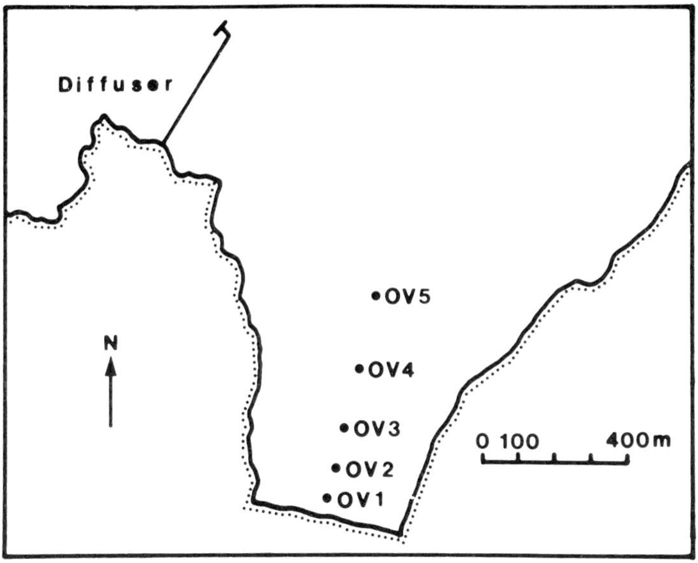

FIGURE 5. Location of sampling stations in Orka Voe.

Shannon-Wiener values in 1985 ranged from 3.22 at OV 1 to 4.92 at OV 5 and Pielou's evenness from 0.63 (OV 1) to 0.81 (OV 5) correlating negatively with percentage mud which was high at inner stations (83.56% at OV 1) but substantially reduced at outer stations (27.06% at OV 5). Similar correlations were established for hydrocarbon (273 ppm total hydrocarbons at OV 1 to 49 ppm at OV 5) and major and trace element data (Table 3).

TABLE 3. Concentration ranges of selected major and trace elements in 3 study areas, 1985.

Element	Sullom Voe	Garths Voe	Orka Voe
% aluminium	2.73-10.54	9.08-11.00	8.68-15.23
% iron	0.66- 4.57	2.39- 5.42	2.46- 6.59
% titanium	0.06- 0.43	0.28- 0.59	0.28- 0.68
ppm copper	13-32	24- 38	18- 40
ppm zinc	1-90	35-107	27-103
ppm lead	9-35	13- 26	16- 32
ppm chromium	14-82	56- 89	54-118
ppm nickel	31-58	36- 59	38- 71

Communities found in mixed sediments are generally more diverse than those found in muddy sediment (Gray, 1984; Yokoyoma and Hayashi, 1980). In an area where hydrocarbon, heavy metal and mud gradients occur together, it is difficult to assign the proportion of the reduced diversity due to each component. This is particularly so in Orka Voe where baseline data is lacking. However, the form of the biological changes are indicative of organic enrichment and the marked reduction in numbers of indicator species in 1985 implies that the communities were under less stress than in 1983 or 1984. However, due to the fluctuating life strategies of opportunistic species, this can only be confirmed by future surveys.

Elevated hydrocarbon concentrations have persisted in Orka Voe from 1981 to 1985. The lack of a clear pattern of weathering/degradation suggests either a chronic and continuous input or remobilisation of material from past input. GC/MS analysis suggests the oil is probably derived from a Middle Eastern crude and possibly represents 'Esso Bernicia' fuel oil that has been trapped in sediments and is being redistributed. There are insufficient data to rule out the diffuser as a source, although it does not appear to be the major source of contamination. Other possibilities include the Crooksetter Burn, which takes drainage water from the terminal to Orka Voe where elevated hydrocarbon values have been recorded (Westwood *et al*., 1986), or seepage from behind the bund or the holding basin.

Major and trace element concentrations were a close reflection of sediment texture. Trace element gradients are unlikely to have been influenced by resuspension of fine material and associated contaminants originating from the diffuser. Equally marked gradients in major elements implies that overall sediment composition at stations OV 1 to OV 3 has been altered by peat and morainic sediment disposal activities when Orka Voe was partially reclaimed. Concentrations of aluminium, iron and titanium were higher in Orka Voe than anywhere else in the survey area (Table 3) suggesting that bund-facing or landfill material was possibly derived from sites remote from Sullom Voe.

4. CONCLUSIONS

In general, Sullom Voe supports a rich and varied fauna. Some localised changes have taken place in areas where chronic input and physical disturbance has occurred. Considering the Sullom Voe terminal handles approximately 45% of North Sea oil, the environmental impact has been minimal.

Station 1 in the inner basin has shown biological differences from other stations since baseline studies were initiated. These differences are due to the partial deoxygenation of the deeper water which occurs in some years. In 1985 aliphatic hydrocarbon concentrations were much higher than in previous years in the inner basin and GC/MS results point to resuspended 'Esso Bernicia' fuel oil. However, there are other minor sources which may also contribute to elevated concentrations.

Communities around the main loading jetties appear largely undisturbed by oil industry activity with the exception of Station 32 at the mouth of Garths Voe. Evidence is not conclusive as to the source of elevated hydrocarbons in this area.

Communities in Garths Voe near the tug jetty have shown signs of disturbance since 1979. In recent years, sediments at stations nearest to the jetty have become coarser but evidence suggests tug movement is not the only cause. Elevated hydrocarbon concentrations have been recorded in Garths Voe, and the observed distribution points to a source in the vicinity of the tug jetty. However, sewage is also brought into this voe *via* 3 burns. The continued presence and changing distribution of indicator species in this area is probably the combination of several inter-related factors.

Elevated hydrocarbon concentrations have been found in sediments in Orka Voe, east of the production and ballast water diffuser. Possible sources of contamination include redistributed 'Esso Bernicia' oil, the Crooksetter Burn, seepage from behind the bund or the holding basin. The diffuser does not appear to be a major source of contamination. Major and trace element data suggests that bund-facing or landfill material may have been derived from areas remote from Sullom Voe. High densities of species indicative of organic enrichment were recorded in 1983 and 1984 in Orka Voe but a marked reduction of these indicator species in 1985 implies the communities are under less stress than previous years. This can only be confirmed by future surveys.

5. ACKNOWLEDGEMENTS

We are grateful to SOTEAG for commissioning the macrobenthic and chemical monitoring programmes as part of the integrated monitoring programme for Sullom Voe and to SVA for funding the work.

Our thanks are due to the staff of the Oil Pollution Research Unit who have helped with all stages of the project over the years and for their helpful comments while preparing this manuscript.

REFERENCES

1. Addy JM: Some studies of benthic communities in areas of oil industry interest. Ph.D. Thesis, University of Wales, 1979.
2. Addy JM: The macrobenthos of Sullom Voe. Proceedings of the Royal Society of Edinburgh 80B, 271-298, 1981.
3. Aurther E: The Terminal. BP Shield International, 12-17, July 1981.
4. Buchanan JB, Sheader M and Kingston PF: Sources of variability in the benthic macrofauna off the South Northumberland coast, 1971-1976. Journal of the Marine Biological Association UK 58, 191-209, 1978.
5. Coakley P, Backe DH and Smith PG: Some observations on the distribution of heavy metals in Sullom Voe. Proceedings of the Royal Society of Edinburgh 80B, 101-111, 1981.
6. Dalby DH: The salt marshes of Sullom Voe. Proceedings of the Royal Society of Edinburgh 80B, 191-202, 1981.
7. Dauvin JC: Impact of Amoco Cadiz Oil Spill on the muddy fine sand *Abra alba* and *Melinna palmata* community from the Bay of Morlaix. Estuarine and Coastal Shelf Science 14, 517-531, 1982.
8. Davies JM, Johnston R, Whittle KJ and Mackie PR: Origin and fate of hydrocarbons in Sullom Voe. Proceedings of the Royal Society of Edinburgh 80B, 135-154, 1981.
9. Department of Energy: Development of the oil and gas resources of the United Kingdom 1985. A report to Parliament by the Secretary of State for Energy, April 1985. HMSO, 82 pp, 1985.
10. Dooley HD: Oceanographic observations in Sullom Voe, Shetland in the period 1974-78. Proceedings of the Royal Society of Edinburgh 80B, 55-71, 1981.

11.Douglas AG, Hall PB, Bowler B and Williams PFV: Analysis of hydrocarbons in sediments as indicators of pollution. Proceedings of the Royal Society of Edinburgh 80B, 113-134, 1981.
12.Eagle RA and Rees EIS: Indicator Species - A Case for Caution. Marine Pollution Bulletin 4(2), pp 25, 1973.
13.Foxton P: The SOTEAG programme of monitoring in Sullom Voe. Proceedings of the Royal Society of Edinburgh 80B, 19-32, 1981.
14.Gray JS: Animal-sediment relationships. Oceanography Marine Biology Annual Review 12, 223-261, 1974.
15.Hiscock K: The rocky shore ecology of Sullom Voe. Proceedings of the Royal Society of Edinburgh 80B, 219-240, 1981.
16.Jones AM and Jones YM: The soft shore environment of Sullom Voe and the north mainland of Shetland. Proceedings of the Royal Society of Edinburgh 80B, 203-218, 1981.
17.Little DI, Howells SE, Abbiss TP and Rostron D: Some factors affecting the fate of estuarine sediment hydrocarbons and trace metals in Milford Haven 1978-82. In: Polutant transport and fate in ecosystems. PJ Coughtrey (ed) British Ecological Society Symposium 1-4 April 1985, in press.
18.Pearson TH: Studies on the Ecology of the Macrobenthic Fauna of Lochs Linnhe and Eil, west coast of Scotland, II. Analysis of macrobenthic fauna by comparison of feeding groups. Vie Milieu, Supplement No. 22(1), 53-91, 1971.
19.Pearson TH, Blackstock J and Douglas A: Chemical and Biological Monitoring Report for Sullom Voe, Orka Voe and Cat Firth, 1984. Report to the Shetland Oil Terminal Environmental Advisory Group, 1985.
20.Pearson TH and Eleftheriou A: The benthic ecology of Sullom Voe. Proceedings of the Royal Society of Edinburgh 80B, 241-269, 1981.
21.Pearson TH and Rosenburg R: Macrobenthic succession in relation to organic enrichment and pollution of the marine environment. Oceanography Marine Biology Annual Review 16, 229-311, 1978.
22.Richardson MG, Dunnet GM and Kinnear PK: Origin and fate of hydrocarbons in Sullom Voe. Proceedings of the Royal Society of Edinburgh 80B, 157-179, 1981.
23.Rosenburg R: Effect of Oxygen Deficiency on Benthic Macrofauna in Fjords. In: Freeland HJ, Farmer DM and Levings CD (eds) Fjord Oceanography. Plenum Press, New York, 499-514, 1980.
24.Stanley SO, Blackstock J, Douglas A, Pearson TH and Taylor J: Chemical and Biological Monitoring for Sullom Voe, Orka Voe and Cat Firth, 1983. Report to the Shetland Oil Terminal Environmental Advisory Group, 1984.
25.Stanley SO, Grantham BE, Leftley JW and Robertson Mrs N: Some aspects of the sediment chemistry of Sullom Voe, Shetland. Proceedings of the Royal Society of Edinburgh 80B, 91-100, 1981.
26.Stanley SO and Pearson TH: Chemical and Biological Monitoring Report of Sullom Voe and Yell Sound 1980. Report to the Shetland Oil Terminal Environmental Advisory Group, 1981.
27.Stanley SO and Pearson TH: Chemical and Biological Monitoring Report of Sullom Voe and Yell Sound 1981. Report to the Shetland Oil Terminal Environmental Advisory Group, 1982.
28.Thistle D: Natural Physical Disturbance and Communities of Marine Soft Bottoms. Marine Ecology Programme Series 6, 223-228, 1981.
29.Tipping E: Humic substances and the surface properties of iron oxides in freshwaters. In: Parker WR and Kinsman DJJ (eds) Transfer Processes in Cohesive Sediment Systems. Plenum Press, New York, 31-46, 1984.

30. Westwood SSC: Macrobenthic Monitoring in Sullom Voe, 1984. Report to the Shetland Oil Terminal Environmental Advisory Group, 1985.
31. Westwood SSC, Little DI and Dodd NM: Chemical and Biological Monitoring in Sullom Voe, 1985. Report to the Shetland Oil Terminal Environmental Advisory Group, 1986.
32. Woodman SSC: Macrobenthic surveys in Sullom Voe 1978-1983, Summary Report. Report to the Shetland Oil Terminal Advisory Group, (1984a).
33. Woodman SSC: Macrobenthic monitoring in Sullom Voe 1983. Report to the Shetland Oil Terminal Advisory Group, (1984b).
34. Yokoyama H and Hayashi I: Zonation and species diversity of smaller macrobenthos in the westernmost part of Wakasa Bay (the Sea of Tango). Journal of Oceanography Society of Japan 36(1), 46-58, 1980.

In addition to the references cited, the following (in-house) limited circulation reports are available:

BIBLIOGRAPHY

1. Addy JM: Sublittoral Macrobenthic Survey in Sullom Voe, Shetland, May 1977. A report to the Shetland Oil Terminal Environmental Advisory Group, 1977.
2. Addy JM: Sublittoral Macrobenthic Survey in Sullom Voe, Shetland, July 1978. A report to the Shetland Oil Terminal Environmental Advisory Group, 1979.
3. Addy JM: Macrobenthic Monitoring in Sullom Voe, Shetland, July 1979. A report to the Shetland Oil Terminal Environmental Advisory Group, 1980.
4. Addy JM: Macrobenthic Monitoring in Sullom Voe, Shetland, July 1980. A report to the Shetland Oil Terminal Environmental Advisory Group, 1981.
5. Addy JM and Griffiths PW: Sublittoral Macrobenthic Survey in Sullom Voe, Shetland, May 1976. A report to the Shetland Oil Terminal Environmental Advisory Group, 1977.
6. Scottish Marine Biological Association: Chemical and Biological Report 1978. A report to the Shetland Oil Terminal Environmental Advisory Group, 1978.
7. Stanley SO and Pearson TH: Chemical and Biological Monitoring Report 1979. A report to the Shetland Oil Terminal Advisory Group.
8. Stanley SO, Pearson TH, Comely CA and Douglas AG: Sullom Voe Survey. A report of work carried out by the Scottish Marine Biological Association between September 1976 and September 1977. A report to the Shetland Oil Terminal Environmental Advisory Group, 1977.

MEASUREMENT OF SCOPE FOR GROWTH AND TISSUE HYDROCARBON CONCENTRATIONS OF MUSSELS (*MYTILUS EDULIS*) AT SITES IN THE VICINITY OF THE SULLOM VOE OIL TERMINAL:- A CASE STUDY

J. WIDDOWS, P. DONKIN, P.N. SALKELD & S.V. EVANS

N.E.R.C., INSTITUTE FOR MARINE ENVIRONMENTAL RESEARCH, PROSPECT PLACE, THE HOE, PLYMOUTH, PL1 3DH, UK.

ABSTRACT

Mussels (*Mytilus edulis*) were used to monitor the degree of oil contamination and the sublethal biological effects of oil in the vicinity of the North Sea oil terminal at Sullom Voe during the period from 1982 to 1985. Spatial and temporal changes in the concentration of aromatic hydrocarbons in the body tissues reflected the location and temporal variations in the number and size of small oil spillages. There was a significant negative relationship between scope for growth and the log10 concentration of aromatic hydrocarbons in the tissues (r=-0.72). A synthesis of data from Sullom Voe and mesocosm oil experiments enabled the former to be placed in a broader context. This not only extended the semi-logarithmic relationship between scope for growth and the tissue concentration of aromatic hydrocarbons but also demonstrated the relatively low to moderate level of oil contamination within Sullom Voe.

INTRODUCTION

The main objective of this field study was to monitor the sublethal biological effects and the degree of hydrocarbon contamination of mussels (*Mytilus edulis*) inhabiting the littoral environment in the vicinity of the North Sea oil terminal in Sullom Voe (Shetland, U.K.) in order to detect any effects of industrial activity in the area. It represents part of a multidisciplinary environmental monitoring programme for the Shetland Oil Terminal Environmental Advisory Group.

The use of mussels as 'indicator species' or 'sentinel organisms' in marine environmental monitoring programmes is now widespread (Goldberg, *et al.*,1978; Farrington, 1983; Davies & Pirie, 1980; Widdows *et al.*, 1981a,b; Lack & Johnson, 1985; Martin, 1985). Measurement of chemical contaminants accumulated within their tissues and sublethal biological effects such as scope for growth, provide a sensitive and quantitative method of assessing changes in environmental quality and pollution impact. The objectives of this paper are: a) to review the combined chemical and physiological measurements carried out in the Sullom Voe area during the period 1982 to 1985, after the oil terminal became operational in 1980; b) to examine the spatial and temporal variation in environmental quality in order to identify those areas that are most affected and to establish whether there has been a gradual deterioration or improvement in environmental quality; and c) to assess the degree to which Sullom Voe has been affected by oil pollution.

Kuiper, J., Van den Brink, W.J. (eds), Fate and Effects of Oil in Marine Ecosystems. ISBN 90-247-3489-4

Fig. 1 Location of sampling sites.
1) Gluss Voe (reference site)
2) Mavis Grind
3) Voxter Voe
4) Calna Taing
5) Scatsta Voe
6) Tanker Jetty 4

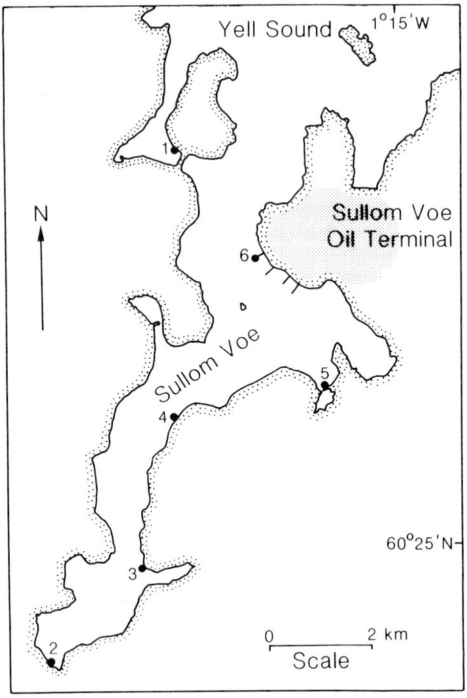

MATERIALS AND METHODS

Mussels (Mytilus edulis) of a standard body size (4.5 cm shell length) were collected from mean low water level at six sites in the region of Sullom Voe (Fig. 1) in July 1982 (sites 1 and 5), July 1983 (sites 1, 2 and 5), August 1984 (sites 1, 2, 5 and 6) and July 1985 (sites 1, 2, 3, 4, 5 and 6).

A mobile laboratory was used to determine the physiological responses (clearance or feeding rate, food absorption efficiency, respiration and excretion rate) of sixteen individuals from each site (for details see Widdows, 1985a), under conditions of ambient temperatures, year to year variation from 11° to 14°C, salinities from 32 to 34°/₀₀, suspended particulates from 0.46 to 0.87 mg l^{-1} and particulate organic matter (POM) from 0.33 to 0.44 mg l^{-1}. Physiological rates were corrected to mass-specific rates for a 'standard mussel' of 1g dry tissue mass using the weight exponent b=0.67 (Widdows et al., 1981). Each rate was then converted to energy equivalents (J g^{-1} h^{-1}) in order to calculate the energy budget and scope for growth, which represents the difference between the energy absorbed from the food and the energy lost via respiration and excretion (Widdows, 1985a, b).

Two groups of five individuals were sampled from each site and their tissues removed from their shells before storing at -20°C. Hydrocarbons were extracted from the tissues by steam distillation and then analysed by high pressure liquid chromatography (HPLC) to give group separation of aromatic hydrocarbons with 2 rings (detected at 225 nm) and aromatics with 3 rings (detected at 254 nm). Quantification was by reference to 2,3-dimethylnaphthalene and 1-methylphenanthrene for 2 and 3 ringed groups respectively.

Expressing tissue hydrocarbon concentration in terms of 2 and 3 ringed aromatic hydrocarbons simply reflects the nature of the analytical procedure which was developed as a rapid and quantitative technique (Donkin & Evans, 1984). It concentrates on the toxic lower molecular weight polyaromatic hydrocarbons which represent a dominant component of the hydrocarbon body burden in oil contaminated mussels.

RESULTS

The results of the annual monitoring of 2 and 3 ringed aromatic hydrocarbons accumulated in the body tissues and the scope for growth (SFG) of *Mytilus edulis* sampled from up to six sites around Sullom Voe are presented in tables 1 and 2.

Gluss Voe (site 1) which lies just outside Sullom Voe represents a reference site with consistently low levels of hydrocarbon contamination between 1982 and 1985 (Table 1). Mussels sampled from Voxter Voe (site 3), towards the southern end of Sullom Voe, were also found to have low levels of hydrocarbon contamination. In contrast, mussels collected from the inner most and southerly part of Sullom Voe at Mavis Grind (site 2) showed a significant level of hydrocarbon contamination. Mussels collected from the near vicinity of the oil terminal, from Scatsta Voe (site 5) and from the tanker loading jetties 3/4 (site 6) showed significant accummulation of hydrocarbons in their tissues, typically an order of magnitude above background concentrations, but with marked temporal variation reflecting the number and size of small oil spillages (details provided by the Sullom Voe Harbour Authority). In 1983 there were relatively few spillages compared to 1982 and 1984, hence the lower body burden values; whereas in April 1985 there was a more significant spillage from 'BP Vision', but this affected the shoreline around Calna Taing rather than nearer the terminal and at site 4 hydrocarbons were accummulated in mussels to tissue concentrations approximately 15-fold above background (site 1).

Table 1. Aromatic hydrocarbon concentration in the body tissues of *Mytilus edulis*. (2 pools of 5 individuals) mean ± range.

Sample site	2+3 ringed aromatic hydrocarbon concentration ($\mu g\ g^{-1}$ wet mass)			
	July 1982	July 1983	Aug. 1984	July 1985
Gluss Voe (1)	0.11±0.01	0.06±0.01	0.08±0.01	0.10±0.01
Mavis Grind (2)	-	1.01±0.11	0.35±0.03	0.51±0.04
Voxter Voe (3)	-	-	-	0.13±0.02
Calna Taing (4)	-	-	-	1.63±0.70
Scatsta Voe (5)	0.92±0.22	0.32±0.06	1.51±0.14⁺	0.23±0.04
Tanker Jetty (6)	-	-	1.37±0.08	0.66±0.07

⁺ Mussels sampled for chemical analysis, in contrast to those sampled for physiological measurement, were collected from a visibly oiled part of the shore.

Table 2. Scope for growth (J g^{-1} h^{-1}) of Mytilus edulis collected from sites in the vicinity of the Sullom Voe oil terminal. (Mean ±S.E.; n=16; Particulate organic matter 0.4 mg l^{-1}).

Sample site		Scope for Growth (J g^{-1} h^{-1})		
	July 1982	July 1983	Aug. 1984	July 1985
Gluss Voe (1)	21.1±2.3	22.8±3.6	21.1±2.0	23.6±1.3
Mavis Grind (2)	-	14.2±7.2	6.3±2.2**	9.6±1.1**
Voxter Voe (3)	-	-	-	17.1±1.6
Scatsta Voe (5)	17.1±1.2	17.0±1.8	19.5±2.0	12.9±1.0*
Tanker Jetty (6)	-	-	10.4±2.4*	7.0±1.6**

Significantly different from Gluss Voe (reference site) at *P<0.05 and **P<0.001.

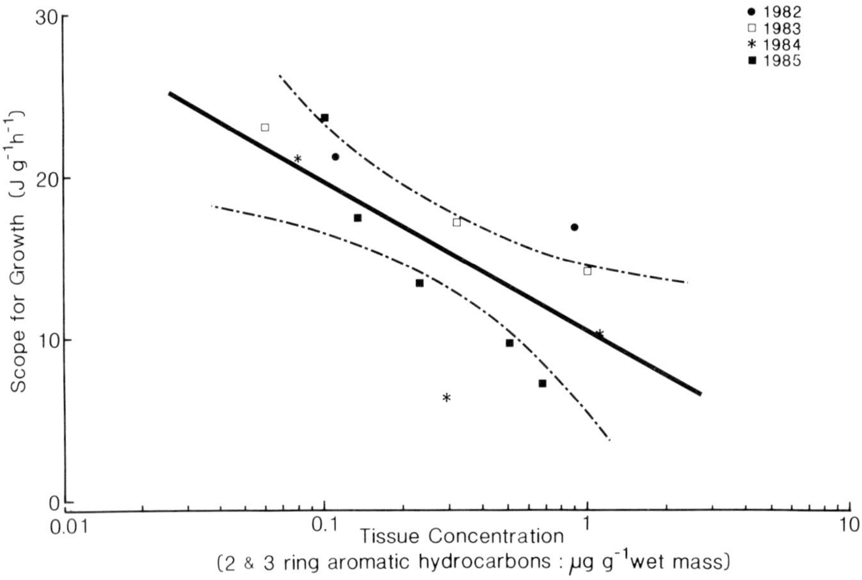

Fig. 2 Relationship between scope for growth (J g^{-1} h^{-1}) and the log concentration of 2 and 3 ring aromatic hydrocarbons in the body tissues of Mytilus edulis sampled from sites around Sullom Voe during the period 1982 to 1985. Scope for growth was standardized for a 1g mussel and ration level of 0.4 mg POM l^{-1}.

The scope for growth of Mytilus edulis collected from the different sites (Table 2) showed an inverse relationship with the degree of hydrocarbon contamination in the body tissues. Mussels from Gluss Voe had consistently higher SFG values (>20J g^{-1} h^{-1}) compared to the contaminated sites at Mavis Grind (site 2), Scatsta Voe (site 5) and the tanker jetties (site 6); no mussels were sampled from Calna Taing (site 4) for physiological measurement in 1985. Mussels sampled from the relatively 'clean' site at Voxter Voe in 1985 also had high SFG values.

The decline in SFG at the more contaminated sites was primarily due to lower clearance or feeding rates, the component of the energy budget which is typically the most responsive to pollution (Widdows, 1985b).

Figure 2 provides a synthesis of the data collected between 1982 and 1985, and illustrates the significant negative relationship (r=-0.72) between SFG and the log10 concentration of 2 and 3 ringed aromatic hydrocarbons in the tissues. The data from Scatsta Voe (1984) were omitted due to the separate sampling of mussels for physiological and chemical measurement; mussels for the latter were collected from a visibly oiled part of the shore. It emphasizes that in order to avoid the potential problem of small scale spatial and temporal variation in environmental contamination, and to achieve a close correspondence between chemical and biological effects measurements it is essential to sample from a pool of animals collected from the same part of the population/shore at the same time. This sampling strategy was carried out in 1985 and resulted in a significant negative correlation (r=-0.93) between SFG and the concentration of aromatic hydrocarbons in the body tissues.

DISCUSSION

A combined chemical and sublethal biological effects monitoring approach, such as that used to assess the pollution effects in Sullom Voe, is able to meet many of the basic requirements of environmental quality assessment and environmental management. Namely, a) to quantify spatial and temporal changes, either deterioration or improvement, in environmental quality and pollution before lethal and damaging effects occur; b) to identify the cause(s) of adverse effects; c) to relate sublethal effects to wider ecological consequences; and d) to assess the degree of pollution, especially in relation to other areas.

The measurement of SFG and hydrocarbon residues in the tissues of Mytilus during the period 1982-1985 has provided information on aspects of spatial and temporal variation in environmental pollution within Sullom Voe. Tissue residue data (Table 1) show that mussels living in the vicinity of the oil terminal (sites 5 and 6) were consistently oil contaminated, with hydrocarbon concentrations typically an order of magnitude above background levels, and that there was a concomitant decline in the SFG or growth potential of mussels at these sites (Table 2). The relationship in Fig. 2 indicates that an order of magnitude increase in the tissue concentrations of 2 and 3 ringed aromatic hydrocarbons can account for an approximately 50% reduction in the growth potential of Mytilus edulis.

As a result of experimental oil exposure and recovery studies, in which concentration-response relationships were established (Widdows et al., 1982, 1985), it is possible to attribute the recorded reduction in SFG of mussels in Sullom Voe to the accummulation of aromatic hydrocarbons within their tissues. Furthermore, other environmental factors which can modify SFG such as temperature, salinity and seston concentration, did not alter sufficiently from year to year or between sites to account for the observed changes in SFG. The concentrations of metals (Cd, Cu, Fe, Mn, Pb and Zn) in the tissues of mussels sampled from the Sullom Voe populations have also been determined (J. Cleary, unpublished data), but none of the metals were accumulated to concentrations that were sufficiently high to produce a toxic effect.

The sampling at additional sites in 1985 demonstrated the absence of a simple pollution gradient along the length of Sullom Voe from the oil terminal (site 6) to Mavis Grind (site 2) at the southern end of Sullom Voe. Mussels from Voxter Voe (site 3) were found to be relatively uncontaminated by hydrocarbons and had high SFG values, whereas mussels from Mavis Grind were consistently found to be moderately contaminated. Evidence suggests that the contamination at the southern end of Sullom Voe (site 2) is due to a localized and gradual seepage of hydrocarbons following the land spillage of diesel oil at a nearby quarry before 1980, and is not the result of oil spills from the terminal.

The temporal variations in SFG and hydrocarbon residues in the tissues of *Mytilus edulis* at sites 5 and 6 near the oil terminal (Tables 1 and 2) indicate that there has been no chronic deterioration in environmental quality and no gradual increase in the level of hydrocarbon contamination in the water column and the littoral environment within Sullom Voe since the terminal became operational. The year to year variations, with higher levels of oil contamination at sites 5 and 6 in 1982 and 1984, appear to be related to the number and size of small spillages (details provided by the Sullom Voe Harbour Authority) during the months preceding the annual sampling in Sullom Voe. However, in April 1985 there was a more significant spillage from 'BP Vision' and due to prevailing wind conditions and currents at the time of the spill the shoreline around Calna Taing (site 4) was the most heavily contaminated region. In contrast, sites 5 and 6 nearer the terminal were less contaminated than 1984, indicating the recovery that can occur during a year.

After quantifying the degree of environmental contamination and the resultant detrimental biological effects, environmental managers ask 'how bad or good are the conditions?' and 'at what level of impact should we start to be concerned and take remedial action?' These aspects can be addressed by presenting the present study in a wider context. A synthesis of data derived from this present field study in Sullom Voe and the mesocosm oil experiments (Widdows et al., 1985) carried out at Solbergstrand (Southern Oslofjord, Norway) serves to place the degree of hydrocarbon contamination recorded in the vicinity of Sullom Voe oil terminal and the associated effects on scope for growth of *Mytilus edulis* into a broader perspective. Figure 3 illustrates the semi-logarithmic relationship (correlation coefficient = -0.87) between scope for growth ($J\ g^{-1}\ h^{-1}$) and the log10 concentration of two and three ringed aromatic hydrocarbons in the body tissues ($\mu g\ g^{-1}$ wet mass), with the triangles representing the data for Solbergstand and the circles representing the data for Sullom Voe (1982-1985). The semi-logarithmic relationship (Fig. 3) demonstrates that:-

a) there is a simple relationship between increasing concentration of aromatic hydrocarbons in the body tissues and declining scope for growth of *Mytilus edulis*;

b) hydrocarbons affect scope for growth over a wide range of aromatic hydrocarbon concentrations in the tissues (ie. three orders of magnitude) without an apparent threshold effect, therefore suggesting a simple mechanism of toxicity based on the loading of the tissues and the absence of any significant physiological adaptation;

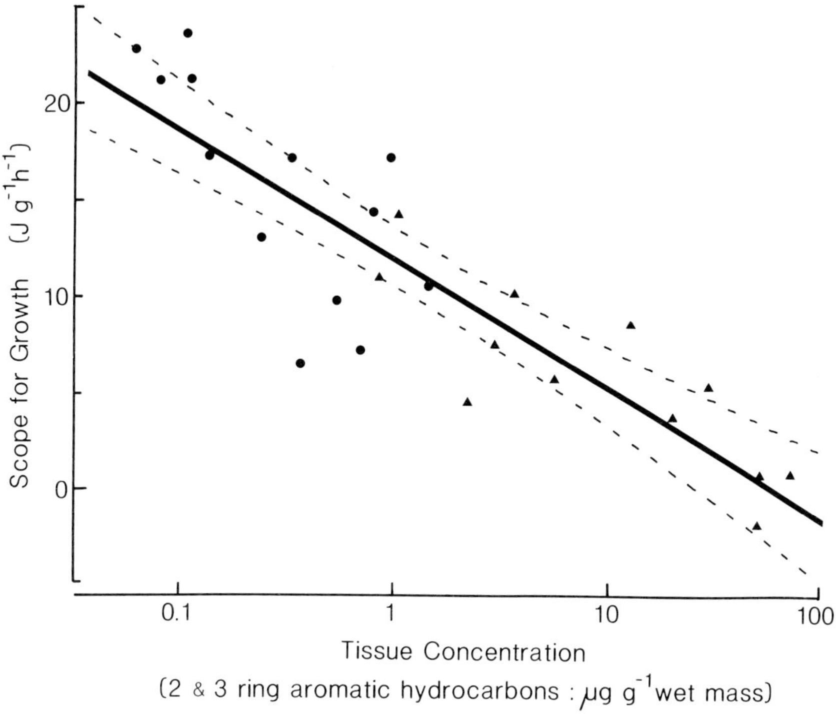

Fig. 3 Relationship between scope for growth and the concentration of 2 and 3 ringed aromatic hydrocarbons in the tissues of *Mytilus edulis*. (Correlation coefficient =-0.87; dashed lines represent 95% confidence limits).
(●) Data from Sullom Voe, Shetlands.
(▲) Data from oil exposure and recovery mesocosm experiments at Solbergstrand, Oslofjord, Norway.

c) mussels from the clean reference site as Gluss Voe have a very low level of hydrocarbon contamination and most sites and mussel populations around Sullom Voe are relatively 'clean'; and
d) the tanker jetty population nearest the source of pollution in Sullom Voe is equivalent to the 'control' population at Solbergstrand on the southern Oslofjord and also sites in the lower Tamar Estuary, S.W. England (unpublished data).

In addition, the mesocosm studies provide important links between effects at the individual, population and community levels of biological organization, thus enabling the prediction of wider ecological consequences from sublethal effects measurements on *Mytilus edulis*. The evidence from these mesocosm studies (Solbergstrand Experimental Station, Progress Report No 5) indicate that the more marked effects of petroleum hydrocarbons, such as impaired reproduction, mortality, reduced recruitment, and a decline in population and community structure occur at concentrations of approximately 10 µg g^{-1} wet mass, which is equivalent to 100 x the background levels recorded in the Shetlands.

There is also a parallel relationship and good agreement between biological effects such as scope for growth, measured at the individual level of organization using an indicator species, and changes in species diversity at the community level. Davies et al.,(1984) have recorded a progressive decline in species diversity (Shannon-Wiener) with a logarithmic increase in oil concentration in the sediment in the North Sea around oil production platforms. Again there is no apparent threshold of effect, but the data suggest that it is necessary to increase the degree of hydrocarbon contamination by several orders of magnitude above background before it is possible to detect a statistically significant decline in species diversity. The parallel relationship between scope for growth and species diversity supports the notion that adverse effects measured at the individual level ultimately manifest themselves at the population and community levels of organisation. Physiological responses therefore serve to complement ecological surveys of community structure, not only by providing an early detection of effects, often before statistically significant changes occur at the community level, but also by providing some insight into the toxicological causes of changes in population and community structure.

In conclusion, the concentrations of hydrocarbons in the body tissues and the sublethal biological responses of mussels indicate that, the degree of oil contamination in Sullom Voe is relatively low to moderate and comparable with sites such as the Tamar Estuary (S.W. England) and the southern Oslo fjord (Norway). There is no evidence of a gradual deterioration of environmental quality in Sullom Voe and there is an apparent recovery to the steady-state condition following any 'short-term' increase in oil contamination and biological impact.

ACKNOWLEDGEMENTS

This work was carried out for the Shetland Oil Terminal Environmental Advisory Group and financed by the Sullom Voe Association Ltd. We thank C.M. Worrall for assistance with the field work.

REFERENCES

1. Davies IC & Pirie JM: Evaluation of a 'Mussel Watch' project for heavy metals in Scottish coastal waters. Mar. biol. 57, 87-93, 1980.
2. Davies JM, Addy JM, Blackman RA, Blanchard JR, Ferbrache JE, Moore, DC, Sommerville HJ, Whitehead A & Wilkinson T: Environmental effects of the use of oil-based drilling muds in the North Sea. Mar. Poll. Bull. 15, 363-370, 1984.
3. Donkin P & Evans SV: Application of steam distillation in the determination of petroleum hydrocarbons in water and mussels (Mytilus edulis) from dosing experiments with crude oil. Anal. Chim. Acta, 156, 207-219, 1984.
4. Farrington JW: Bivalves as sentinels of coastal chemical pollution: The Mussel (and Oyster) Watch. Oceanus 26, 18-29, 1983.
5. Goldberg ED, Bowen VT, Farrington JW, Harvey G, Martin JH, Parker PL, Riseborough RW, Robertson W, Schneider E & Gamble E: The Mussel Watch. Environ. Conserv. 5, 101-125, 1978.
6. Lack TJ & Johnson D: Assessment of the biological effects of sewage sludge at a licensed site off Plymouth. Mar. Poll. Bull. 16, 147-152, 1985.
7. Martin M: State Mussel Watch: Toxics surveillance in California. Mar. Poll. Bull. 16, 140-146, 1985.

8. Widdows J: Physiological measurements. In: The Effects of Stress and Pollution on Marine Animals by Bayne et al. New York: Praeger, 1985a.
9. Widdows J: Physiological responses to pollution. Mar. Poll. Bull. 16, 129-134, 1985b.
10. Widdows J, Bakke T, Bayne BL, Donkin P, Livingstone DR, Lowe DM, Moore MN, Evans SV & Moore SL: Responses of *Mytilus edulis* on exposure to the water accommodated fraction of North Sea oil. Mar. Biol. 67, 15-31, 1982.
11. Widdows J, Bayne BL, Donkin P, Livingstone DR, Lowe DM, Moore MN & Salkeld PN: Measurement of the responses of mussels to environmental stress and pollution in Sullom Voe: a base-line study. Proc. Roy. Soc. Edin. 80B, 323-338, 1981a.
12. Widdows J, Donkin P & Evans SV: Recovery of *Mytilus edulis* L. from chronic oil exposure. Mar. Environ. Res. 17, 250-253, 1985.
13. Widdows J, Phelps DK & Galloway W: Measurement of physiological condition of mussels transplanted along a pollution gradient in Narragansett Bay. Mar. Environ. Res. 4, 181-194, 1981b.

OIL-BASED DRILLING FLUIDS: ARE THEY AN ENVIRONMENTAL RISK?

JERRY F. PAYNE*, L. FANCEY*, J. KICENIUK*, U. WILLIAMS*, ANVER RAHIMTULA**, S. KHAN**, JIM OSBORNE***
*DEPT. OF FISHERIES AND OCEANS, SCIENCE BRANCH, P.O. BOX 5667, ST. JOHN'S, NEWFOUNDLAND, CANADA, A1C 5X1
**MEMORIAL UNIVERSITY OF NEWFOUNDLAND, ST. JOHN'S, NEWFOUNDLAND, CANADA,
***ENVIRONMENTAL PROTECTION SERVICE, ST. JOHN'S, NEWFOUNDLAND, CANADA

KEYWORDS
Drilling fluids, polycyclic aromatic hydrocarbons, toxicity, mixed-function oxygenases, biological monitoring, chick-embryo bioassay.

INTRODUCTION

The use of oil-based drilling fluids has been discouraged in hydrocarbon exploration and production in the marine environment but these drilling fluids are presently being used to a considerable degree in the United Kingdom and Norwegian sectors of the North Sea. Field studies in the North Sea have demonstrated only localized impacts around individual drilling sites, including those sites where 'toxic' diesel oils were employed as base oils in drilling fluids (1-2). Yet the question of disposal of cuttings contaminated with oil from drilling fluids remains somewhat controversial (3-5), including in Canada where oil-based fluids have been permitted for use on an experimental basis in the Arctic and off the East Coast.

Many of our studies over the past several years on the sublethal effects of petroleum hydrocarbons, are generally applicable (6-10) but we have also begun to investigate the potential toxicity of oil-based drilling fluids, which are derived from petroleum (11) but have elevated levels of aliphatic and reduced levels of aromatic hydrocarbons. Prudhoe Bay crude oil (PBCO) was taken as a 'model' hydrocarbon mixture and the effects of aliphatic versus aromatic fractions on cellular energy functions and toxicity in the sensitive chick-embryo bioassay was studied in detail. Results supported the hypothesis that paraffin or aliphatic hydrocarbon mixtures have an overall low toxicity potential compared with diesel oil. Evaluation of the toxicity of some low-aromatic base oils in the chick embryo bioassay as well as their mixed-function oxygenase (MFO) enzyme induction potential in fish, further supported this hypothesis. Also, on the basis of observations on the MFO enzyme induction potential of drilling fluid base oils, we suggest that this particular index can be a useful forerunner for delineating areas of effect/no effect around petroleum development sites using either diesel- or mineral oil based- drilling fluids.

EFFECT OF AROMATIC VERSUS ALIPHATIC FRACTIONS OF PETROLEUM HYDROCARBONS ON THE BIOENERGETIC FUNCTIONS OF MITOCHONDRIA

The mitochondrial inner membrane possesses the energy transducing functions, and oxidative phosphorylation in mitochondria is responsible for supplying over 95% of the total energy requirements of eukaryotic cells. Respiratory control index (RCX) and P/O ratios are generally used as indicators of mitochondrial function (12). However, state 3 respiration (rate of oxygen uptake in the presence of adenosine diphosphate and substrate) and state 4 respiration (rate of oxygen uptake in the presence of

Kuiper, J., Van den Brink, W.J. (eds), Fate and Effects of Oil in Marine Ecosystems. ISBN 90-247-3489-4

substrate only) may also serve as indicators of mitochondrial function (13). One of the possible causes of morbidity in animals exposed to hydrocarbons is decrease in respiration rate and energy production. Thus one of our main objectives was to examine if crude oil or its aliphatic or aromatic sub-fractions affected the mitochondrial bioenergy-related processes such as respiration and electron transfer. Also included in the study was an investigation of the effects of the polar fraction which contains the sulfur, oxygen and nitrogen substituted congeners (NOS). Possible sites of toxic action were evaluated with polarographic and enzymatic techniques using rat liver mitochondria.

The effect of various concentrations of PBCO and its aliphatic, aromatic and NOS fractions on β-hydroxybutyrate -dependent respiration rates and respiration control ratios (RCR) of rat liver mitochondria are shown in Figure 1. Addition of dimethyl sulfoxide (DMSO) extracts of PBCO or its aromatic fraction caused a concentration-dependent inhibition of state 3 respiration and an increase in state 4 respiration. The NOS fraction was effective only at high concentrations while the aliphatic fraction was ineffective. The aromatic fraction was clearly the most effective with 50 micrograms inhibiting β-hydroxybutyrate-dependent state 3 respiration by approximately 70%. Figure 1 lists only the results obtained with β-hydroxybutyrate but succinate dependent respiration was also sensitive to inhibition by PBCO or its aromatic fraction.

In addition, the effect of PBCO and its aliphatic, aromatic and NOS fractions on β-hydroxybutyrate dehydrogenase activity were also determined. As indicated in Fig. 2, enzyme activity was inhibited significantly by PBCO and the aromatic fraction while the aliphatic fraction had little effect. Similar results were obtained in studies on succinate dehydrogenase. Effects on mitochondrial adenosine triphosphatase were also investigated. This particular enzyme is stimulated by electron chain uncouplers or by destruction of the mitochondrial inner membrane. A stimulation of about 5 fold was observed with 50 micrograms of the aromatic fraction while the aliphatic and polar fraction were much less effective.

It can be concluded from this series of studies that, compared with complex mixtures of aliphatics, aromatics can be expected to have a greater potential to adversely effect cellular energy functions.

EVALUATION OF AROMATIC VERSUS ALIPHATIC FRACTIONS OF PETROLEUM HYDROCARBONS IN THE CHICK EMBRYO BIOASSAY

The chick-embryo bioassay is one of the most sensitive tests available for evaluating the toxicity potential of hydrocarbons. Recent investigations have shown that application of microliter quantities of crude oil to the egg shell results in drastically reduced embryo survival (14-16), with toxicity being directly related to aromatic hydrocarbon content (more specifically polycyclic aromatic hydrocarbon content). The embryo is especially sensitive during the earlier stages of development, around day 7. This period corresponds to the stage of organogenesis including liver differentiation and the appearance of hepatic MFO enzyme activities. Figure 3 shows the effect of application of various amounts of PBCO and its aliphatic, aromatic and polar fractions on chick embryo mortality. The

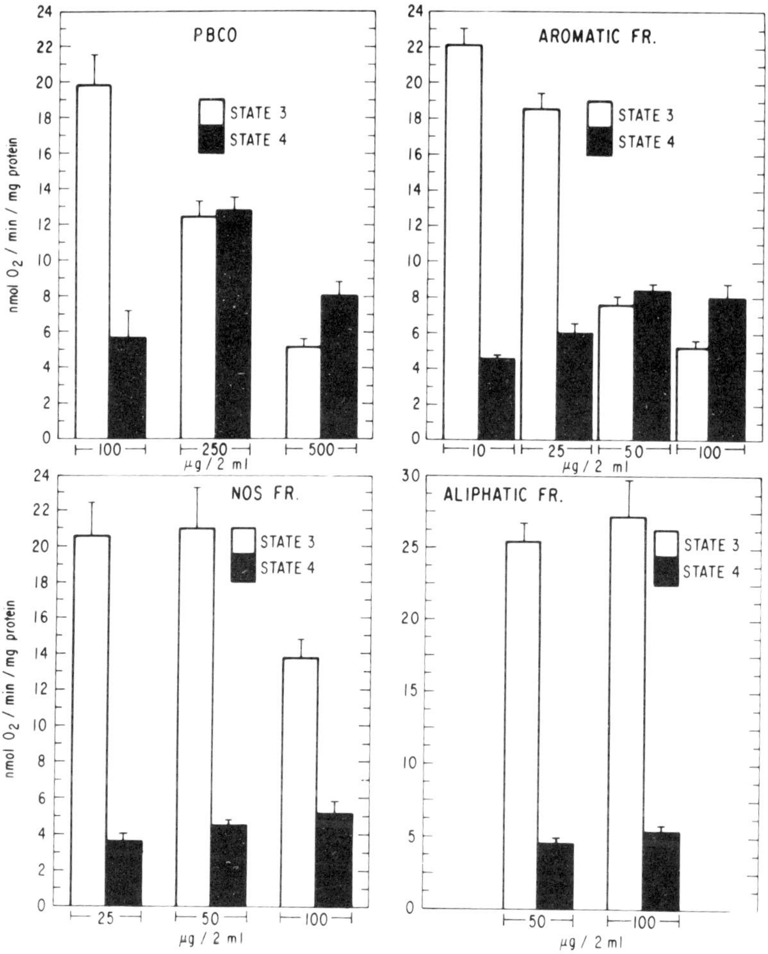

Figure 1. Effect of DMSO extracts of PBCO or its fractions on respiration rates during oxidation of β-hydroxybutyrate by mitochondria. The fractionation of PBCO was carried out on a silica/alumina column. DMSO extracts were made by vigorously vortexing equal volumes of DMSO and the appropriate fraction followed by centrifugation. Mitochondria were isolated from livers of male Sprague-Dawley rats (200-250 g) as described by Sottocasa et al.(17) using an isolation medium composed of 250 mM sucrose, 3.4 mM Tris and 1 mM EGTA (pH 7.4). β-hydroxybutyrate supported respiration rates of isolated mitochondria in state 3 and state 4 were determined polarographically as described by Estabrook (12) using an oxygen monitor equipped with a Clark oxygen electrode. The activity of β-hydroxybutyrate dehydrogenase was measured as described by Delafield and Doudoroff (18) using β-hydroxybutyrate as substrate and mitochondria solubilized with Triton X-100 (0.05%).

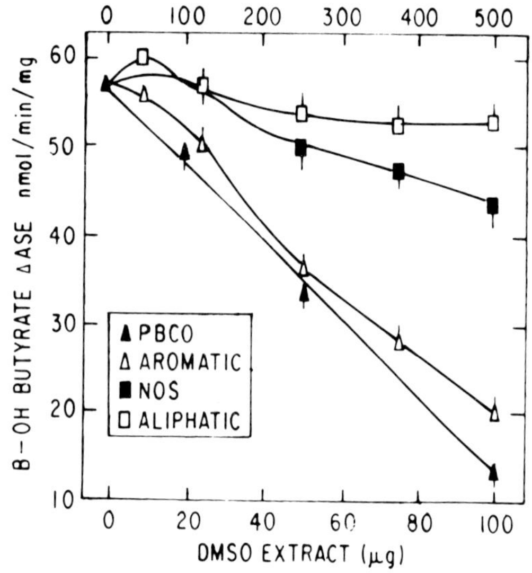

Figure 2. Effect of varying amount of a DMSO extract of PBCO or its fractions on mitochondrial β-hydroxybutyrate dehydrogenase activity. Activity is expressed as nmoles of NAD^+ reduced per min per mg protein. Each point represents the mean ± S.E. of 4 independent observations. The numbers along the abscissa at the top and bottom refer to micrograms of DMSO extracts of PBCO or its fractions (respectively) added per ml of the assay mixture.

7-day-old embryos were shown to be very sensitive and both the crude and aromatic fraction resulted in increased embryo mortality with increasing dosage. The aromatic fraction was especially toxic having an approximate LD-50 around 0.4 microliters. By contrast the aliphatic fraction demonstrated an extremely low level of toxicity. In essence the acute toxicity effects observed in the chick-embryo bioassay reflected the results obtained in the energy related studies-namely the aliphatic fractions demonstrated a lower toxicity potential.

EVALUATION OF VARIOUS DRILLING MUD BASE OILS IN THE CHICK EMBRYO BIOASSAY

In this study we examined the toxicity of a variety of drilling mud base oils as well as oil obtained from the Shell blowout off the East Coast of Canada and a diesel oil, on 12 day old chick embryos. Hydrocarbon analysis of the various base oils was carried out by high performance liquid chromatography which can resolve the relative abundance of 2, 3 and 4 ringed aromatics. Application of increasing amount of the blowout and diesel oils to the surface of fertile eggs resulted in increasing embryo death. In contrast the base oils displayed little toxicity. Application of up to 20 microliters of oil Tradename Safver 1, 3 and 5 resulted in the death of only 1 embyro out of 30 while application of 20 microliters of Safver oil number 4 resulted in the mortality of 4 out of 30 embryos. In general, results corroborated the hypothesis that base-oils enriched in aliphatics are relatively non-toxic.

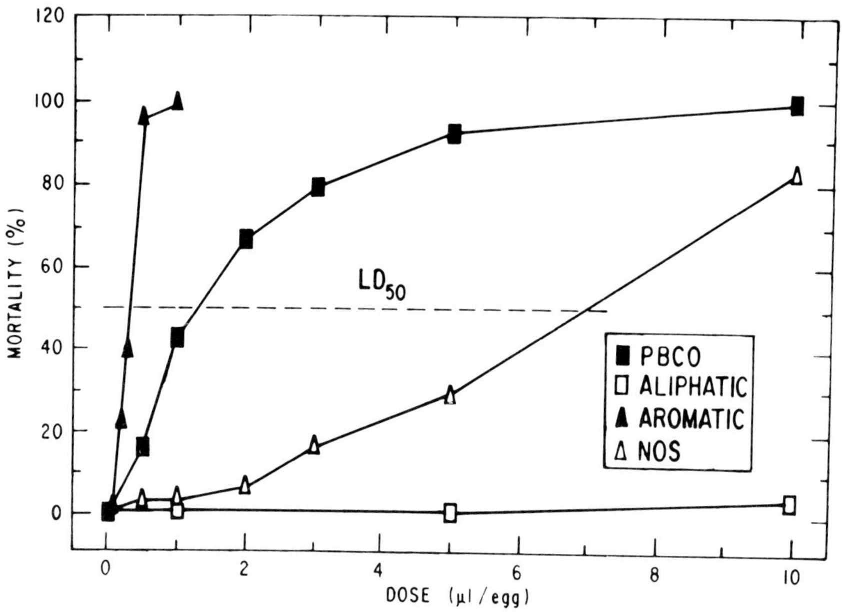

Figure 3. Toxicity of PBCO or its aliphatic aromatic and NOS fractions on the chick embryo. Groups of fertile eggs (n = 30) were spotted with the indicated amount of PBCO or its aliphatic, aromatic and NOS fractions on day 7 of development. Hydrocarbon samples were applied with a microliter syringe just below the airspace of upright eggs as described by Albers (15). Eggs were incubated at 37.5°C in commercial incubators and candled once every two days and opened on day 15. All dead embryos were weighed to determine the age of death.

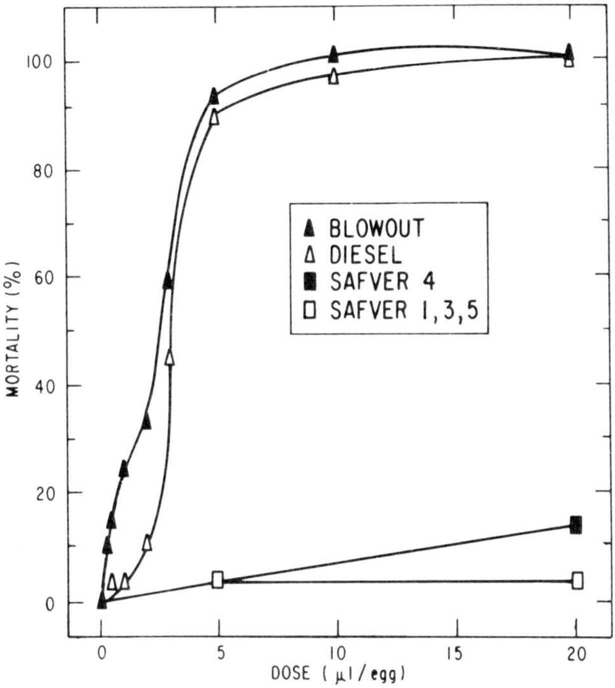

Figure 4. Toxicity of various drilling fluid base oils compared with diesel and Shell blowout oil in the chick embryo assay. The procedure was similar to that described in Fig. 3. On day 7, 0-20 microliters of the various oils were applied by a microliter syringe just below the airspace. Control and treated eggs were candled every other day up to day 19 in incubators, at which time they were opened and scored for mortality. Chemical analyses of the drilling mud base oils was carried out by high performance liquid chromatography on a Waters Energy Analysis Column. Diesel was shown to be enriched in diaromatics and triaromatics compared to Safver 1. Safver 4 was essentially free of both diaromatics and triaromatics while Safver 5 contained some diaromatics but was essentially free of triaromatics. (Safver 5 was somewhat unique however in that it also contained relatively high levels of mono aromatics which chromatographed unipolar with benzene.) Further details concerning the chemical analysis can be found in Payne et al. (19).

EVALUATION OF VARIOUS DRILLING MUD BASE OILS FOR MFO INDUCTION POTENTIAL IN FISH

The induction of MFO enzymes has been validated on a number of occasions in the field as a sensitive index of low level hydrocarbon exposure (reviewed by Payne (20)) including sites in the North Sea where diesel based drilling muds were used (21). We carried out studies to determine the MFO enzyme induction potential of various base oils compared with diesel oil. A relatively high level of MFO enzyme induction was observed in fish injected with diesel oil. Male and female fish were assessed and the levels of induction in both kidney and liver tissues were comparable. Diesel oil was also effective in inducing a low level of enzyme activity in heart tissues (data now shown). By comparison, two of the base oils (Tradename: Safver 1 and Safver 5) were less effective in inducing liver enzymes in either male or female fish, while kidney enzyme levels were not significantly changed by treatment. No significant differences in enzyme activity were noted in either kidney or liver tissues of male and female fish treated with the third base oil (Tradename, Safver 4). For comparison purposes it should also be emphasized that the dose used in all the synthetic oil treatments was double the dose used in the diesel oil exposure. Overall our results on injection trials with base oils together with those of Leaver et al. (22) and Addison et al., (23), who investigated induction in flatfish exposed to oil based drilling mud cuttings, indicate that cuttings containing diesel oil substitutes can be expected to have a lower MFO enzyme induction potential. Thus both larger doses and or longer exposure periods will likely be required before induction occurs in fish around petroleum development sites contaminated with diesel oil substitutes.

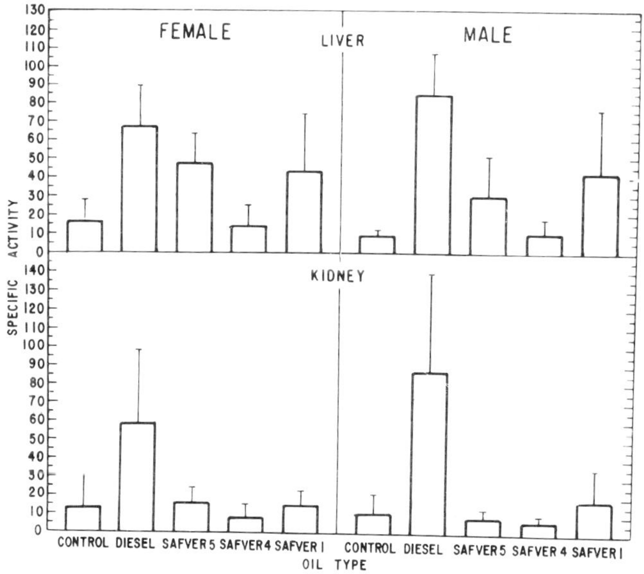

Figure 5. Induction of ethoxyresorufin dealkylase in fish exposed to drilling mud base oils. Rainbow trout (*Salmo gairdneri*) were injected intraperitoneally with either 100 μl of diesel oil or 200 μl of base oil (Tradename Safver 1, Safver 4 and Safver 5) and exposed for 2 weeks at 1°C in flow through aquaria (60 liter). One unit of activity is equivalent to the production of one picomole of resorufin per mg of protein per min. Enzyme differences were evaluated by means of Student's t-test and two-way analysis of variance (adapted from Payne et al. (19))

MFO ENZYME INDUCTION AS A POTENTIAL MONITORING INDEX AROUND HYDROCARBON DEVELOPMENT SITES USING OIL-BASED DRILLING FLUIDS

The observation that drilling fluid base oils have a residual capacity for MFO enzyme induction is of interest from two perspectives. Firstly, the decreased potential for MFO enzyme induction is in general agreement with an overall reduced toxicity potential. Secondly, we have recently established that the induction of MFO enzymes can be expected to appear in fish well in advance of harmful biological effects. Figure 6 demonstrates the percentage change in various indices found in flounder chronically exposed for 4 months to various levels of polycyclic aromatic hydrocarbons (PAH) from petroleum. The higher sediment levels represent PAH concentrations expected around oil rig sites. As indicated, MFO enzyme induction was observed at much lower levels of sediment hydrocarbon contamination and can be expected to appear well in advance of effects of biological significance, making it an attractive monitoring index for delineating zones of "effect/no effect".

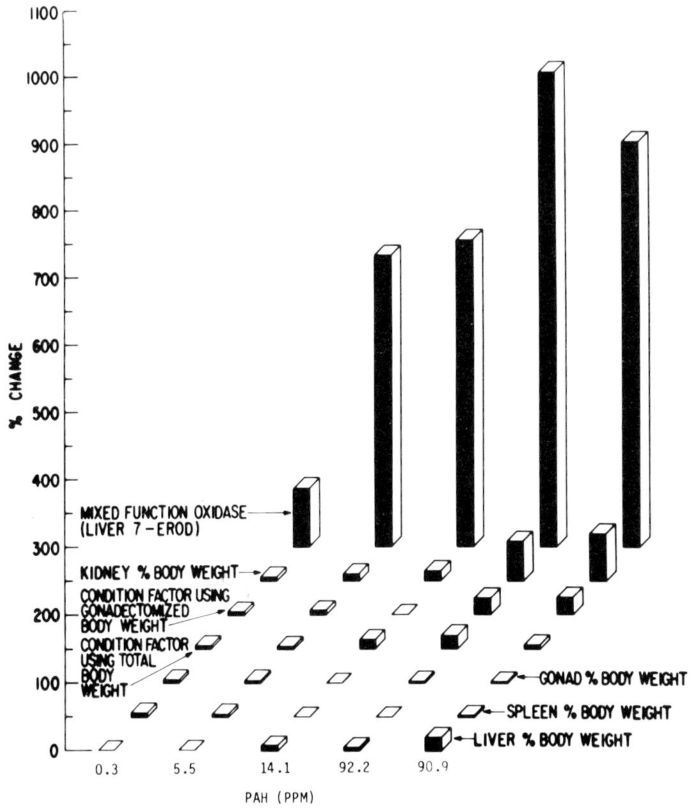

Figure 6. Percentage change in various somatic indices in flounder chronically exposed for four months to various levels of PAH from petroleum.

SUMMARY

Studies were carried out to obtain a better appreciation of the toxicity potential of the new drilling fluid base oils compared with diesel oil, which were traditionally used as base oils in drilling muds. Investigations included studies on base oils per se as well as on model aliphatic and aromatic fractions from petroleum which form the base of different oils. The sensitive chick embryo bioassay was used to evaluate acute toxicity while effects on mitochondrial processes were investigated in order to assess potential perturbations on a critical biological function-namely cellular respiration rate and energy production. Also investigated was the MFO enzyme induction potential of different base oils compared with diesel oil. Overall, base oils are indicated to have a low toxicity potential and it would appear that any concerns related to the usage of 'toxic' diesel oil should be minimized by substitution of low-aromatic base oils in drilling fluid formulations. Evidence was also presented demonstrating that MFO enzyme induction can be expected to act as an early warning response well in advance of the appearance of more serious biological effects. As such, MFO enzyme induction can be a useful tool for delineating areas of potential "effect-no-effect" on fish populations around petroleum development sites using oil based drilling fluids.

REFERENCES

1. Davies, J. M., J. M. Addy, R. A. Blackman, J. R. Blanchards, J. E. Ferbrache, D. C. Moore, H. J. Sommerville, A. Whitehead & T. Wilkinson, 1984. Environmental effects of the use of oil-based drilling muds in the North Sea. Mar. Pollut. Bull. 15: 363-370.
2. Matheson, I., P. F. Kingston, C. S. Johnston & M. J. Gibson, 1986. Statfjord Field environmental study. Report of the Norwegian State Pollution Control Authority/Statfjord Unit Joint Research Project on Oil Based Drilling Fluids: Cleaning and Environmental Effects of Oil Contaminated Drill Cuttings. Trondheim, Norway, Feb. 1986.
3. Engelhardt, F. R., H. Hall, R. J. Paterson & D. C. Strong, 1983. Oil-based Drilling Muds in the North Sea: A Perspective. Environmental Protection Branch Technical Report No. 3, Canada Oil and Gas Lands Administration, Energy Mines and Resources Canada and Indian and Northern Affairs Canada.
4. Hutcheson, M. S., P. L. Stewart, R. Odensc, B. F. Fowler, & D. Green, 1984. Report on development of Toxicity Testing Guidelines for Oiled Cuttings. Atlantic Oceanics Company Ltd. Dartmouth, Nova Scotia, Canada.
5. Bakke, T., N. W. Green, K. Naes & A. Pedersen, 1986. Drill cuttings on the sea bed: field experiment on benthic recolonization and chemical changes in response to various types and amount of cuttings. Report of the Norwegian State Pollution Control Authority/Stratfjord Unit Joint Research Project on OIl Based Drilling Fluids: Cleaning and Environmental Effects of Oil Contaminated Drill Cuttings. Trondheim, Norway, Feb. 1986.
6. Payne, J. F., 1977. Mixed function oxidases in marine organisms in relation to petroleum hydrocarbon metabolism and detection. Mar. Pollut. Bull. 8: 112-116.
7. Payne, J. F., J. W. Kiceniuk, W. R. Squires & G. L. Fletcher, 1978. Pathological changes in a marine fish after a 6-month exposure to petroleum. J. Fish. Res. Board Can. 35: 665-667.

8. Payne, J. F., R. Maloney & A. Rahimtula, 1979. Are petroleum hydrocarbons an important source of mutagens in the marine environment? 1979 Oil Spill Conference, American Petroleum Institute Publication No. 4308: 553-536.
9. Payne, J. F., J. W. Kiceniuk, R. Misra, G. Fletcher & R. Thompson, 1983. Sublethal effects of petroleum hydrocarbons on adult American lobsters (Homarus americanus). Can. J. Fish. Aquat. Sci. 40(6): 705-717.
10. Khan, R. A. & J. Kiceniuk, 1984. Histopathological effects of crude oil on Atlantic cod following chronic exposure. Can. J. Zool. 62(10): 2038-2043.
11. Thoresen, K. M. & A. A. Hinds, 1983. A review of the environmental acceptability and the toxicity of diesel oil substitutes in drilling fluid systems. Report of the IADC/SPE 1983 Drilling Conference, New Orleans, Louisiana, Feb. 1983.
12. Estabrook, R. W., 1967. Mitochondrial respiratory control and the polarographic measurement of ADP:O ratios. In R. W. Estabrook & M. E. Pullman (eds.) Methods in Enzymology, Vol. X, Academic Press, New York: 41-47.
13. Masana, O., M. Takaaki, I. Fumio, E. Kohei, S. Ritsue, M. Tadamichi & I. Bunji, 1983. Classification of potentially toxic chemicals based on their effects on mitochondrial respiration. Physiol. Chem. Phys. Med. NMR 15: 229-232.
14. Ellenton, J. A., 1982. Teratogenic activity of aliphatic and aromatic fractions of Prudhoe Bay crude and fuel oil No. 2 in the chicken embryo. Toxicol. Appl. Pharmacol. 63: 209-215.
15. Albers, P. H., 1978. The effect of petroleum on different stages of incubation in bird eggs. Bull. Environ. Contam. Toxicol. 19: 624-630.
16. Hoffman, D. J., 1978. Embryotoxic effects of petroleum hydrocarbons in mallard ducks and chicks. Toxicol. Appl. Pharmacol. 46: 183-190.
17. Sottocasa, G. L., B. Kuylensteirna, L. Ernster & A. Bergstrand, 1967. An electron transport system associated with the outer membrane of liver mitochondria. J. Cell. Biol. 32: 415-438.
18. Delafield, F. P. & M. Doudoroff, 1969. β-hydroxybutyrate dehydrogenase. In J. M. Lowenstein (ed.) Methods in Enzymology, Vol. XIV, Academic Press, New York: 227-231.
19. Payne, J. F., L. Fancey, J. Kiceniuk, U. Williams, J. Osborne & A. Rahimtula, 1985. Mixed-function oxygenases as biological monitors around petroleum hydrocarbon development sites: potential for induction by diesel and other drilling mud base oils containing reduced levels of polycyclic aromatic hydrocarbons. Mar. Environ. Res. 17: 328-332.
20. Payne, J. F., 1984. Mixed function oxygenases in biological monitoring programs: review of potential usage in different phyla of aquatic animals. In G. Personne, E. Jaspers & C. Claus (eds.) Ecotoxicological Testing for the Marine Environment Vol. 1, State Univ. Ghent and Inst. Mar. Scient. Res., Bredene, Belgium: 625-655.
21. Davies, J. M., J. S. Bell & C. Houghton, 1984. A comparison of the levels of hepatic and hydrocarbon hydroxylase in fish caught close to and distant from North Sea oil fields. Mar. Environ. Res. 14: 23-45.
22. Leaver, M. J, J. M. Davies & D. Raffaelli, 1985. The induction of hepatic aryl hydrocarbon hydroxylase activity in juvenile plaice by exposure to oil based drilling mud cuttings. Mar. Environ. Res. 17: 149.

23. Addison, R. F., K. Doe & A. Edwards, 1984. Effects of oil based drilling mud cutting on winter flounder (*Pseudopleuronectes americanus*): *absence of acute toxicity or mixed function oxidase induction.* Can. Tech. Rep. Fish. Aquat. Sci. 1307: 19 p.

PRACTICAL METHODS FOR MONITORING ROCKY SHORE COMMUNITIES

K. HISCOCK

Oil Pollution Research Unit, Field Studies Council,
Orielton Field Centre, Pembroke, Dyfed, U.K.

1. INTRODUCTION

The shoreline is often the recipient of the most severe impacts from oil spills or effects of low level chronic pollution from effluents. Techniques for monitoring changes in rocky shore communities, including effects of oil pollution, have in part been derived from descriptive survey methods and in part designed for specific purposes. Both approaches may suffer from difficulties if selected methods are inappropriate to monitoring aims or are used on shore types different to those for which they were intended. Reliance on well established but unsuitable techniques has occurred in several instances. There has also been criticism of apparently simple methods by idealists who may imagine that all rocky shores are extensive unbroken platforms at one tidal level, when in practice they are usually assemblages of bedrock, boulders, pools, crevices and sediments at different tidal heights. Also, techniques designed in the office can be highly impactical when the worker reaches the shore.

Techniques available for survey and monitoring of rocky shore communities were described and discussed at a workshop organised by the Field Studies Council and sponsored by British Petroleum International. The conclusions of that workshop (Hiscock, 1985) together with the results of more recent studies are used to present a summary of practical methods for monitoring rocky shore communities.

2. PRACTICAL CONSIDERATIONS

Rocky shore communities are easy to monitor effects of oil pollution because:

- Plants and animals live on the surface and can be surveyed or sampled *in situ* without disturbance of the habitat.
- The range of species present is fairly small.

However, there are also difficulties in studying rocky shores or interpreting data from surveys because:

- The period of time when the tide is out is short and may not occur in daylight.
- Shores are often very broken and include many different habitats and therefore assemblages of species.
- Shore communities are very dynamic.
- Shore communities are heterogeneous even on apparently homogeneous surfaces.
- Species living on the shore live in a stressful environment and do not generally reveal subtle effects of oil pollution.
- Data generated by surveys is often unsuitable for statistical analysis.

Kuiper, J., Van den Brink, W.J. (eds), Fate and Effects of Oil in Marine Ecosystems. ISBN 90-247-3489-4

One of the most important considerations when planning a monitoring strategy is to take account of the shore type to be surveyed. Each provides different problems for the use of various survey techniques. The following main types are considered:

- Extensive unbroken rock platforms or sea walls surveyed at one tidal level. (Eminently suitable for quantitative survey producing results amenable to statistical analysis.)
- Broken bedrock including gullies, overhangs, rock platforms, rock pools, etc. (Require careful instruction on habitats/levels to be surveyed/excluded and care in ensuring accurate site relocation. Quantitative techniques can be used with care.)
- Stable boulder shores which include surfaces ranging from horizontal to overhanging surfaces. (Very difficult to use quantitative techniques though semiquantitative methods can work well.)
- Under boulder communities. (Probably very sensitive to oil spills but not previously monitored. Quantitative or semiquantitative techniques on specific boulders possible. Care required to prevent damage during survey.)
- Shingle or cobbles which may be mobile. (Quantitative survey possible on fairly flat substratum if consideration given to patchiness of communities. Semiquantitative methods work well.)
- Rock pools. (Probably sensitive to oil spills but not previously monitored. Difficult to survey quantitatively and semiquantitative methods best.)

Also the community present will affect the techniques used. Thus, on shores dominated by algae, photography will not be a useful technique if target species are obscured. Counting will be required for some species (such as gastropods) whilst a measure of percentage cover will be appropriate for foliose or encrusting species. However, the measures obtained from these two techniques are not suitable for treatment together in statistical analyses. Techniques available include:

- Visual assessment by experienced scientists.
- Species listing (with or without abundance notations).
- Recording abundance of species at marked sites.
- Quantitative recording using marked areas at selected vertical heights.
- Quantitative recording by replicated random sampling.
- Manipulative studies/settlement plates/transplants.
- Morphological, physiological, cell biology or reproductive studies.

For species listing and recording abundance, it is likely that all conspicuous species will be included. For quantitative or experimental studies, only selected species will most likely be studied.

Survey objectives may include:

- Description of gross change at a site.
- Detection of a gradient from a point source.
- Description of subtle change in populations.
- Detection of sublethal effects.

3. CONCLUSION

The report of the Rocky Shore Survey and Monitoring Workshop (Hiscock, 1985) is essentially a catalogue of the techniques available for surveys and monitoring as well as an appraisal of the advantages, disadvantages and suitability of the methods which were described. At the conclusion of the meeting, a table was prepared which summarised the suitability of different techniques for different objectives. Practicality in relation to different shore types was not considered but is included here in Table 1.

TABLE 1. Summary of survey techniques considered by the author to be appropriate and practical for different monitoring objectives on different shore types. 1 = Highly recommended, 2 = Applicable but not highly recommended, 3 = Not applicable, * = Over-detailed. Based on part on the conclusions of the Rocky Shore Survey and Monitoring Workshop (Hiscock, 1985).

OBJECTIVE	SHORE TYPE	VISUAL ASSESSMENT	SPECIES LISTING	ABUNDANCE OF SPECIES AT MARKED SITES	QUANTITATIVE RECORDING IN MARKED AREAS	QUANTITATIVE RECORDING BY RANDOM SAMPLING	MANIPULATIVE STUDIES/ SETTLEMENT PLATES/ TRANSPLANTS	MORPHOLOGICAL/ PHYSIOLOGICAL/CELL BIOLOGY/ REPRODUCTIVE STUDIES
Description of gross change	Unbroken platforms	1	1	1	1	2*	3	3
	Broken rock	1	1	1	1	3	3	3
	Stable boulders	1	1	1	3	3	3	3
	Under boulders	1	1	1	2	3	3	3
	Shingle and cobbles	1	1	1	1	2*	3	3
	Rock pools	1	1	1	3	3	3	3
Detection of a gradent from a point source	Unbroken platforms	2	2	2	1	1	1	2*
	Broken rock	2	2	2	1	3	1	2*
	Stable boulders	2	2	1	3	3	1	2*
	Under boulders	2	2	1	2	3	1	2*
	Shingle and cobbles	2	2	1	2	2	1	2*
	Rock pools	2	2	2	3	3	1	2*
Description of subtle change in populations	Unbroken platforms	3	3	3	1	1	2	2
	Broken rock	3	3	3	1	1	2	2
	Stable boulders	3	3	2	3	3	2	2
	Under boulders	3	3	3	1	3	2	2
	Shingle and cobbles	3	3	3	1	2	2	2
	Rock pools	3	3	3	3	3	2	2
Detection of sublethal effects		3	3	3	3	3	3	1

REFERENCES

1. Hiscock K: Rocky Shore Survey and Monitoring Workshop. May 1st to 4th 1984. Published by British Petroleum International Ltd, London. 136 pp. 1985.

A SUMMARY OF RESEARCH AND MONITORING STUDIES CARRIED OUT IN NORTH SEA OILFIELDS BY THE OIL POLLUTION RESEARCH UNIT 1973-1986

D. LEVELL

Oil Pollution Research Unit, Field Studies Council,
Orielton Field Centre, Pembroke, Dyfed, U.K.

The Oil Pollution Research Unit (OPRU) has carried out 41 macrobenthic monitoring studies in 20 North Sea oil and gas fields over the past 13 years (see summary Table). These studies have included measurement of benthic sediment parameters such as particle size and hydrocarbon and selected metal concentrations to aid in the interpretation of biological data and for the characterisation of the behaviour and fate of organic and inorganic contaminants.

These surveys have been contracted by the oil industry and completed within strict time and cost constraints. Nevertheless, the wide range of the studies has provided an opportunity to examine some alternative strategies and methodologies for marine benthic monitoring.

The effectiveness of various sampling configurations in detecting and delimiting zones of impact has been investigated, including large-scale regular grids (84 stations - Beatrice 1981, 72 stations - Thistle 1982), radiating transects (Ekofisk), two intersecting transects (Beryl 'B' - 1985) and, more recently in response to government legislation, a single transect aligned with the residual water movement. Varying levels of sample replication have also been employed to assess the degree of field variability and to determine minimum acceptable sample number.

When selecting monitoring strategies, a clear understanding and definition of the aims of the work is vital. On the basis of our studies, it is recommended that ideally, any initial descriptive or 'baseline' survey of a new area should consist of a large number of single sample stations covering as wide an area as feasible (usually out to 5 km or more from platform locations) in order to detect natural environmental and macrobenthic variations. These should be combined with a central 'core' of stations near to the focus of attention and one or more peripheral stations which are likely to remain undisturbed, from which a suitable number of replicate samples should be obtained. The 'core' and reference stations enable a more complete and confident description of the communities and sediment conditions, and would form the basis for future monitoring surveys. This approach is particularly recommended in areas of strong environmental gradients or heterogeneity.

Subsequent monitoring generally requires the detection of trends rather than the determination of absolute values at given locations, and this can permit reduction of sample number compared with the 'baseline'. However, in areas of critical importance such as the definition of boundaries between zones of differing levels of impact, for example, between undisturbed and slightly disturbed zones, it has been shown (in the relatively homogeneous sediments surrounding Beryl 'B') that 5 biological, 3 hydrocarbon, 1 metals and 1 grain size sample are the minimum sampling requirements. Sample replication is required to determine not only total

Kuiper, J., Van den Brink, W.J. (eds), Fate and Effects of Oil in Marine Ecosystems. ISBN 90-247-3489-4

quantitative values at these critical regions, but also the qualitative nature of parameters such as total sediment hydrocarbons composed of varying proportions of, for example, diesel or kerosine base oils, or weathered crude oil.

Investigation into the influence of sieve mesh size on the sensitivity and interpretation of biological monitoring has shown that a moderate increase in sensitivity in the definition of some species distributions resulted from the analysis of 0.5 mm combined with 1.0 mm sieve fractions (Beatrice 1981 and Beryl 1985). This was particularly so with *Pholoe inornata*, *Ophryotrocha* spp. and ?*Raphidrilus* sp., three small opportunist polychaete species which occur in distinct zones surrounding some long-established platforms discharging oil based mud contaminated drill cuttings. Relatively little new information on the trends of biological community parameters and the majority of species distributions resulted from the combined data from 0.5 and 1.0 mm sieves when compared with the analysis of the 1.0 mm fraction alone. This small increase in sensitivity for some species is not felt to be justified by the disproportionate increase in laboratory analytical commitment necessary to process the smaller size fraction.

The data gathered by OPRU whilst monitoring the discharges in oilfields developed using water based, diesel based or alternative base oil drill muds have contributed, along with other monitoring agencies and contractors, to the description and definition of zones and scale of disturbances found in a variety of North Sea oilfield developments (Davies *et al*., 1984).

Several oilfields have now reached the peak of developmental drilling, and the decline in drilling and associated discharge volumes will provide an interesting opportunity to study and assess the degradation rates of sediment hydrocarbons and any concomitant macrobenthic community recovery. Areas currently of particular interest are:

1. The determination of the causative agents of the environmental impact of drilling discharges, i.e. direct physical smothering and sediment disturbance *vs*. organic enrichment *vs*. toxicological impact of drill mud components.
2. The relative impact of the newer low toxicity oil based muds compared with diesel based muds and the relative longevity of effects.
3. The continual problem of the definition of local background or undisturbed conditions.
4. The effects of the increasingly significant quantities of production water and associated chemical discharges.
5. The inconsistency and incompatibility of different analytical methodologies.

ACKNOWLEDGEMENTS

The contribution of numerous OPRU staff, both past and present is gratefully acknowledged. OPRU wishes to express thanks to the Oil Companies listed below who have shown an enlightened attitude to environmental monitoring, and enabled more than just the obligatory minimum survey work to be performed.

	Oil or Gasfield	Operator	Dates surveyed
1.	Amethyst	Britoil plc	1986
2.	Auk	Shell (U.K.) Expro.	1977
3.	Beatrice	Britoil plc	1977,1980,1981,1982,1983,1985
4.	Beryl	Mobil North Sea Ltd	1983,1985
5.	Block 11/25	Shell (U.K.) Expro.	1985
6.	Buchan	BP Pet. Dev. Ltd	1980,1982
7.	Cleeton	BP Pet. Dev. Ltd	1986
8.	Cyrus	BP Pet. Dev. Ltd	1986
9.	Eldfisk	Phillips (Norway)	1984
10.	Ekofisk	Phillips (Norway)	1973,1975,1977,1981,1984
11.	Forties	BP Pet. Dev. Ltd	1975,1978,1981,1983,1984
12.	Indefatigable	Shell (U.K.) Expro.	1986
13.	Leman	Shell (U.K.) Expro.	1986
14.	Magnus	BP Pet. Dev. Ltd	1979
15.	Maureen	Phillips (London)	1979,1981,1983
16.	Ravenspurn	BP Pet. Dev. Ltd	1986
17.	Sean	Shell (U.K.) Expro.	1986
18.	Thistle	Britoil plc	1982,1985,1986,1986
19.	Tor	Phillips (Norway)	1984
20.	Ula	BP Pet. Dev. Ltd	1984

BIBLIOGRAPHY. The following publications include data generated by the above monitoring surveys.

1. Addy JM, Levell D and Hartley JP: Biological monitoring of sediments in the Ekofisk Oilfield. In: Proc. conference on Assessment of Ecological impact of oil spills. American Institute of Biological Sciences, Arlington, Va. 514-539, 1978.
2. Hartley JP: Biological Monitoring of the Seabed in the Forties Oilfield. In: Proc. of Conference on Ecological Damage Assessment. Soc. of Petroleum Industry Biologists, Arlington, Va. 215-229, 1979.
3. Hartley JP: Methods for monitoring offshore macrobenthos. Mar. Poll. Bull. 13(15), 150-154, 1982.
4. Hartley JP: The benthic ecology of the Forties Oilfield (North Sea). J. Exp. Mar. Biol. Ecol. 80, 161-195, 1984.
5. Hartley JP and Ferbrache J: Biological monitoring of the Forties Oilfield (North Sea). In: 1983 Oil Spill Conference (Prevention, Behaviour, Control, Cleanup), San Antonio, Texas, U.S.A. 407-414, 1983.
6. Hartley JP and Bishop JDD: The macrobenthos of the Beatrice Oilfield, Moray Firth, Scotland. Proc. Roy. Soc. Edin., in press.
7. Bishop JDD and Hartley JP: A comparison of the fauna retained on 0.5 mm and 1.0 mm meshes from benthic samples taken in the Beatrice Oilfield, Moray Firth, Scotland. Proc. Roy. Soc. Edin., in press.
8. Addy JM, Hartley JP and Tibbetts PJC: Ecological effects of Low Toxicity Oil-based Mud Drilling in the Beatrice Oilfield. Mar. Poll. Bull. 15(12), 429-436, 1984.
9. Davies J, Addy J, Blackman R, Ferbrache J, Moore D, Somerville H, Whitehead A and Wilkinson T: Environmental effects of oil-based mud cuttings. Mar. Poll. Bull. 15, 363-370, 1984.

Kuiper, J., Van den Brink, W.J. (eds), Fate and Effects of Oil in Marine Ecosystems. ISBN 90-247-3489-4

SECTION IV
RISK MANAGEMENT AND RISK ANALYSIS

SEABEL
HAZARD IDENTIFICATION AND RESPONSE DECISION FOR CHEMICAL SPILLAGES AT SEA

Ir. P. Bockholts
TNO, Division of Technology for Society, Department of Industrial Safety
Apeldoorn, the Netherlands

1. INTRODUCTION

Chemicals are commonly used products in our today's life. Applications are widely spread, resulting in an extensive transport; throughout the world large quantities of chemicals are shipped over oceans, seas and other waterways. This transport includes certain dangers to people and the environment. International regulations apply and are agreed upon by many countries, with the aim to reduce the risks to a minimum. These preventive measures do not exclude any accident and they occur. Some recent cases are listed in table 1.

Emergency response facilities are therefore required in order to minimize the consequences and damage in case of accidents. With accidents we mean: pollution of sea and shore, harm to the health of people, effects on fishing grounds, shipping trade, the marine environment etc.

In order to act adequately in case of accidents due to potential spillages of dangerous substances and to prevent calamities, many countries have established an emergency management organization.

In the Netherlands such an emergency management organization for combating calamilities in case of serious accidents on the North Sea is a subdivision of the North Sea Directorate of the Ministry of Works.

Kuiper, J., Van den Brink, W.J. (eds), Fate and Effects of Oil in Marine Ecosystems. ISBN 90-247-3489-4

Type of incident	Name of ship	Date of incident	Chemical(s) involved/ amount	Main threat	Final measure(s)
Sinking of ship after collision	Olaf	July 1986	Fly ash	Bio-accumulation of heavy metals	Salvage of ship and its cargo
Collision of fishing vessel with tanker	Orleans	January 1986	Oil	Fire on tanker	Natural dispersion (wind force > 7)
Collision of ferry with vessel	Mount Louis	August 1984	Uranium hexafluoride	Toxic gas; threat to human health	Salvage of 30 tanks
Loss of lorry with drums from vessel	Dama Optima	January 1984	Dinoseb	Marine toxicity	Some drums could be traced, some are still missing
Collision	Katina	June 1982	Heavy fuel oil 1650 m^3 oil containing 45% water	Oil pollution; threat to sealife 150 m^3 washed up on the Dutch coast	1440 m^3 has been recovered by mechanical removal techniques
Collision in foggy weather	Stanislaw Dubois	April 1981	Caustic soda, calcium carbide, 465 tonnes	Explosion; reaction between calcium carbide + seawater gives explosive acethylenegas	Ship has finally been scuttled because no permission was given to enter port
Loss of 51 chlorine cylinders	Sindbad	December 1979	Fluid chlorine pressure 2-6 bar; one cylinder contains 400 m^3 gas	Spontaneous burst of cylinder due to corrosion. In worst case inhalation can lead to death	Some cylinders have been destroyed by experts at open sea; 13 are still missing
Polymerization reaction in tanker in the harbour	Stolt Busan	March 1979	Styrene	Fire, explosion; threat to human health	Cooling with water; Addition of toluene and inhibitor
Escape of oil in sea during well-testing	Penrid 58	January 1979	Oil with high specific weight and great viscosity	Oil pollution of the sea	33 tonnes oil/water emulsion mechanically removed
Run aground	Amoco Cadiz	March 1978	Oil	Oil pollution; threat to sealife and coast	Mechanical removal
Loss of tank cars from ferry during storm		November 1977	Ethylene oxide	1 tank car washed ashore	Tank car was removed from beach
Oil-gas blow out at platform	Bravo	April 1977	Oil-gas mixture	Blow out of 20.000 tonnes during 12 days	Oilslick completely disintegrated by natural weathering
Leaking container that exploded later	Franciszek Lubrzycki	March 1977	42 barrels of sulphur dichlorine	Toxic and corrosive gases; immediate threat for humans, esp. lungs and eyes	Transport of container to open sea, where contents of drums were spouted out

Table 1. Some accidents in which the Ministry of Works has been involved. Information source: TNO Incident Database FACTS.

The North Sea Directorate has great experience with combating oil pollution and spends great effort in developing new techniques for combating spilled chemical substances in particular for the North Sea. The EC also wishes to promote the development of new tools for combating chemical spillages at sea. It is within this scope that TNO is developing a decision support system for combating of chemical spillages at sea. The decision support system is called "SEABEL". In this system today's knowledge on estimating the physical effects of chemical spills by simulation models and the knowledge on available combating techniques will be implemented. The system will become available in two versions: a computer program and a Handbook.

2. OBJECTIVES

Today's hazard identification and response decision-making for chemical spills at sea is based on experts' best knowledgde and judgement. However, the matter is recognized as a complex one, where many different elements play a role; e.g. behaviour and properties of chemicals, the variety in actual situations at the spot of the accident. Also legal and insurance aspects have to be considered in certain cases.

For some of the incidents the (potential) threat for either people (harbour workers, salvors etc.) or for the environment was so great that the Emergency Situation came into operation. Table 1 may illustrate among others the main threat of the incident and the measures that were taken to get control of this threat.

The information system is a tool for decision makers because more different aspects can be considered than an expert can do and predictions or conclusions can be drawn more accurately and consequently than an expert can do. It may be expected that particularly in cases where time pressure is felt, a decision support system will have its advantages. In addition to this, the system can be used to investigate other cases with certain similarities in various ways. It can also be used for training and education purposes.

Three basis questions are to be answered by decision makers for emergency response:

1. What is the actual and predicted hazard.
2. How urgently are decisions to be drawn.
3. What measures are suitable in terms of success and required effort.

The answer to the first question should be a comprehensive picture of the actual danger and its potentials in terms of harm to people, property loss, and environmental damage and consequences.

The second question is important since it is decisive in the case immediate actions are required or not and whether more time is available for further detailed investigations of the accident and the preparation of the most suitable response method.

The third question deals with the selection of appropriate measures based on available methods, responsibility, technical feasibility, expected success, required effort and costs.

3. HAZARD IDENTIFICATION AND EMERGENCY RESPONSE

The decision support system can be divided into two main parts:

* The first part deals with the hazard identification.
* The second part of the system deals with the selection and estimation of obtainable success of individual measures and their weighing.

The hazard identification results in a quantitative and qualitative presentation of the effects of a release of a chemical substance. Depending on the type of incident (dispersion, fire, explosion, etc.), time related concentration curves, heat radiation curves and overpressure curves are calculated for a particular area. On the basis of these calculations the damage to persons, biological life, properties and the environment is estimated and the potential danger is indicated. The output of the hazard identification part is input for the response decision part.

The nature and gravity of the (possible) effects and consequences of a release of a chemical substance are the first indications for measures to be taken.

The location of the incident, weather conditions, time pressure and other limiting conditions influence the decision of the measures to be taken. If the selection of measures yields more possibilities, precluding each other, these measures have to be weighed one against another in an evaluation process. The amount of reduction of damage to people, biological life, properties and environment could be a graduated scale for the assessment of the possible measures.

4. SEABEL STRUCTURE

SEABEL will consist of four modules each covering a separate function. The four modules are logically linked but can also be used independently:

The functional models are:

Module I: accident diagnosis, dealing with data gathering from the accident;

Module II: effect diagnosis, dealing with simulation of the effects of spilled chemicals;

Module III: hazard identification, dealing with the estimated effects related to the actual area of the accident;

Module IV: decision support module, dealing with the selection and evaluation of the measures to be taken.

Module I covers the information gathering procedure. It will supply the user with information collection in a structured way in order to avoid omissions. For this purpose a set of checklists is available from which the relevant ones will be presented on the screen. Furthermore a set of databases is available in the module in order to search data on chemicals, equipment and recommended data for not obtainable ones. The output of the module is a list of all relevant items on the described accident including all gathered data. The output is a compilation of relevant checklists. The user will have 3 basic options at his disposal:

* data entry on selected and on the screen presented checklists.
* data retrieval in the various databases.
* data review for verification and data entry progress checking.

Additional functions are:

* references to competent information sources
* user instructions.

The ultimate aim is that module I contains all information that is available about the accident. In the subsequent modules this information is used for prediction of the hazard identification and emergency response decision support.

Module II covers the simulation of the effect of the spilled chemicals. Various mathematical models are available in SEABEL with which the behaviour of the chemicals can be described.

The following models will be included in the present version:

* releases of gases and liquids
* heat radiation
* vapour cloud explosions
* dispersion of neutral and heavy gases
* oil slick behaviour.

Depending on the chemicals that are to be considered one or more of these models are used in order to calculate phenomena like amount of spilled material, the travel of the chemicals in the atmosphere and on the water surface, peak pressures of explosions and heat production of fires. For other phenomena models can be included at a later stage. It is obvious that the data for the effect calculations are obtained from the previous module; they are partially dependent on the actual accident and partially obtainable from databases in the SEABEL system.

The results of the calculations give a realistic impression of the expected consequences of the accident in terms of physics. The results need now to be merged with the location in order to estimate the real hazards for people, property and environment. This is handled in the next module.

Module III merges the estimated effects of the spilled material with location-dependent information. The information will be available in a geographic database. This database contains data on traffic lines, fixed objects like platforms etc., cables, pipelines, dumping areas, wrecks and shoreline. The function of merging results in a graphical presentation of the harmed area. A presentation of data that represent the hazards for people, property and environment can then be given. At this stage the hazard identification is complete and emergency response can be selected and evaluated. This part is handled in module IV.

Module IV . All information for a hazard identification is now completed and the decision making for emergency response can be processed. The elements to be considered are summed up and related to each other.

The decision to be taken consists of the following stages:

1) selection of measures to be evaluated;
2) evaluation of selected measures.

In the first stage all applicable measures, based on the description of the incident, are selected. In the second stage the measures are evaluated using a certain set of judging rules.

In order to be able to make the selection, certain data must be available with respect to the applicability of measures under various circumstances . A database with all available measures is therefore included in SEABEL. Each measure is described in terms of application, consideration, limitation and conditions for use. The effectiveness of each measure is related to the "do nothing" option.

Parameters that are used for the evaluation are number of victims, property damage, environmental damage, costs of measure, value of cargo, effectiveness in terms of repression and prevention, compulsory, recommended or possible, legal and insurance aspects. The effects of the measures on the calamity are examined not only in the short term but also in the long(er) term. Finally the measures that have been selected are to be evaluated in comparison with each other in order to make the right choice.

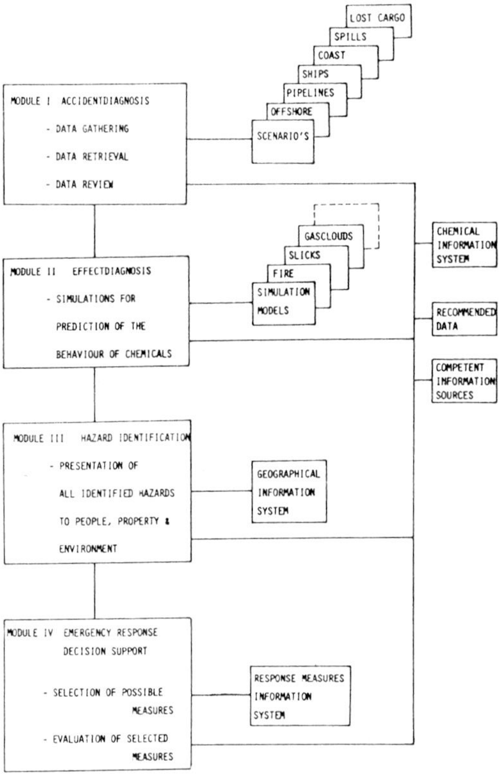

Structure of the SEABEL system.

5. CONCLUSIONS

Although the project that is described here is not yet fully operational, some conclusions can be drawn:

* The decision support system is a frame in which today's knowledge on the combating of chemical spillages at sea is embedded. New developments can be added at any convenient time.
* The connection with already existing tools like databases, models, checklists etc. into a frame with a very modular structure substantially enhances the power and capabilities of the individual systems.
* The system will be a powerful tool for decision makers, but will never replace the experts; it is a tool on behalf of the experts
* Apart from use in operational circumstances, systems like SEABEL can be used for training purposes. Cases can be defined in order to train operational personnel as well as management. Experience can be achieved in aspects of time lapses, escalation mechanisms, obtainable success.

OIL SPILL RISK MANAGEMENT: THE STRATEGIC RESPONSE CASE

H.N. PSARAFTIS
Department of Ocean Engineering (Room 5-211)
Massachusetts Institute of Technology,
Cambridge, MA. 02139
USA.

1. INTRODUCTION AND OUTLINE OF THE PAPER

The purpose of this paper is to examine some important aspects of oil spill risk as they relate to the strategic problem of providing adequate response capability to cope with oil spills that may occur in a certain geographical area. The paper explains the rationale of the methodology developed and discusses an illustrative application in the U.S. New England region.

Managing the risk of oil spills includes developing strategic response plans for what should be done ahead of time so that oil spills that may occur in the future can be responded to in an acceptable way. The strategic aspect of the response problem involves the making of the following types of decisions: (a) where should one locate adequate response capability? (b) what should be the levels of emergency response equipment that should be stockpiled for this purpose? (c) what types of equipment should be stockpiled? and (d) what should be the policy of allocation of such response capability among points or zones of high oil spill risk?

Strategic response decisions typically involve planning horizons of considerable duration (eg. 5 to 15 years). Since those decisions have to be made before actual oil spill incidents in the area of interest occur, the strategic planner has to base those decisions on information on the risk of oil spills in that area. There are two main dimensions of oil spill risk, and both should be explicitly considered by the planner: The first is the probabilistic behavior of oil spill occurrence, and the second concerns the economic and environmental consequences of oil pollution damage. Of course, not all oil spill response decisions are strategic in nature. Tactical (or operational) decisions deal with actions that are taken after the occurrence of a particular spill, and are in themselves an important part of the oil spill response problem. However, a detailed consideration of such decisions is beyond the scope of this paper, which focuses on the strategic aspect of the problem (for detailed analytical modeling of tactical and operational decisions see Ziogas [10], Psaraftis and Ziogas [4], and Demis [1]).

The first dimension of oil spill risk concerns the probabilistic behavior of oil spill occurrence. It turns out that this behavior exhibits certain peculiarities that cannot be ignored in the strategic response problem. For instance, it is generally accepted that the distribution of the volume of an individual spill is extremely skewed (more on this in Section 2.1). In that respect, it becomes meaningless to consider statistics such as average spill size, let alone base strategic response decisions on such statistics. Another probabilistic feature that is relevant in this problem is that the simultaneous occurrence of two or more spills in the same area is very unlikely (see Section 2.2). This means that in designing a

Kuiper, J., Van den Brink, W.J. (eds), Fate and Effects of Oil in Marine Ecosystems. ISBN 90-247-3489-4

strategic response system one need not worry too much about the possibility that equipment stockpiled in one location may have to be used for two spills that occur in the same vicinity at the same time.

The second dimension of oil spill risk concerns the consequences of oil spills. Despite the general consensus that oil spill damage assessment is not a precise art, in Section 2.4 we briefly report on recent progress toward quantifying the damage costs of an oil spill under a variety of senarios.

In addition to information on oil spill risk, solving the strategic response problem involves assumptions regarding the criterion according to which a good (or optimal) system must be chosen (see Section 2.3). In this study we have assumed that our objective is to minimize the expected sum of system costs (costs to create and maintain the response system) and damage costs (as described above). We also include "policy stipulations" that force the system to respond to spills up to a certain volume within a prespecified time (see Section 2.5).

In Section 3 of the paper we present a formal list of inputs to and decision variables of our model, and also briefly outline our solution methodology.

Our New England illustrative application (section 4) seems to indicate, among other things, that (a) it is generally less cost-effective to invest in expensive large-scale cleanup technology that would be used only rarely for large (> 50,000 gallons) spills, than to acquire capability to cope with smaller spills that occur on a daily basis, and (b) equipment to combat small spills should be geographically dispersed across many small local response sites, whereas capability to respond to large spills should be consolidated at one or a few large-scale strike centers.

We conclude in Section 5 with comments on the results of this work.

2. PROBLEM STRUCTURE AND MODELING PHILOSOPHY

2.1 Large Variability of Spill Volumes

The probabilistic peculiarities of oil spill occurrence have been reported by various researchers who analyzed oil spill statistics in the past - among others, by Devanney and Stewart [2]. As mentioned earlier, the distribution of the volume of an individual spill is extremely skewed. To state one example, the average spill size of all spills in and around U.S. waters in 1982 was reported by the U.S. Coast Guard (USCG) to be 2,290 gallons [8]. However, 55% of these spills were at least 20 times smaller than this average, while 61% of all oil spilled was spilled in just two spills, each of which was over 4,000 times the above average [8]. It becomes therefore obvious that in managing the risk of future spills, one should devise a method that explicitly takes into account the large variability of spill volumes, rather than their average values.

2.2 Non-Simultaneity of Spills

It can be also reasonably assumed that a simultaneous occurrence of two (or more) spills which would cause queueing or congestion in the use of the response resources of a particular geographical area is very unlikely (although not impossible). The general validity of the non-stimultaneity assumption was tested in TSC [7], where the probability of one overlap of two or more spills of 50,000 gallons or more within a year and within the entire U.S. was estimated to be of the order of 0.01 to 0.001 for a typical spill duration of 5 days. The same study indicated that there were only about 20 spills in the above size range each year during the 1974-77 period.

2.3 Criteria for the Strategic Problem

In this paper we shall assume that the goal in the strategic response problem is to minimize the expected sum of response system costs and the costs due to damages from spills that may occur in the area, the latter costs multiplied by a user-specified "weight".

The consideration of both system costs and damage costs as part of the problem objective makes sense intuitively, because any response system requires funds and one would like to know not only how much a system would cost but also how much damage that system would avert. However, it is clear that such an objective function implies not only risk neutrality on the part of the decision-maker, but equally importantly, that oil spill damage costs can be evaluated with some confidence. We have assumed risk neutrality both for analytical convenience, and because very little or nothing has been reported to date regarding the risk preference structure of "society" regarding oil spills (the possible exception is the recent work of a Norwegian research group who seem to have obtained some insights on this issue - see Fredrikson [3]). At the same time, and for reasons similar to those discussed earlier regarding expected volumes, we shall have to introduce constraints that somehow capture the decision-maker's risk aversion, particularly when it comes to very large, very rare, catastrophic spills. This will be discussed in section 2.5.

2.4 Treatment of Damages

Regarding the evaluation of damages, despite the general consensus that "oil spill cleanup and damage assessment are not, and are never likely to be precise arts" [9], some progress has been made in this area in recent years, and related efforts are continuing. In this paper we shall take advantage of recent related work at MIT, the purpose of which has been to quantitatively evaluate the damage costs of an oil spill under a variety of scenarios. Such an approach takes an input spill-specific information (location, size, sea state, wind, oil type), area-specific information (inventory of environmental and economic resources), and information on the response, and produces estimates of damages, broken down in several categories (value of lost oil, organisms, beaches, marshes, recreation, etc). A detailed description of the damage assessment algorithms is beyond the scope of this paper and can be found elsewhere (see for instance Demis [1]). This paper assumes that these damages can be predicted as a function of several spill parameters and of the response to the spill and describes how such information can be used in the strategic decision-making process. This approach exhibits considerable flexibility with regard to possible future improvements in damage assessment.

Regarding now the damage weight, its role is twofold: First, it can be used to represent how much the decision-maker is willing to pay in system costs in order to reduce damage costs by $ 1 (and, in that respect, a high value of that weight increases the relative importance of damage costs vis-a-vis system costs). Second, the weight can be used to perform sensitivity analysis on the value of damages, which, as mentioned before, is never likely to be precisely known.

We should finally mention that in the strategic response planning context, total damage costs can be grouped into two components, the damage "due to unsatisfied demand," and the damage "due to delay and equipment inefficiency." These two components are defined and discussed in Section 3.3.

2.5 The "Benign Neglect" Response and Policy Stipulations on Response Times

Our model assumes that a "benign neglect" response to a spill is a *permissible* response. Of course, this assumption is patently false in the real world, where "benign neglect," or, essentially, "doing nothing," is likely to be not only politically unacceptable, but also illegal under existing regulations. Nevertheless, and since oil pollution combat funds are by no means unlimited, we feel that one cannot reject such an assumption *a priori*, particularly if a "benign neglect" response can be justified on a cost/benefit basis (if the latter is the case, one can essentially rename that response as "surveillance and monitoring" so as to make it politically more palatable). Our model will give us the flexibility to examine the merits of such an option.

At the same time, there are constraints that the decision-maker may wish to impose so as to *guarantee* some minimum level of response. An example of such constraints is the directive issued by President Carter to the USCG in 1977 "respond adequately to a 100,000 ton spill within 6 hours," provided of course that the phrases "respond adequately" and "within 6 hours" are unambiguously defined. In this paper we shall have the option to consider similar "policy stipulations," which will be seen to translate into probabilistic constraints on response times. Such policy stipulations can be considered to reflect the decision-maker's risk aversion in an *indirect* way, not only in terms of insuring against the adverse consequences of a catastrophic spill in a particularly vulnerable area, but also in cases the risk the decision-maker is averse to is a "political" risk, that is, the risk of the system being unable to handle a very large spill.

3. THE MODEL

3.1 Inputs

The model inputs are the following (inputs with indices i,j,e, and k in parentheses are considered for all possible values of these indices):

J = Known and finite set of oil spill "risk points", that is, points in the area of interest where oil spills are likely to occur. These can be offshore drilling locations, points along tanker traffic lanes, etc.

f(j) = Oil spill frequency of occurrence (spills/year) at risk point j of set J.

K = Known and finite set of oil spill volumes (to discretize continuous spill volume distribution). Usually we take K = 3.

v(j,k) = Discretized oil spill volume of index k (k belongs to K) at risk point j of set J.

T(j,k) = Duration of oil spill volume v(j,k) (in hours).

p(j,k) = Probability that spill at j will have a volume equal to v(j,k).

I = Known and finite set of candidate stockpiling facilities in the area of interest.

FC(i) = Fixed cost to open facility i of set I (equivalent annualized value, \$/year).

E = Known and finite set of equipment types.

a(i,e) = Acquision cost of equipment type e of set E stockpiled at facility i of set I, per unit of capability stockpiled. Includes manning and maintenance (in \$/gallon of capability/year).

OP(e,j) ="Coefficient of operability" of equipment type e of set E if deployed at risk point j of set J. This is the probability that weather conditions at j will allow operation of that equipment type (between 0 and 1).

RE(e) ="Coefficient of recovery efficiency" of equipment e of set E, given equipment is operable. This is the fraction of oil volume encountered by the device that is actually recovered (between 0 and 1).

d(i,e,j)=Dispatching time of equipment type e of set E from facility i of set I to risk point j of set J (in hours).

TC(i,e,j)= Transportation cost of equipment type e of set E from facility i of set I to risk point j of set J, per unit of capability transported (in \$/gallon of capability).

b(i,e,k)=On-scene operational cost of equipment type e of set E from facility i of set I to a spill of volume v(j,k), per unit of capability (in \$/gallon of capability).

DP(j,k) ="Damage potential" of a spill of volume v(j,k) at j of set J. This is the damage cost per unit volume of nonrecoverable oil if a spill of volume v(j,k) occurs at risk point j of set J (in \$/gallon). Input furnished by the damage assessment subroutine.

w(j,k) ="Damage rate" of a spill of volume v(j,k) at j of set J. This is a derived input, equal to DP(j,k)/T(j,k) (in \$/gallon/hr).

W =User-specified "damage weight" (positive). This tells us how damage costs are weighed vis-a-vis system costs. If W=1, these are weighed equally.

In addition to the above standard inputs, there are also several optional inputs, in case policy stipulations on response times are imposed. These optional inputs are the following for each risk point j of set J:

a(j) =User input between 0 and 1 (non-dimensional).

R(j) =User input greater than 0 (in hours).

Q(F(j)) =Derived input. This is the "F(j)-fractile" of the spill volume distribution at risk point j of set J, that is, Q(F(j)) = CDF^{-1}(F(j)) where CDF^{-1} is the inverse of the cumulative distribution function of the discretized spill volume at j.

A strategic response plan satisfies an (F, R) stipulation if it guarantees a response capability of at least Q(F(j)) to every risk point j of set J, and, if all equipment comprising such capability can be dispatched to j in no more than R(j) hours.

3.2 Decision Variables

The model's decision variables (or outputs) are the following:

y(i) = 1 if candidate location i is opened, zero otherwise.

u(i,e) = Response capability of equipment of type e stockpiled at facility i (in gallons)

r(i,e,j,k) = Response capability of equipment of type e stockpiled at i which is designated to respond to a spill of volume v(j,k) at risk point j(in gallons).

We note here that the capability u(i,e) of an equipment package of type e of set E located at facility i of set I is defined as the maximum volume of spill that that package is capable of fully recovering in the hypothetical situation in which (a) the package can operate on scene for the entire duration of the spill and (b) its recovery efficiency is equal to 1.0. Of course, neither (a) or (b) will hold in general, but the model takes this into account via inputs d(i,e,j), OP(e,j) and RE(e), as defined earlier (for a complete analysis of how these inputs are considered see the model's mathematical formulation in Tharakan [6] and Psaraftis et al [5]).

3.3 Objective Function Components

As stated in Section 2.3, in the strategic response problem we wish to minimize the expected sum of response system costs and the costs due to damages from spills, the latter multiplied by a user-specified damage weight.

The objective function in this problem can be written as follows:

$$Z = Z(1) + Z(2) + Z(3) + Z(4)) + W(Z(5) + Z(6)).$$

where terms Z(1) to Z(4) are components of the system response cost, and components Z(5) and Z(6) are components of the damage cost. Specifically, those components are defined as follows (all costs are per annum and in dollars):

Z(1) = Cost to open new facilities.
Z(2) = Cost to acquire and maintain capability.
Z(3) = Expected cost to mobilize and transport equipment to the spill site.
Z(4) = Expected cost of on-scene cleanup.
Z(5) = Expected cost of damage "due to unsatisfied demand" (see below).
Z(6) = Expected cost of damage "due to delay and equipment inefficiency" (see below).

All six components are known functions of (a) the problem inputs defined in Section 3.1 and (b) the decision variables defined in Section 3.2 (see Psaraftis et al [5] for more details). The problem is to identify a combination of decision variables so that Z is minimized.

The definition of Z(5) and Z(6), the two damage cost components, merits some discussion. Z(5) is the damage cost that certainly occurs whenever the total on-scene capability (after all equipment comprising such capability has arrived on scene) is <u>less</u> than the volume of the spill (hence the term "unsatisfied demand"). But even if there is no unsatisfied demand, the fact that equipment cannot arrive on scene instantaneously, and the fact that in general there are losses due to equipment inefficiencies, would make some additional oil escape. The damage due to that oil is represented by Z(6). A prompt response plan would generally result in a low Z(6), whereas a full-scale mobilization of equipment would generally result in a low Z(5). From a strategic response perspective, differentiating damage costs according to this scheme is important and can provide further insights on the merits of a particular response plan.

In Tharakan [6] and in Psaraftis et al [5], the strategic response problem was formulated as a mixed integer programming optimization problem. To date, we have been able to reasonably handle problems involving up to 8 facilities, up to 5 equipment types at each facility, up to 19 risk points and up to 3 spill volumes at each risk point. The code was written in FORTRAN and implemented on a VAX 11/782. Running times for typical problems have been no more than 3-4 CPU minutes per run of the strategic model. This suggests that even larger problems can be easily tackled, particularly if the codes implementing the model are further refined.

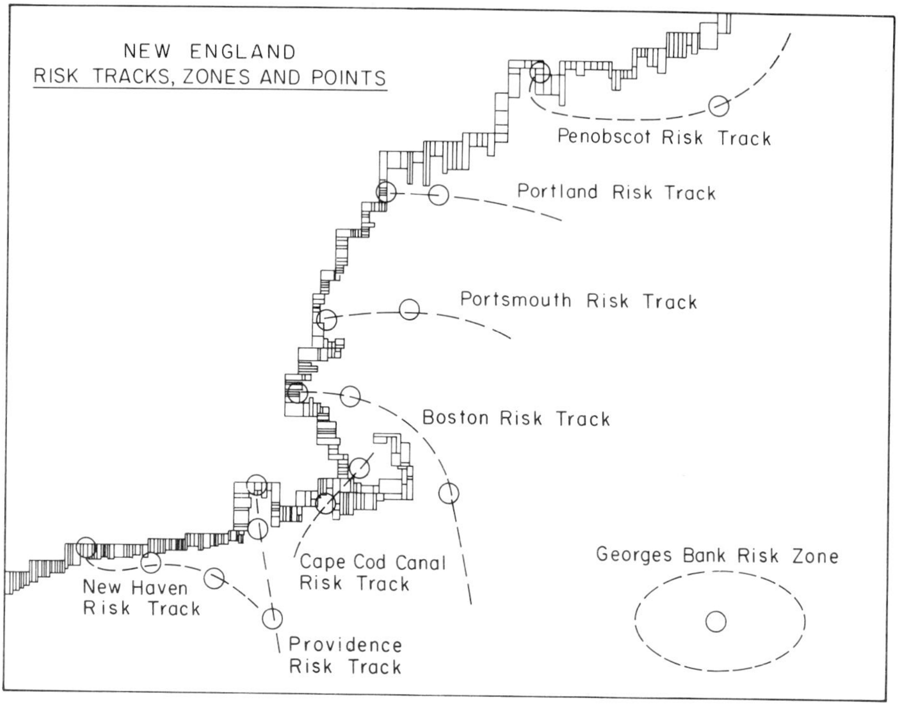

FIGURE 1.

j	Location	f(j) (spills/year)
1*	Penobscot Bay, ME (harbor)	7.00
2	Penobscot Bay, ME (offshore)	0.04
3*	Portland, ME (harbor)	55.00
4	Portland, ME (harbor)	0.35
5*	Portsmouth, NH (harbor)	6.76
6	Portsmouth, NH (offshore)	0.04
7*	Boston, MA (harbor)	60.00
8	Boston, MA (offshore # 1)	0.19
9	Boston, MA (offshore # 2)	0.19
10*	Providence,RI (harbor)	66.75
11	Providence, RI (offshore # 1)	0.21
12	Providence, RI (offshore # 2)	0.21
13*	New Haven, CT (harbor)	44.25
14	New Haven, CT (offshore # 1)	0.14
15	New Haven, CT (offshore # 2)	0.14
16	Cape Cod Canal, MA (offshore # 1)	0.06
17	Cape Cod Canal, MA (offshore # 2)	0.06
18*	Georges Bank, MA (operational)	56.00
19	Georges Bank, MA (blowouts)	0.02

TABLE 1: Risk points and frequencies. Asterisks denote "small" spills.

4. AN ILLUSTRATIVE EXAMPLE

This section presents an illustrative application of our model in the New England area. The main features of the scenario that we examined are the following (see Psaraftis et al [5] for complete details):

(1) There are a total of 19 risk points (see Figure 1). Their locations and frequencies are displayed in Table 1. Frequencies have been derived using information on oil throughputs in the area and the methodology outlined in Devanney and Stewart [2]. The risk points can be divided into two major categories,"small" spills (up to about 50,000 gallons) and "large" spills (50,000 gallons and above). Spill volume distributions have been discretized to 100, 10,000 and 10,000 gallons for small spills and to 100,000, 1,000,000 and 10,000,000 gallons for large spills. The corresponding probabilities (p(j,k)'s) have been set to 0.895, 0.0855 and 0.0195 for harbor spills, 0.989, 0.01 and 0.001 for platform operational spills, 0.25, 0.60 and 0.15 for tankship spills and 0.61, 0.32 and 0.07 for platform blowout spills. Notice the dramatic frequency difference between small and large spills.

(2) There are a total of 8 candidate equipment stockpiling facilities: Portland, ME; Portsmouth, NH; Boston, Woods Hole and Provincetown, MA; Providence, RI; and New Haven, CT.

(3) We assume a total of 4 candidate equipment types, all in the mechanical removal category. Categories 1 and 2 are essentially associated with small spills in protected areas, while categories 3 and 4 consist of heavy-duty equipment, typically used in large high-seas spills. Category 2 (4) differs from category 1 (3) only in acquisition costs, efficiency and operability, all of which are higher for category 2 (4) than for category 1(3).

j	DP(j,1)	DP(j,2)	DP(j,3)
1*	10.88	7.88	16.30
2	1.39	1.08	1.06
3*	59.20	68.39	51.60
4	1.34	1.18	0.98
5*	10.04	3.67	1.39
6	2.22	1.28	0.99
7*	89.34	91.28	75.90
8	1.32	1.19	1.03
9	1.15	1.12	1.04
10*	66.74	79.87	60.19
11	2.33	1.40	1.08
12	0.96	0.96	1.02
13*	6.87	7.06	41.75
14	1.59	1.03	0.98
15	0.92	0.92	0.94
16	1.14	1.04	0.97
17	1.11	1.45	11.47
18*	0.85	0.85	0.85
19	0.85	0.90	0.95

TABLE 2: Damage potential DP(j,k): Damage per unit volume of non-recoverable oil from a spill of volume index k at risk point j ($/gallon). Asterisks denote "small" spills.

(4) Finally, Table 2 displays the values of the "damage potentials" DP(j,k), obtained by separately running the Damage Assessment subroutine. It is interesting to observe that small spills generally exhibit significantly higher damage potentials than large spills, quite probably because these spills occur much closer to shore and are likely to impact the environmental and economic resources along the coast more severely.

4.1 "Large" Spill Analysis

In Psaraftis et al [5], twelve of the "large" spill problem were examined. Due to space limitations, we give here a flavor of these results:

(1) In our "basic scenario" the damage weight is equal to 1.0, there are no fixed costs to open a facility, and no response time stipulations, and there is a "standard" mobilization delay of 3 hours at each facility, plus a 3-hour deployment delay at each of the spill locations. Not surprisingly, the optimal solution under this scenario is a "benign neglect" response, with expected damages totalling $ 4.651 million per year, all in the form of unsatisfied demand.

(2) The "benign neglect" policy of the basic scenario remains optimal even if the damage weight is doubled. However, if the weight becomes 5.0 we begin to see 1 million gallons of capability being stockpiled at Woods Hole. Since such capability can only respond to spills up to 1 million gallons, it results in an expected damage of $3.935 million pear year, mainly in unsatisfied demand. Such capability averts only $ 0.716 million per year in damages vis-a-vis the "benign neglect" policy, at an additional expected system cost of $ 2.428 million per year. The equipment type chosen among the four candidates is # 3, probably because its low acquisition costs offset the benefits realized by a higher efficiency and operability (the latter are features of technology # 4).

(3) Raising the weight even further (to 10.0 initially and then to 15.0) produces some expected and some unexpected results. The fact that total system capability increases to a level capable of handling the maximum spill size (10 million gallons) is certainly predictable. If W=10.0, ten million gallons of capability (again of type 3) are stockpiled in Woods Hole and there is no unsatisfied demand. What is less predictable is what happens when W is raised to 15.0. First, two additional facilities are established, one in Provincetown and one in Penobscot, taking away a small fraction of capability from Woods Hole, with the total capability remaining constant at 10 million gallons. Second, while Woods Hole continues to be equipped with type 3 equipment, the two new facilities introduce technology # 4, which is more expensive, yet more efficient and reliable. One can explain this by the observation that the only way to further reducing damages from those when W=10 is by reducing Z(6) (Z(5) is already zero). But Z(6) is the damage due to delays and inefficiencies in the response. Delays can be reduced by "spreading out" the capability to three locations instead of one. Inefficiencies can be reduced by using equipment # 4 instead of # 3.

(4) Still, it is not immediately clear why both technologies are needed when W=15.0. We can shed more light into this issue only by looking at the entire response array r(i,e,j,k) (not shown here). This investigation reveals the following facts (see also Figure 2): While the 10-million gallon spills are responded to from all 3 facilities (using both types of technology), the 1-million gallon spills (which are more frequent) are only

responded to from Provincetown and Penobscot (which are equipped with technology # 4 only). Finally, the smallest spills in the range (100,000 gallons) are responded to either from Provincetown (risk points 9, 11, 12, 14, 15, 16, 17 and 19) or from Penobscot (risk points 2, 4, 6, 8), using technology # 4 only (see Figure 2). In other words, equipment type # 3 (stocked at Woods Hole) is acquired only to respond to the massive, 10-million gallon spills. Since such technology is used relatively rarely, it is preferable from type # 4 which would cost more to acquire. By contrast, the higher efficiency and reliability of type # 4 make it more desirable to respond to the smaller but more frequent 1-million and 100,000 gallon spill, with no massive outlays.

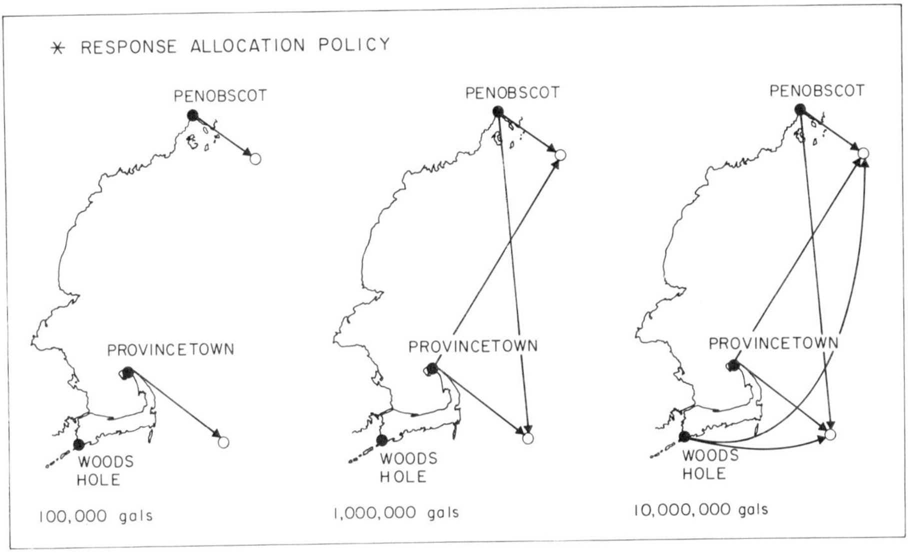

FIGURE 2

4.2 "Small" Spill Analysis

In Psaraftis et at [5], seven variants of the "small" spill problem were examined. Excerpts of these results follow:

(1) By contrast to the "large" spill case, here one needs a very low (close to zero) damage weight to make a "benign neglect" policy optimal. Expected damages then total $5.09 million per year, all due to unsatisfied demand. If the damage weight is set equal to one, the optimal policy is to stockpile capability that would leave no unsatisfied demand (if we assume zero mobilization and on-scene deployment delays, zero fixed costs and no response stipulations). Expected damages are then reduced to $1.067 million per year, all due to delay and equipment inefficiency, and expected system costs are $1.531 million per year. All but one facilities are opened.

(2) The allocation policies as well as the technological "mixes" that are recommended for the small spill problem are far more intricate than their equivalent in the large spill problem. A further investigation reveals that geographically, each facility typically responds to all volumes generated at risk points in its vicinity, and may also respond to more

distant locations but only to larger volumes at those locations. Indeed, each 100-gallon spill is typically responded to only from one location (usually, but not always, from its closest facility). By contrast, each 1,000 gallon spill is responded to by more locations (as many as four in this case), while even more facilities (as many as seven here) may be designated to respond to each 10,000-gallon spill (see Psaraftis et al [5] for more details).

5. CONCLUSIONS

Managing the risk of oil spills is a complex and multi-faceted problem. In this paper we have focused on strategic response as one of the weapons in the oil spill risk manager's arsenal. The discussion of the previous section has given a flavor of the potential of the model as a flexible tool for analysis of options before policy makers.

One should be cautious in generalizing these results to other geographical areas. For instance, an application of this model to the U.S. West Coast might very well lead to different conclusions, since, by contrast to New England, the wind in the West Coast typically blows towards the shore - and this may increase the "damage potentials" of large spills dramatically.

In spite of such caveats, we believe that the results of the New England application of this model provide strong evidence that, over the range of scenarios examined, it is generally more cost-effective to invest in "small-spill" response capability (i.e. less than 50,000 gallons) than to acquire expensive large-scale equipment that would be used only rarely (and, if used, would probably avert far less damages than its cost could justify). For the large spill case one would require more than $12.5 million per year in system costs (most of which in equipment acquisition and maintenance funds that would have to be disbursed with certainty) to bring the expected value of damages from $ 4.7 to about $ 2.3 million per year. It might be worthwhile to investigate whether these same response system funds could be better utilized in a "revolving fund" (or insurance) scheme of direct compensation of damages, with minimal (i.e. "surveillance and monitoring") response. A portion of these same funds could also be used to strengthen the response system for small spills.

A second finding of the New England application concerns the siting patterns of response resources for these two categories of spills. We saw that small-spill capability is typically dispersed geographically, while large-spill capability is typically consolidated at one or a few "strike centers". Of course, departures from this pattern could occur under alternative assumptions regarding the scenario examined. In an actual implementation of the model, a significant effort should be spent to determine which scenario is the most accurate representation of the real world.

ACKNOWLEDGEMENTS

Support for the MIT Oil Spill Model has been provided by the National Oceanic and Atmospheric Administration's Office of Sea Grant, the U.S. Coast Guard, the U.S. Navy, the Commonwealth of Massachusetts, the Doherty Foundation, JBF Scientific Corporation, the Spill Control Association of America, Texaco and Petro-Canada. Opinions expressed in the paper are solely those of the author. The computations of this paper have been supported by the Department of Ocean Engineering at MIT.

REFERENCES

1. Demis D.J. "Oil Spill Management: The Damage Assessment Model and the Spatial Allocation of Cleanup Equipment," unpublished S.M. thesis, MIT Department of Ocean Engineering, June 1984.
2. Devanney, J.W., III, R.J. Stewart. "Bayesian Analysis of Oil Spill Statistics," Marine Technology 11, 365-382 1974.
3. Fredrikson, G.F.W. "Cost/Benefit Analysis of Norwegian Oil Spill Preparedness," Report 82 02 22-2, Central Institute for Industrial Research, Oslo, October 1983.
4. Psaraftis, H.N., B.O. Ziogas. "A Tactical Decision Algorithm for the Optimal Dispatching of Oil Spill Cleanup Equipment," Management Science 31, 1475-1491 1985.
5. Psaraftis, H.N., G.G. Tharakan, A. Ceder. "Optimal Response to Oil Spills: The Strategic Decision Case", Operations Research, 34, No. 2, 1986.
6. Tharakan, G.G. "The Complementary Locations on a Network," unpublished ScD dissertation, MIT Department of Ocean Engineering, May 1982.
7. TSC. "Deployment Requirements for U.S. Coast Guard Pollution Reponse Equipment," Transportation Systems Center, Report to the U.S. Coast Guard 1979.
8. USCG. "Polluting Incidents In and Around U.S. Waters: Calendar Year 1981 and 1982," U.S. Coast Guard, Commandant Instruction M16450.2E, 1982.
9. White, I.C., J.A. Nichols. "The Cost of Oil Spills," Proceedings, Oil Spill Conference, San Antonio, TX. 1983.
10. Ziogas, B.O. "Management of Oil Spill Cleanup," unpublished PhD dissertation, MIT Department of Ocean Engineering, February 1982.

ENVIRONMENTAL IMPACT ASSESSMENT PROCEDURE FOR OIL POLLUTION IN THE DUTCH WADDEN SEA - THE F3 PIPELINE EXAMPLE

R.P.W.M. JACOBS

Nederlandse Aardolie Maatschappij B.V., Environment Department,
P.O. Box 28000, 9400 HH Assen, The Netherlands

ABSTRACT

The framework used for environmental impact assessment for oil pollution in the very complex and highly sensitive environment of the Dutch Wadden Sea is presented. It consists of a four-step procedure, employs a systematic approach , and requires intermediate judgements and decisions. The results have been used to arrive at appropriate contingency measures in case of an F3 pipeline leakage to set priorities for protective and clean-up spill response actions, and to select the preferred and rejectable spill response methods in specific habitats.

INTRODUCTION

In 1974 the Nederlandse Aardolie Maatschappij B.V. discovered exploitable amounts of oil and gas in the F3 block in the northern part of the Dutch continental sector of the North Sea. The oil was found to be a very light type of crude, in fact a condensate. A two-phase pipeline from a production platform to the onshore treating facilities (circa 230 km) was considered necessary to transport the hydrocarbons (see Figure 1). Part of the pipeline (i.e. 23.5 km) would cross the Dutch Wadden Sea, a shallow coastal intertidal sea between the Wadden Islands and the mainland. This area is known for its ecological, cultural and scenic values; moreover, a major part of the sea is a nature reserve (see Figure 2) and in 1984 the entire sea was designated a protected wetland and added to the list falling under the Convention on wetlands of international importance (Convention of Ramsar 1971).

In order to be fully prepared in the unlikely event of oil spillage from the F3 pipeline and to be able to take the appropriate measures to combat the spilled oil, a well-prepared and approved contingency plan was considered essential for optimal protection of the vulnerable habitats and sensitive biota in the Wadden Sea. The framework of that plan consisted of the identification and prediction of the impact of spillage incidents. To assess the type and magnitude of any environmental change that might result from spillages, a systematic approach was used. The subsequent application of this procedure in the plan has improved the general confidence in the identified reasonable spill responses and the decisions to be made in spill situations, which have been laid down in the contingency plan.

In assessing oil spill requirements the oil industry is making increasing use of environmental impact assessment as a tool to arrive at optimal prevention and reduction of impact (see e.g. Wilkinson and McEwan, 1983; Conover et al., 1985). Although a large number of procedures have been developed for and used in environmental impact assessment, they all have specific principles in common. Actual approaches and objectives usually depend on specific situations, which differ with the environment and the project concerned. The aim of this paper is to provide detailed information

Kuiper, J., Van den Brink, W.J. (eds), Fate and Effects of Oil in Marine Ecosystems. ISBN 90-247-3489-4

on the application of the preferred procedure used for the actual F3 situation. It may serve as an example of impact assessment in large coastal intertidal seas. The sensitivity of such environments to oil pollution is generally recognized (Gundlach and Hayes, 1978; Baker, 1983; Anon., 1985) and similar habitats in estuaries and embayments have been included in other contingency plans (e.g. Wilkinson and McEwan, 1983).

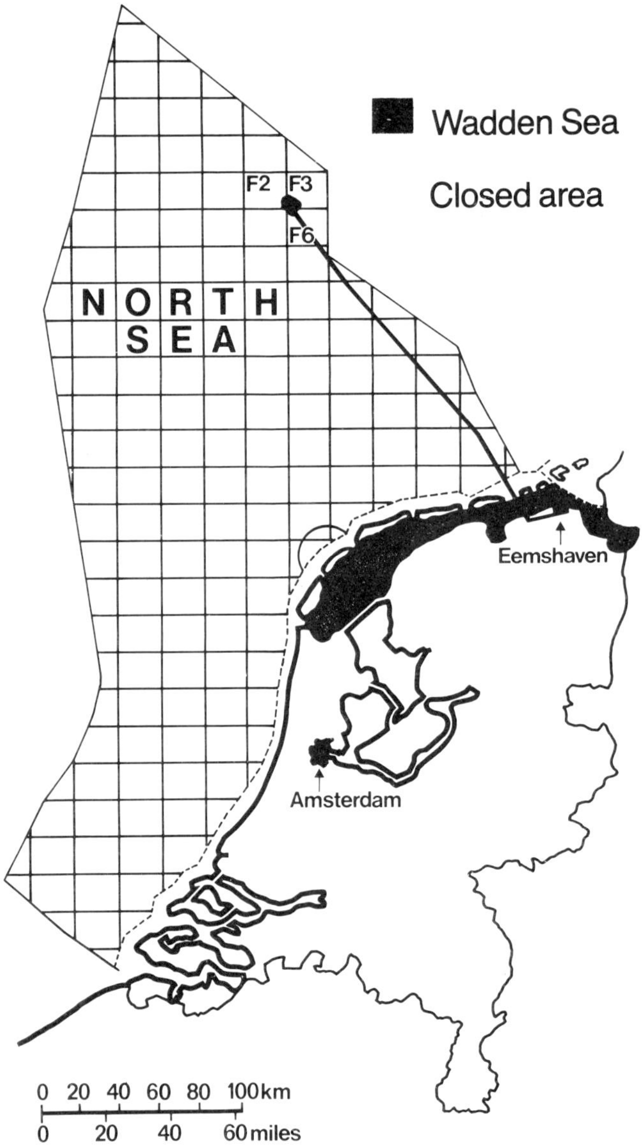

FIGURE 1. Situation of the F3 oil and gas field and the projected pipeline for transport of hydrocarbons to the shore.

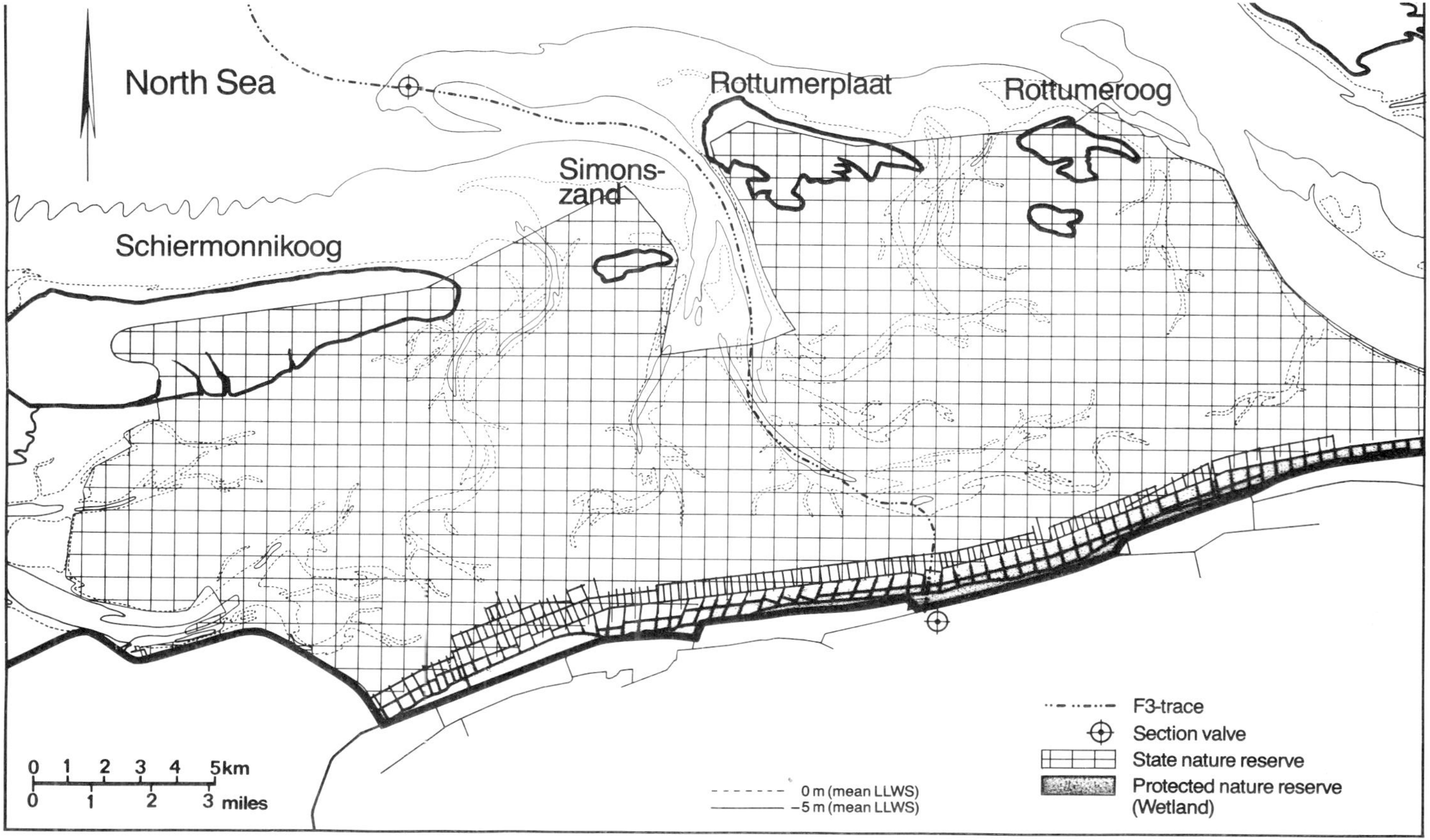

FIGURE 2. Projected route of the F3 pipeline in the Wadden Sea. The part between the two section valves is about 23.5 km.

IMPACT ASSESSMENT PROCEDURE

A four-step process was used to assess risk and environmental impact and to give response decisions to those assessments, evaluating all the available scientific information along with the values and judgements that can be brought to bear.

The four steps, i.e. risk analysis, identification of area at risk, identification of hazard and impact evaluation, are described below. Each step requires that the necessity to proceed along the set lines be evaluated.

Risk analysis

The analysis of the possibility and probability of a pipeline spillage is usually used to classify the actual risk and to justify balanced decisions on risk response. Very early in the development phase of the project the crossing under the Wadden Sea was considered reason enough to take all reasonably possible and economically feasible measures to prevent any chance of a spillage:

- pipeline routing: preferably in the deepest areas, i.e. the main channels (see Figure 2);
- burying of the pipeline at least 2 m below the sediment surface;
- special integrity provisions: thickness of the pipe wall, anticorrosive coating, concrete coating, cathodic protection;
- automatic protective devices for over- and under-pressure situations;
- section valve stations with automatic low pressure switches at both ends of the Wadden Sea (see Figure 2);
- situation of the offshore section valve station on an artificial island to safeguard the valve;
- regular maintenance, control and monitoring of pipeline route and integrity, process and protective devices.

Moreover, special measures will limit the amount of oil that could enter the Wadden Sea in case of a leakage:

- sweeping the pipeline with spheres such that oil is transported in slugs of max. 700 m3 each;
- distance between two spheres 27 km, so that always only one slug is present in the Wadden Sea section;
- regular surveying of the pipeline route with aircraft.

Based on the above, the situation was judged and it was concluded that the incident probability was negligibly small, thus in effect an "insignificant risk". However, as an insignificant risk is not equivalent to an absolute "zero risk", any discussion about performing a risk analysis and about the possible acceptability of any, even the smallest, risk was considered pointless in view of the sensitivity of the area. For this reason a response was developed to abate even the smallest risk, without considering costs and benefits of risk reduction or risks of alternative routes.

Identification area at risk

With the aim to optimise the required actions in case of a spillage and to delineate the appropriate remedial response for specific habitats, the area at risk needed to be delimited. The type of data gathered included information on (1) the amount of oil that might be spilled and the physical and chemical characteristics of the oil and, (2) hydrographical and meteorological characteristics of the pipeline route and the surrounding area.

The special engineering designs guarantee that in case of a pipeline rupture the maximum amount of oil lost would be 700 m3. However, this would be oil under pipeline conditions, i.e. under high pressure. As a consequence of a subsequent drop to atmospheric pressure, 60% of this volume would immediately disappear due to flashing, so that only 280 m3 would be found on the water surface. This is the maximum volume of oil that would have to be dealt with.

Compositional data on the reservoir fluid were available. Starting from these characteristics the composition of the spilled oil after flashing was modelled. Moreover, the physical properties of this fluid were determined. These data were used in computer calculations (TRANSSPILL programme, Ministry of Transport and Public Works, North Sea Directorate) on the fate

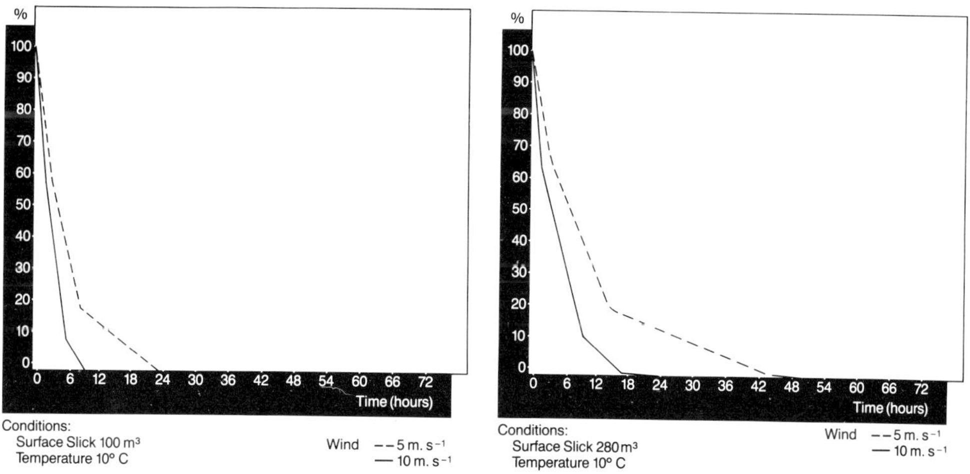

FIGURE 3. Examples of joint effects of evaporation and dispersion on slick volume, indicated as a residual percentage (calculated with the TRANSSPILL computer model). The upper figure refers to an initial surface slick with a volume of 100 m3. The lower figure concerns a pipeline rupture, spilling 700 m3 oil and leaving 280 m3 on the water surface after flashing.

of the oil at sea. The relative contributions of various weathering processes such as dispersion, evaporation, dissolution and spreading were thus obtained. As a result thereof the residual volumes of oil and the thickness of the slick in relation to time could be estimated. A few examples are presented in Figure 3.

In order to be able to define the extent of any influence of a spill, the spatial boundaries of the area at risk first had to be determined. Basic data required were the physical conditions of the environment, i.e. hydrographical (currents, tides, etc.), geomorphological (sediment, erosion, etc.) and meteorological (temperature, wind, etc.). The inherent

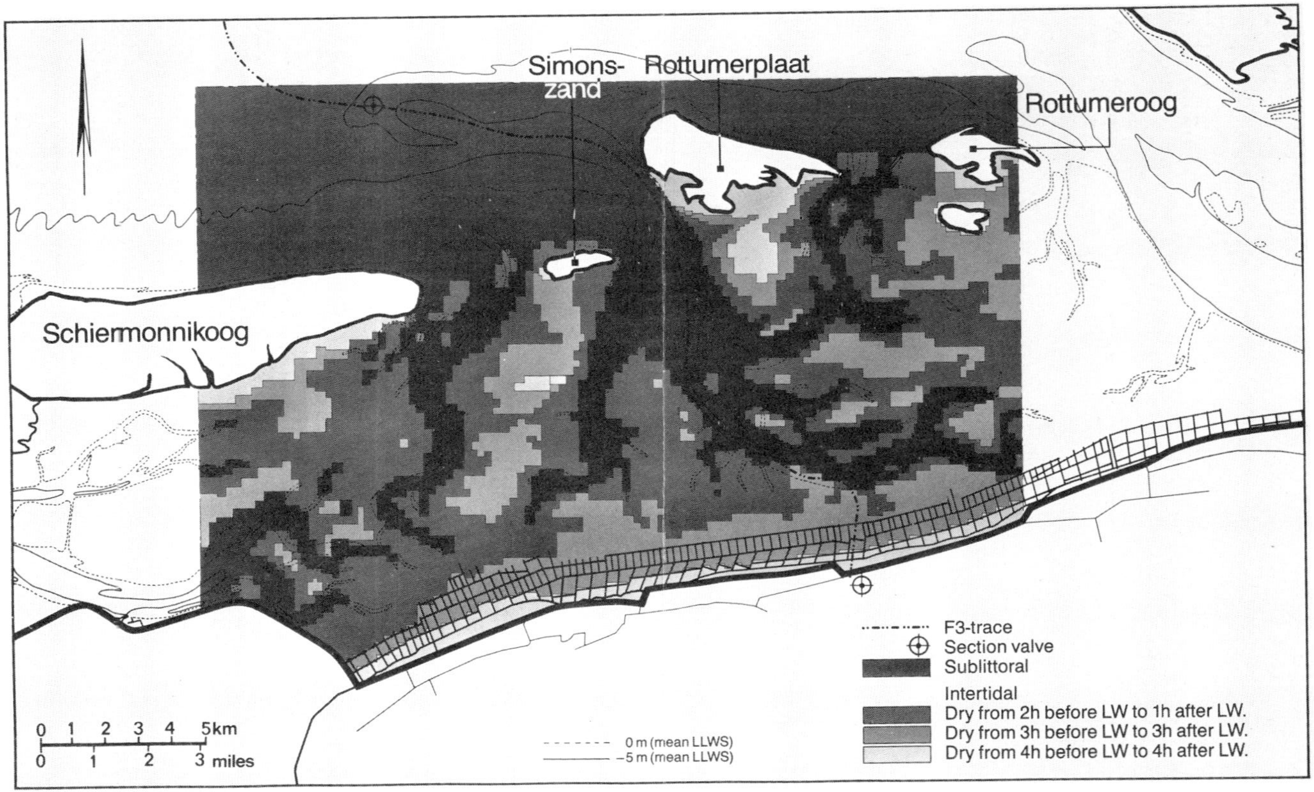

FIGURE 4. Area covered by the computer (STOLOM) programme used for spill trajectory simulations for environmental impact assessment and to be used in predicting slick movements in case of a real spillage. In this figure a part of the area information used in the programme is also presented and this concerns the dry areas in relation to the tidal cycle.

extreme complexity of the Wadden Sea, both in space and in time, required detailed site-specific information. However, a mass of data was available on currents and water levels in relation to the tidal cycle (e.g. Rijkswaterstaat, 1975). These data formed, after some adaptation, the area information for computer simulations of slick trajectories. The programme used was STOLOM (Short Term On-Line Oil Movement Model) from Shell Internationale Petroleum Maatschappij B.V. (see Wright et al., 1986). Part of the data set used is shown in Figure 4, local currents (direction and velocity) not being presented. For specific simulations data on spill location, moment in the tidal cycle and wind (direction and velocity) still had to be added. With STOLOM slick movements have been simulated for a variety (of combinations) of conditions, starting from several spill locations. This has resulted in trajectories over the first 12 hours after

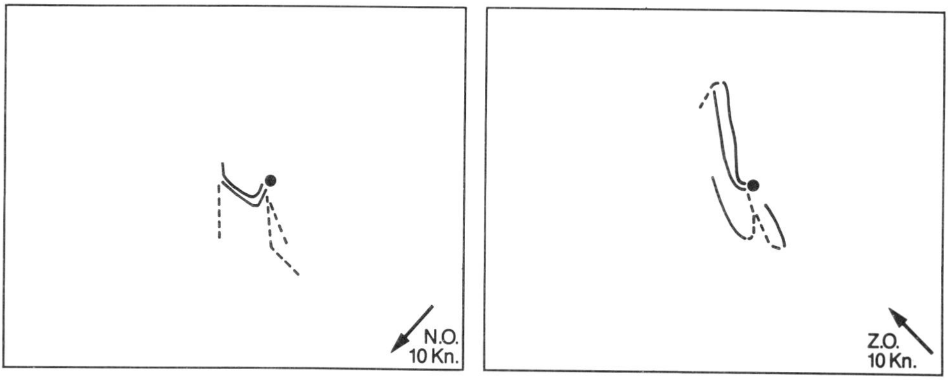

FIGURE 5. Examples of simulated slick trajectories during the first 12 hours after a spillage. Spill location about halfway between the valve stations; different wind directions (left north-east, right south-east); moment of spillage variable in the tidal cycle (continuous lines: at high water or receding tide; broken lines: at low water or incoming tide). A specific line indicates the movements of the centre of the slick in relation to time and stops when the oil hits land (at low water); eventually the slick can move on with the next incoming tide.

a spill. These show only the movements of the centre of the slick and no volumetric calculations. However, combined with the results of the time-dependent fate of the spilled oil (TRANSSPILL results) it was possible to predict not only the trajectories, but also the amounts of oil that would strand. A few examples are presented in Figure 5. From all the simulations it has been concluded that the actual area at risk is restricted to a strip of about four kilometres wide along both sides of the pipeline; thus, the actual area at risk is much smaller than the Wadden Sea sector indicated in Figure 4.

At the end of this step in the impact assessment procedure a confirmation was required of potential adverse effects to justify the next step. Hence, the estimated amounts of spilled oil that various habitats could receive were considered. It appeared that these amounts, particularly in vulnerable habitats, could be such that hazard identification was required.

Identification of hazard

For an identification of the potential harm posed by a spillage it was

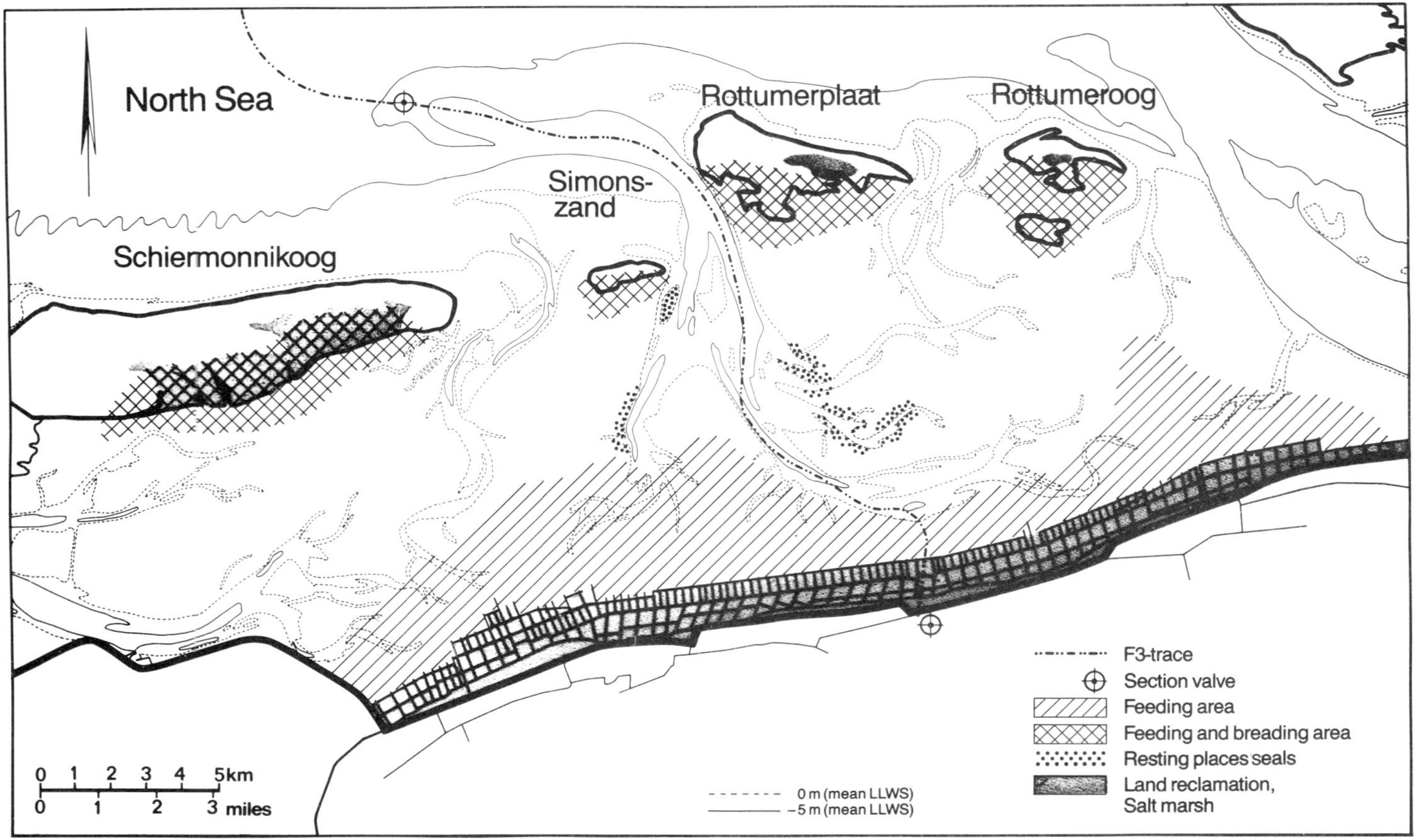

FIGURE 6. Part of the area information used in the impact assessment procedure: some characteristic habitats and important areas for seals and birds.

necessary to establish the biological responses and tolerances in local organisms and populations subjected to the stress of F3 hydrocarbon exposure. This included hazard assessment for pelagic and benthic organisms, birds, intertidal vegetation, etc.

Results from field and laboratory studies can be used to predict effects of oil spills on organisms. Laboratory studies on organisms known to occur in coastal areas have indicated that oil and petroleum hydrocarbons may be quite toxic to the flora and fauna. Effects and effect concentrations have been compiled and summarized by Connell and Miller (1981), Bergman (1982), Miller (1982) and Anon. (1985) for different taxonomic groups. Miller (1982) concluded that for most adult or mature organisms a lethal response can be expected in the range of 1-100 mg/l soluble aromatic derivates. The range is indicative of species-specific sensitivity, which is a remarkable but general phenomenon. Even within one taxonomic group differences may exist with regard to the effects of the water-soluble fraction of a particular oil. Size and developmental stages of animals and plants influence the sensitivity markedly, early life stages (e.g. larvae) being generally more susceptible than the eggs and adults of the same species. Moreover, large differences exist in toxicity of different oils and oil types, generally depending on the aromatic content.

Recorded and other information on different oils was collected, critically examined and validated with regard to their physical, chemical and ecotoxicological characteristics. This was compared with the information available on F3 oil after flashing and by means of extrapolation the ecotoxicological properties could be assessed. Based on the deduced information, it is expected that the water-soluble fraction of F3 oil is slightly toxic to mature aquatic organisms (according to the GESAMP (GESAMP, 1982) classification) and moderately toxic to early life stages. The toxicity of the parent material, however, is less (about a factor five). With regard to fate and behaviour it has been concluded that following a spillage the oil will float on and evaporate from the water surface, but will also partly dissolve (particularly the aromatic hydrocarbons) and disperse therein. The modelled rate of disappearance from the water surface is illustrated in Figure 3; the actual evaporation is estimated to be about 50-75% within 12 hours, depending on the temperature and wind conditions. Oil may penetrate sediment after stranding to a depth of a few centimetres, the actual depth being largely determined by the local water table. However, hydrocarbons may be transported to greater depth as a consequence of bioturbation. Concentrations in the sediment will be dependent on the amount of oil stranded, and the texture and the particle size distribution, with estimated maximum concentrations of 500 mg/kg. F3 oil is expected to be readily biodegradable; the rate depends largely on local environmental conditions and is high under aerobic and low under anaerobic conditions.

These results confirmed a potential for causing adverse effects following a spillage, thus the quantitative and qualitative dimensions of the situation required detailed consideration, i.e. the next step: impact evaluation.

Impact evaluation

This step in the assessment process focussed on weighing the environmental effects posed by the identified hazard, and the exposure to it. Firstly, this required an inventory of the physical vulnerability and biological sensitivity of the ecosystems to the F3 oil in the area at risk. For specific ecosystems (salt marshes, seagrass beds, intertidal flats, etc.; see Figure 6) the characteristics of environmental units (water column, sediment) and the occurrence of taxonomic categories (microbiota to mammals and algae to higher plants) were compiled. For some species it was necessary to go beyond species level to life history stages (e.g. fish eggs and larvae) or occurrence in a particular season (e.g. feeding and breeding of birds, reproductive periods). Important areas of concern were identified by this process. The information base used has been derived from existing information in the published literature (e.g. Anon., 1976; Bergman, 1982; Wolff, 1983), government and industry "grey literature", and unpublished data files.

Secondly, the environmental information was evaluated together with the information on identified hazard and actual exposure possibilities, in order to identify and qualify potential effects and, as a part thereof, the ecosystem recovery (time).

The identified potential impacts were subsequently compared with experimental impact studies carried out in the Wadden Sea (e.g. Farke et al., 1985; Kuiper et al., 1985, 1986) and showed reasonably good agreement. The evaluation resulted in a rating of ecosystem susceptibility, species sensitivity and habitat vulnerability. Based thereon priorities have been determined for spill response actions, i.e. for protection and clean-up, so that first of all the most serious impacts could be reduced or eliminated. Although such action is aimed at removal of oil, this may cause further damage. Cleaning of particular habitats, for instance salt marshes, is not only very difficult, but can also cause erosion. Weighing of any additional adverse effects from specific spill responses against possible alternatives or the do-nothing option has led to a selection of preferable methods for combating oil in the habitats concerned

CONCLUSIONS

The value of the applied approach has been many fold:

- By modelling fate and behaviour of spilled oil it was possible to demonstrate that only a small sector in the Wadden Sea is at risk in the case of an F3 pipeline spillage.
- Even in the restricted area at risk, the environment appeared so diverse and many parts so sensitive to oil pollution that the necessity of diverse response devices was demonstrated.
- Appropriate response methods for different areas could be identified.
- Oil response capability, including type and amount of equipment required and location of storage bases for quick response to areas most at risk, could be identified.

The results were used in the preparation of the F3 contingengy plan (NAM, 1985).

This approach is equally applicable to all types of oil spills in the Wadden Sea and to all parts thereof. However, uncertainties with regard to spill characteristics (e.g. location, quantity and type of oil) will diminish the quality and adequacy of the response actions required, so that an optimum action plan, such as that achieved for the F3 pipeline, would probably be hard to attain.

The use of sophisticated modelling in assessing fate and behaviour of oil spilled at sea, in delineating impact boundaries and in predicting slick trajectories has proven its central role in impact assessment. In addition, usage of systematic approaches serves as a base for establishing the framework that facilitates practical decision making, being essential for managing oil spills in sensitive and complex environments like the Wadden Sea.

ACKNOWLEDGEMENTS

As the impact assessment procedure formed an integrated part of the F3 contingency plan, this paper has only been possible thanks to the discussions and contributions of the whole team during the preparation of the plan. I would, therefore, like to thank J. Verhagen, D. Hilberink, J. van Buuren and T. Verhoeff.

REFERENCES

Anonymous: Waddenzee: Natuurgebied van Nederland, Duitsland en Denemarken. Harlingen, Landelijke Vereniging tot Behoud van de Waddenzee; 's-Graveland, Vereniging tot Behoud van Natuurmonumenten in Nederland, 368 pp, 1976.

Anonymous: Oil in the Seas: Input, Fates and Effects. Washington, D.C.: Nat. Acad. Press, 601 pp, 1985.

Baker JM: Impact of oil pollution on living resources. The Environmentalist 3 (Suppl. 4): 1-48, 1983.

Bergman M: Gedrag, bestrijding en biologische effecten van olie in estuariene gebieden. 1. Literatuuroverzicht. RIN-rapport no. 82/18, 420 pp, 1982.

Connell DW and Miller GJ: Petroleum hydrocarbons in aquatic ecosystems - behaviour and effects of sublethal concentrations: Part 1 and 2. CRC Critical Reviews in Environmental Control 11: 37-104, 105-163, 1981.

Conover SAM, Strong KW, Hickey TE and Sander F: An evolving framework for environmental impact analysis. II. Applications. J. Environ. Mgmt 21: 359-374, 1985.

Farke H, Wonneberger K, Gunkel W and Dahlman G: Effects of oil and a dispersant on intertidal organisms in field experiments with a mesocosm, the Bremerhaven caisson. Mar. Environ. Res. 15: 97-114, 1985.

GESAMP: The evaluation of the hazards of harmful substances carried by ships. Rep. Stud. GESAMP 17: pag. var., 1982.

Gundlach ER and Hayes MO: Vulnerability of coastal environments to oil spill impacts. MTS Journal 12: 18-27, 1978.

Kuiper J, Van het Groenewoud H, Admiraal N, Hoornsman G, Van der Meer M, Schulting FL and Verkoelen PJMD: Lot en effecten van olie vrijgekomen bij een gesimuleerde breuk van de F3 pijpleiding in Model Plankton Ecosystemen. Report MT-TNO no. 85/162, 149 pp, 1985.

Kuiper J, Van het Groenewoud H, Admiraal N, Scholten M, De Wilde PAWJ, Van Moorsel G, Dekker R, Wolff WJ and Brouwer C: The influence of dispersants on the fate and effects of oil in model tidal flat ecosystems. Final report Oil Pollution Experiments with special reference to the use of Finasol OSR5 (OPEX 1984-1986). Report MT-TNO no. R86/182, 94 pp, 1986.

Miller GJ: Ecotoxicology of petroleum hydrocarbons in the marine environment. J. Appl. Toxicol. 2: 88-98, 1982.

NAM: Calamiteitenbestrijdingsplan F3 pijpleiding Waddenzee. Draaiboek en Appendix. Assen, Nederlandse Aardolie Maatschappij B.V., 1985.

Rijkswaterstaat (Directie Groningen; Meet- en Adviesdienst Delfzijl): De waterbeweging in het stroomgebied van de Zuid-oost Lauwers, op het Uithuizerwad en in het Ra. Rapport no. 75-1, 1975.

Wilkinson TG and McEwan R: An environmental impact study and pollution contingency plan for an ecologically sensitive area in the North Sea. Proceedings 1983 Oil Spill Conference. Washington D.C.: Amer. Petr. Inst., publ. no. 4356, pp 135-141, 1983.

Wolff J(ed.): Ecology of the Wadden Sea. Three Volumes. Rotterdam, Balkema, 1983.

Wright RAD, Wilkinson TG and Sommerville HJ: Prediction of oil spill movement by micro-computer. Wat. Sci. Tech. 18: 357, 1986.

INTERNATIONAL ACCEPTANCE OF SHORELINE SENSITIVITY MAPPING AS AN INTEGRAL SPILL-RESPONSE COMPONENT

Erich R. Gundlach
E-Tech, Inc.
70 Dean Knauss Dr.
Narragansett, RI
02882-1443 USA

Bart J. Baca
Coastal Science & Engineering, Inc.
P.O. Box 8056
Columbia, SC 29205 USA

This poster session presents a series of maps that portray oil-sensitive environments and wildlife in a color-coded format that has received wide international acceptance as necessary for spill-response planning and response. Some of the areas where this system of sensitivity mapping has been applied include Kuwait, South Africa, Indonesia, Panama, Nigeria, and parts of western Europe and North America (both Canada and the United States).

Key components of this mapping system include:

a) characterization and ranking of shoreline types in relation to potential oil-spill damage. Shorelines are color-coded and ranked on a scale of 1 to 10, with 10 being the most sensitive. Typically, marshes, mangroves, and sheltered tidal flats are the most sensitive.

b) a description of biological components including shoreline species, areas of wildlife concentrations (e.g., rookeries, nesting sites, etc.), and species of special concern (e.g., endangered or protected status). Seasonal data and protected status are indicated graphically.

c) presentation of sites of particular socio-economic importance, including such areas as coastal parks, recreational facilities, marinas, etc., and

d) primary response strategies to deal with a most- likely spill incident. This includes boom placement sites, boat launching ramps, nearest landing strips and major airports, etc.

The presentation of the above information is graphical, using symbols on a series of base maps. A short explanatory text is included to aid interpretation. Format varies somewhat between countries and different regions. This is especially true when dealing with dealing with different climates; as from Alaska to tropical Panama or arid Kuwait.

Future trends indicate that the usefulness of such maps will extend further through computer integration and formatting. The most recent oil-spill models are able to simulate oil deposition and removal from different types of shorelines (see Gundlach, Reed, and Spaulding; this conference). The shoreline characterization provided by these sensitivity maps conforms easily to the parameters needed for these models.

Kuiper, J., Van den Brink, W.J. (eds), Fate and Effects of Oil in Marine Ecosystems. ISBN 90-247-3489-4

A COMPUTER SIMULATION MODEL DEPICTING OIL IMPACT AND REMOVAL WITHIN COASTAL ENVIRONMENTS

Erich R. Gundlach, E-Tech, Inc.
Mark Reed and Malcolm Spaulding, Applied Science Associates, Inc.
70 Dean Knauss Drive
Narragansett, RI 02882-1443

This poster session describes the development of an oil-spill model that realistically simulates the active deposition and removal of oil along various types of shorelines. The model, developed for predicting oil spill impacts along the Alaskan Bristol Bay coastline, contains five primary components:

(a) an offshore transport submodel;
(b) a nearshore/surfzone transport submodel;
(c) a shoreline interaction submodel;
(d) a chemical weathering submodel; and
(e) a suspended sediment submodel being developed.

The offshore trajectory model brings the oil into the nearshore/surfzone environment. The distribution of the oil will vary with environmental conditions (e.g., wind speed and direction, the current field, sea-surface elevation, etc.).

The nearshore/surfzone transport submodel takes the oil from offshore areas and computes its movement shoreward or alongshore (or offshore again). In preparation for a sediment interaction model being developed, a component for predicting alongshore sediment transport is also included.

The shoreline interaction submodel is entirely new and consists of three primary aspects: (1) characterization of the shoreline into one of seven geomorphic types, each having particular characteristics relative to oil deposition and removal; (2) application of an oil-removal coefficient that reflects an oil-removal rate for each shoreline type and varies with tidal and wave conditions, and (3) application of an oil-holding capacity that realistically limits the amount of oil able to be deposited on each particular shoreline segment (as during supertanker spills). Removal rates are calculated according to a first-order decay function:

$$M_i = M_{io}e^{-K_f t}$$

where: M_i = mass of oil within each beach segment i.
M_{io} = mass of oil originally deposited on the beach.
K_f = the removal rate coefficient.
t = time in days since original deposition.

The seven geomorphic classes and proposed rate coefficients are as follows. The higher the value, the more rapid is oil removal.

Kuiper, J., Van den Brink, W.J. (eds), Fate and Effects of Oil in Marine Ecosystems. ISBN 90-247-3489-4

Exposed rocky shores	K_f=0.90-0.99
Sheltered rocky shores	0.01-0.05
Eroding peat scarps	
low <1m waves	0.10-0.20
high >1m waves	0.70-0.80
Sand beaches (<1m waves)	
beach face	0.20-0.20
backshore	0.10-0.15
Sand beaches (>1m waves)	0.50-0.60
Gravel beaches (<1m waves)	
beach face	0.10-0.20
backshore	0.05-0.10
Gravel beach (>1m waves)	0.40-0.50
Tidal flats	0.90-0.99
Marshes	0.001-0.01

The other key component relating to oil deposition on the shoreline is the application of an oil-holding capacity. This capacity limits the amount of oil able to be deposited on the shoreline, varying with shoreline slope, oil viscosity (e.g., thickness), and oil incorporation into the beach. As an example, during a large spill with onshore winds, a certain amount of oil will be deposited but the excess will (depending on nearshore conditions) be transported along shore and possibly onto adjoining beaches.

The fourth submodel integrated into this model enables the simulation of oil weathering in all three environments: offshore, nearshore/surfzone and onshore. Three primary weathering components are included: evaporation, dispersion, and emulsification. Through the use of a modular format, weathering algorithms developed by different researchers can be inserted depending on model requirements.

The final component of this overall model is the inclusion of a parameter reflecting oil interactions with suspended particulate matter (currently under development). This submodel will be integrated within the offshore and nearshore/surfzone transport models.

In summary, the model will compute nearshore wave environment, currents, and sediment transport for each shoreline segment as a function of time. Advection, spreading, emulsification, evaporation, dispersion, and shoreline deposition/removal will be calculated at each time step for up to 90 days after the spill. Throughout the run, the dynamic mass balance of oil can be quantified for the sea surface, water column, sea floor, atmosphere, and the shoreline. The sum of all individual components equals the total mass of oil introduced across the model boundaries up to that time. In addition, the oil mass will be spatially distributed into three compartments for each shoreline segment: beach, surf zone, and nearshore zone.

THE ALVENUS OIL SPILL ALONG THE TEXAS COASTLINE

Erich R. Gundlach
E-Tech, Inc.
70 Dean Knauss Drive
Narragansett, RI 02882-1443

On 30 July 1984, the tanker Alvenus suffered a major rupture along its bow which caused the loss of approximately 10,000 tons of heavy Venezuelan crude oil. This loss was the third largest in United States' waters following the Santa Barbara and Argo Merchant oil spills. This poster session will focus on the cleanup methods and the role of coastal processes influencing both oil transport and the movement of beach sands.

The British-registered Alvenus was in transit to deliver 14,952 gallons of crude oil to Conoco facilities in Lake Charles, Louisiana, when the accident occurred 18 km south of Calcacieu Pass, Louisiana. The rupture caused the loss of roughly 1.2 million gallons of Mery crude and 1.6 million gallons of Pilan crude. The very high viscosity of these oils added greatly to the difficulty of cleaning up the spill. Strong winds and currents rapidly carried the oil toward the low-lying sand beaches of the Texas coast.

Major impacts of oil occurred on Galveston Island, Texas, on 6 August. On the beaches, oil formed a continuous swash over 30 m wide and up to 4 cm thick. Altogether, 22 km of fine-grained sand beaches and 16 km of seawall were affected. Cleanup of the seawall was relatively straightforward utilizing high-pressure spraying with a small amount of fine sand mixed in. Cleanup of the beaches was most effectively carried out by road graders working in groups of six, each pushing the oil higher up the beach until reaching a common collection berm. Numerous passes were made to remove the thickest accumulations. An analysis of coastal processes, including wave action, currents, and sand movement, indicated that oil previously deposited on the nearshore/surfzone bottom, was actively transported shoreward by wave action (much like a conveyor belt). In this manner, previously cleaned beaches appeared very oily again the next following day. Cleanup of the beaches lasted five weeks whereas oil removal from the seawall took several months.

Nearly 100,000 cubic meters of oily sand were removed from the beaches. Heavily oiled material was taken to a certified inland sanitary landfill site; moderately oiled material was left on the island and used as landfill for a city park; lightly oiled material was graded along the upper part of the beach to add to dune stability. Even with this extensive removal of beach sand, beach erosion could not be determined on the short-term because of continued accretion following a major hurricane which occurred the year before.

Kuiper, J., Van den Brink, W.J. (eds), Fate and Effects of Oil in Marine Ecosystems. ISBN 90-247-3489-4

Scientific names of organisms mentioned in the texts

References are to the author(s) of the paper in which the scientific name is mentioned.

Abra alba	Dixon; Westwood et al
Abra nitida	Westwood et al
Acartia tonsa	Girling
Amphiura filiformis	Dixon
Apistobranchus tullbergi	Westwood et al
Arenicola marina	Dekker & van Moorsel
Artemisia maritima	Scholten, Leendertse & Blaauw
Asterionella glacialis	Scholten, Kuiper et al
Augeneriella lagunari	Dicks & Westwood
Avicennia marina	Dicks & Westwood
Balaena glacialis	Engelhardt
Balaena mysticetus	Engelhardt
Balaenoptera acutorostrata	Engelhardt
Balaenoptera physalus	Engelhardt
Callorhinus ursinus	Engelhardt
Capitella capitata	Dicks & Westwood; Westwood et al
Capitomastus minimus	Westwood et al
Cerastoderma edule	Dekker & van Moorsel
Chaetoceras debilis	Scholten, Kuiper et al
Choromytilus meridionalis	Mason
Corbula gibba	Westwood et al
Corophium volutator	Dekker & van Moorsel
Crangon crangon	Bardot & Castaing; Dixon
Crassostrea gigas	Page et al
Dugong dugon	Engelhardt
Emiliania huxleyi	Skjoldal & Thingstad
Enhydra lutris	Engelhardt
Erignathus barbatus	Engelhardt
Eschrichtius robustus	Engelhardt
Festuca rubra	Scholten, Leendertse & Blaauw
Globicephala melaena	Engelhardt
Halichoerus grypus	Engelhardt
Halimione portulacoides	Scholten, Leendertse & Blaauw
Halodule wrightii	Thorhaug & Marcus
Harpinia antennaria	Dixon
Hydrobia ulvae	Dekker & van Moorsel
Juncus gerardii	Scholten, Leendertse & Blaauw
Lagenorhynchus acutus	Engelhardt
Lobodon carcinophagus	Engelhardt
Lutra lutra	Engelhardt
Macoma balthica	Dekker & van Moorsel
Mediomastus fragilis	Westwood et al

Kuiper, J., Van den Brink, W.J. (eds), Fate and Effects of Oil in Marine Ecosystems. ISBN 90-247-3489-4

Megaptera novaeangliae	Engelhardt
Monodon monocerus	Engelhardt
Mya arenaria	Gilfillan et al
Mysella bidentata	Dixon; Westwood et al
Mytilus edulis	Widdows et al; Gilfillan et al; Parker et al
Navicula spec.	Scholten, Kuiper et al
Nereis diversicolor	Dekker & van Moorsel
Nitzschia closterium	Scholten, Kuiper et al
Odobenus marinus	Engelhardt
Ophryotrocha spec.	Levell
Orcinus orca	Engelhardt
Owenia fusiformis	Westwood et al
Pectinaria koreni	Westwood et al
Perna perna	Parker et al
Phaeocystis poucheti	Scholten, Kuiper et al
Phoca groenlandica	Engelhardt
Phoca vitulina	Engelhardt
Pholoe inornata	Levell
Plagiogramma spec.	Scholten, Kuiper et al
Platichthys flesus	Reiersen & Fugelli
Polydora spec.	Westwood et al
Porphyra spec.	Parker et al
Prorocentrum spec.	Scholten, Kuiper et al
Pseudocuma longicornis	Dixon
Puccinellia maritima	Scholten, Leendertse & Blaauw
Raphidrilus spec.	Levell
Rhizophora mangle	Gilfillan et al
Rhizosolenia delicatula	Scholten, Kuiper et al
Salmo gairdneri	Payne et al
Scalibregma inflatum	Westwood et al
Scolelepis fuliginosa	Westwood et al
Skeletonema costatum	Scholten, Kuiper et al
Spartina alterniflora	Gilfillan et al
Spartina anglica	Scholten, Leendertse & Blaauw
Syringodium filiforme	Thorhaug & Marcus
Temora longicornis	Scholten, Kuiper et al
Thalassia testudinum	Thorhaug & Marcus
Thyasira flexuosa	Westwood et al
Triglochin maritima	Scholten, Leendertse & Blaauw
Tubificoides benedeni	Westwood et al
Turritella communis	Westwood et al
Tursiops truncatus	Engelhardt
Ursus maritimus	Engelhardt
Zalophus californianus	Engelhardt